Câmpean, Deliiski, Bedelean

Wood Drying and Steaming

Mihaela Câmpean, Nencho Deliiski, Bogdan Bedelean

Wood Drying and Steaming

HANSER

Print-ISBN: 978-1-56990-998-0
E-Book-ISBN: 9798950184338

Bibliographic information of the German National Library:
The German National Library lists this publication in the German National Bibliography; detailed bibliographic data are available on the Internet at http://dnb.d-nb.de.

AI-based systems were used for the translation of this work as well as for the creation of alternative text descriptions for images and illustrations. Responsibility for the content, as well as the final editorial review and approval, rests solely with the author(s) and the responsible editors.

Vilshofener Straße 10 | 81679 München | *info@hanser.de*
www.hanserpublications.com
www.hanser-fachbuch.de
Editor: Frank Katzenmayer
Production management: Frauke Schafft
Cover concept: Marc Müller-Bremer, www.rebranding.de, München,
Cover design: Thomas West
Cover picture: stock.adobe.com/luigidezotti
Typesetting: in-house
Printed and bound by Elanders Waiblingen GmbH, Waiblingen
Printed in Germany

Table of Contents

Preface

This book is part of the series on wood as a material, published by Carl Hanser Verlag, Munich, Germany. This book series is continuously expanding, and the individual volumes are intended to reflect the current state of knowledge in their respective fields.

It is a great honor for me to be part of this project. My thanks go to Prof. Peter Niemz as coordinator of this series. He has always supported me in my endeavors and firmly believed in me and in my ability to complete this project, and so this book came into being, intended for both practice and teaching.

The thermal treatment of wood by drying and steaming is of particular economic importance with regard to the production of quality products under sustainable conditions. It is our duty as researchers and educators in the field of wood processing to meet this responsibility. High-quality thermal treatment enables the saving of considerable amounts of wood material and energy. For this purpose, knowledge of both the theoretical foundations of drying and steaming and their correct application is indispensable. This was the guiding thread that led me in developing this book: the linking of theory and practice, in order to draw readers' attention to the small details that make the difference between quality and reject material – details that are theoretically grounded and have been continuously refined through practical experience and research.

My thanks also go to the two co-authors, Prof. Nencho Deliiski from the University of Forestry Sofia (Bulgaria) and Dr. Bogdan Bedelean from the Transilvania University of Brasov (Romania), who have significantly raised the level of this book. In Chapter 2, Section 2.2.2.9, and Chapter 5 (B. Bedelean) as well as Chapter 10 (N. Deliiski), it is shown that the theoretical foundations are the starting point for modeling the complex wood drying and steaming processes, which are increasingly gaining importance today. Dr. Sergiu Georgescu also made a special contribution in preparing the figures for this book.

I would also like to express my sincere gratitude to my father, Dr. Ion Marinescu. From him I have learned everything I know about the thermal treatment of wood. He has always encouraged and supported me. Since my studies and later during the writing of my dissertation, he guided me with great patience in the field of wood drying and passed on to him his passion for this important process. He also always oriented himself toward close cooperation between science and practice.

And it was again my father who initiated our participation in the IUFRO wood drying conferences from 1997 onwards. These events gave me the opportunity to establish international contacts – among those with whom I was in contact on my journey from doctoral student to my current position, and from whom I have learned a great deal. These include Prof. Thomas Trübswetter (1935–2008), Prof. Helmuth Resch, Prof. Stavros Avramidis, Prof. Alfred Teischinger, Prof. Tom Moren, Prof. Patrick Perre, Prof. Yves Fortin, Prof. Margot Sehlstedt-Persson, Prof. Hans Welling, Prof. Lena Antti, Prof. Robert Nemeth, Prof. Ales Straze, and many others. I dedicate this book to all of them and thank them wholeheartedly for helping me become a better lecturer for my students. I hope that this book will be helpful to German-speaking students and production specialists worldwide.

I would like to thank Mr. Frank Katzenmayer and Ms. Christina Kubiak from Carl Hanser Verlag for the excellent collaboration on the manuscript preparation and publication.

Prof. Dr. Mihaela Campean

Transilvania University of Brasov, Romania

March 2026

1 Objective and Economic Significance of Wood Drying

Wood drying is the technological process in which water is removed from the material (wood). The main objective is to achieve a moisture content that corresponds to the practical service conditions of the wood component in question (Table 1.1).

Table 1.1 Target Final Moisture Content for Wood Components Depending on Application (based on [1–3])

Application		Target wood moisture content (final moisture content after drying) [%]
Structural solid timber	Submerged in water (e.g., bridges)	> 30
	Weathered components	18 – 20
	Components protected from weathering:	
	In well-ventilated, unheated rooms	15 ± 3
	In well-ventilated, weakly heated rooms in winter	12 ± 3
	In well-ventilated, well-heated rooms in winter	9 ± 3
Exterior cladding, balconies, soffits of roof overhangs, packaging, pallets, etc.		15 ± 2
Barrels		14 – 16
Windows, exterior doors		13 ± 2
Ship and boat building		12 – 15
Glued laminated timber construction		12 ± 2

Application	Target wood moisture content (final moisture content after drying) [%]
Firewood	11 - 16
Furniture, floors, parquet, interior doors, wall paneling in centrally heated rooms	10 ± 2
Sawdust (for briquettes and pellets)	8 - 12
Ceiling paneling in centrally heated rooms, glued laminated timber for the interior of office buildings and houses	8 ± 2
Aircraft construction	7 - 10
Veneers, chips	6 - 12
Musical instruments	6 - 8

As a hygroscopic material, wood has the ability to adjust its moisture content by releasing or absorbing water/water vapor from its surroundings in order to reach equilibrium with its environment.

The target equilibrium moisture content (*EMC*) depends on the temperature and the relative humidity of the drying medium, as well as on the wood species. The relationship between the three parameters is expressed graphically by means of so-called sorption isotherms. The graph created by Loughborough and Keylwerth for Sitka spruce (Figure 1.1) is still used today for the approximate determination of the equilibrium moisture content as a function of the ambient state parameters for all wood species.

The equilibrium moisture content that wood seeks to attain in an environment with a temperature of 20 °C and a relative humidity of 45 % is 8.3 %.

In reality, for example, yew and black locust have a lower equilibrium moisture content than spruce or beech, and some tropical wood species also often reach lower equilibrium moisture contents due to their high extractive content. The equilibrium moisture content of wood can also be reduced by thermal treatment (e.g., the ThermoWood process) or by impregnation with a salt-containing solution [5].

The hygroscopicity of wood is utilized during drying by selecting and controlling the drying parameters (air temperature and air humidity) such that the resulting equilibrium moisture content is always below the desired final moisture content of the wood. Thus, the wood is continuously stimulated to release moisture.

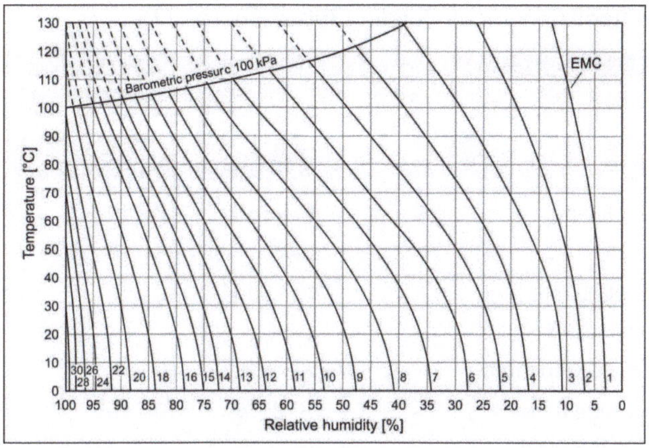

Figure 1.1 Hygroscopic equilibrium of sitka spruce as a function of temperature (source: [4])

The reduction of wood moisture content during drying is also accompanied by changes in the shape and size of the wood component. It is said that the wood "works". Since all objects made of wood should exhibit a certain dimensional stability, the installation of wet or insufficiently dried wood components is a serious mistake that leads to undesirable damage, because the wood also attempts to adapt its moisture content to the surrounding environment once installed. This leads to deformations, cracks, loosening of joints, peeling of lacquer films, veneers, melamine coatings, etc.

To prevent these undesirable effects, the wood must be dried as close as possible to the in-service moisture content. For this purpose, the so-called service classes were introduced [6]:

- service class 1 (SC 1) – for wood equilibrium moisture contents of 5 to 15 %, mostly not > 12 %, corresponding to a temperature of 20 °C and a relative humidity of the surrounding air that exceeds 65 % for only a few weeks per year;
- service class 2 (SC 2) – for wood equilibrium moisture contents of 10 to 20 %, corresponding to a temperature of 20 °C and a relative humidity of the surrounding air that exceeds 85 % for only a few weeks per year;
- service class 3 (SC 3) – for wood equilibrium moisture contents of 12 to 24 %, corresponding to climatic conditions that lead to higher wood moisture contents than those specified in service class 2.

Target wood moisture contents depending on the typical climatic conditions in various building types are given in Table 1.2.

Museums are to be regarded as heated buildings, churches as enclosed buildings with periodically high moisture exposure, so that the typical in-service moisture content for furniture and other wooden objects specific to these building types can be taken from Table 1.2.

Table 1.2 Wood Moisture Contents in Various Building Types (source: [7])

Building type		Typical climate (temperature [°C] / relative humidity [%])	Wood moisture content [%]
Low moisture exposure	Heated buildings	18 – 23 °C / 40 – 55 %	8 – 10
	Sports halls	20 °C / 40 – 50 %	8 – 10
	Air-conditioned indoor swimming pools	27 – 30 °C / 50 – 60 %	8 – 10
Moderate moisture exposure	Closed halls, unheated	3 – 22 °C / 55 – 80 %	10 – 16
	Open storage halls, without wet stored goods	5 – 22 °C / 55 – 85 %	10 – 19
	Ice sports halls, unheated	5 – 9 °C / 60 – 80 %	12 – 17
	Agricultural buildings (e.g., cold-air livestock barn)	0 – 23 °C / 60 – 85 %	12 – 19
High moisture exposure	Closed riding halls, irrigated	0 – 23 °C / 60 – 90 %	14 – 20
	Composting facilities	up to 40 °C / up to 100 %	> 24

Since the in-service moisture content of wood is in all cases significantly lower than the moisture content of freshly felled wood, drying processes are encountered in all wood-processing sectors:

- in sawmills (for drying sawn timber and parquet strips);
- in bentwood factories (for drying bent furniture components);
- in factories and workshops that manufacture musical instruments or sporting goods from wood;
- in veneer factories (for drying veneers);
- in the wood-based panel industry (for drying wood chips and wood fibers);
- in the packaging industry;
- in other small wood-processing industries (for drying ice cream or coffee sticks, matches, toothpicks, pencils, etc.).

In addition to controlled moisture reduction and the associated prevention of dimensional and shape changes, the drying of wood components also contributes to other improvements in properties, such as:

- the increase of mechanical strengths;
- better gluability;
- enabling various surface treatments;
- protection against fungal and insect attack (these find favorable conditions for development at wood moisture contents between 20 % and 100 %);
- the reduction of transport costs by reducing the weight of the wood.

Therefore, the drying of wood is required as a mandatory operation at the beginning of every wood-processing process. This underlines the economic importance of this process, as the duration and quality of drying directly influence production costs and also the quality of the final wood products.

Understanding the behavior of wood in its interaction with the drying environment and the rules that must be observed at each stage of the technological process are essential prerequisites for shortening the drying time and limiting damage to the material caused by cracks, deformations, color changes, and other specific defects.

References

[1] Trübswetter, T. (2009): *Holztrocknung. Verfahren zur Trocknung von Schnittholz – Planung von Trocknungsanlagen*. 2nd ed., Carl Hanser Verlag, Munich, Germany

[2] Marinescu, I. (1979): *Uscarea lemnului*. Editura Tehnică, Bucharest, Romania

[3] Franke, B.; Franke, S.; Schiere, M.; Müller, A. (2019): *Praxisleitfaden – Beurteilung der Holzfeuchte für Tragwerke und Brücken für die Planung, Errichtung und Nutzung.*. 2nd ed., Berner Fachhochschule, Switzerland

[4] Kollmann, F. F. P. (1951): *Technologie des Holzes und der Holzwerkstoffe. Band 1*. 2nd ed., Springer Verlag, Berlin/Göttingen/Heidelberg

[5] Niemz, P.; Sonderegger, W. (2021): *Holzphysik. Eigenschaften, Prüfung und Kennwerte*. 2nd ed., Carl Hanser Verlag, Munich, Germany

[6] DIN EN 1995-1-1 (Eurocode 5): Bemessung und Konstruktion von Holzbauten

[7] Informationsdienst Holz (2022): *Holzschutz bei Ingenieurholzbauten. Holzbau Handbuch, Reihe 5, Teil 2, Folge 1*. 2nd ed., Studiengemeinschaft Holzleimbau e. V., Wuppertal, Germany

2 Wood and the Drying Medium

Drying is characterized by a continuous interaction between the material to be dried (in this case wood) and the drying medium. Therefore, the factors influencing the drying process can be divided into two major categories (Figure 2.1).

INFLUENCING FACTORS - WOOD DRYING	
Wood Properties	**Properties of the Drying Medium**
• Wood species (structure)	• Type (gaseous / liquid)
• Bulk density	• Temperature
• Wood moisture content	• Pressure
• Equilibrium moisture content	• Absolute humidity (moisture content)
• Shrinkage and swelling	• Relative humidity
• Chemical composition	• Density and specific volume
• Thermal properties	• Enthalpy
• Electrical properties	• Velocity and flow type

Figure 2.1 Classification of factors influencing wood drying

Their influence largely determines the behavior of wood during drying, both in terms of drying time and drying quality and the associated costs. Knowledge of these characteristics facilitates understanding of the principles underlying the development of drying schedules and drying control.

2.1 Important Wood Properties

To understand the drying behavior of wood, it is necessary to know the characteristics of its structure, its physical and mechanical properties, and its chemical composition.

In contrast to other materials, wood is an organic material, inhomogeneous and anisotropic, whose properties vary not only with respect to growing conditions, but also from tree to tree of the same wood species and even within the same tree, along the height and cross-section, and depending on growth or structural defects, etc.

This great variability of wood properties also influences the drying behavior of this material (see Appendix 1), which requires adaptation of the applied drying schedule depending on the wood species, the initial moisture content, and the dimensions of the wood to be dried.

2.1.1 Anatomical Structure

Wood is a complex material consisting of millions of plant cells firmly intergrown with one another. These have various shapes, sizes, and arrangements within the wood mass, and are responsible for three important functions: support (fibers), conduction (tracheids), and storage (vessels, parenchyma).

Softwood consists of 95% axial tracheids. These are elongated cells oriented parallel to the stem axis. The tracheids of earlywood have large bordered pits on the walls, which serve the exchange of liquid from cell to cell. Latewood tracheids have fewer, narrower, more slit-shaped pits. In the living tree, they ensure the conduction of water and nutrients from the roots to the crown. They retain their conducting function even after the tree has been felled, thus facilitating the movement of water in the wood during drying. For this reason, softwoods are considered easy-to-dry wood species (see also Appendix A).

On the other hand, hardwood has a more complex structure consisting of various cell types: fibers – with thick cell walls capable of providing the structural support function, vessels and rays – which ensure water conduction but are extremely thin-walled, and parenchyma cells – in which reserve substances are stored. The alternation of strong and thin-walled anatomical elements makes the drying of hardwoods, especially dense hardwoods, more difficult. Drying of these wood species is frequently accompanied by drying defects such as cracks and distortions, especially when they exhibit both wide and narrow wood rays (as is the case, for example, with oak, hornbeam, or black locust).

Among hardwood species, those with a ring-porous structure (Figure 2.2 a), e.g., oak, ash, black locust, chestnut, present greater drying difficulties than species with a diffuse-porous structure (Figure 2.2 b), e.g., beech, birch, sycamore maple, even if they have similar densities. The less homogeneous structure tends to crack and warp during drying. Therefore, in the case of these wood species, less severe drying schedules must be applied, which leads to an extension of the drying time.

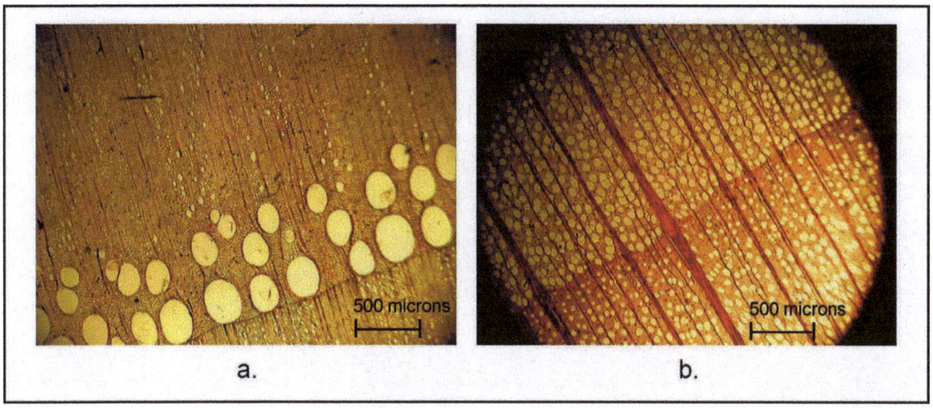

Figure 2.2 Pore arrangement in hardwood: a) ring-porous structure (oak) – cross-section, 40× magnification; b) diffuse-porous structure (beech) – cross-section, 40× magnification (source: [1], with permission)

Trees in the temperate climate zone add one growth ring each year during the warm seasons. This consists of two distinct zones: earlywood – which is formed in spring, and latewood – which develops in summer. The widths of the two zones and their ratio influence drying, because the wide-lumen earlywood is a good conductor of water, while the narrow-lumen latewood has a greater density and dries more slowly. The difference between the two annual ring zones plays an important role in the drying of resinous softwoods and ring-porous hardwoods.

Another characteristic of wood structure that influences its drying is the presence of heartwood (not present in all wood species). Heartwood is known to be formed only in certain woody plants and only from a certain age of the tree onward. Heartwood consists of inactive lignified cells that no longer participate in the conduction of water and the transport of nutrients. The fact that water conduction occurs only through the sapwood is significant for wood drying insofar as the removal of water from it proceeds more easily than from the heartwood. Therefore, species without heartwood, e.g., birch, alder, willow, dry easily and without stress. The only aspects to be considered in their case are the high initial moisture content, which requires a longer time for the removal of free water, and the risk of moisture pockets in the wood, which can lead to uneven drying.

The heartwood of softwoods contains more resins, fats, and tannins, which, as a result of oxidation processes, are the cause of darkening [2]. The situation is more complex in hardwoods. In these, tyloses (Figure 2.3) as an inner lining of the cell cavities. These are fine membranes that seal the vessels airtight. Formation occurs preferentially in ring-porous heartwoods.

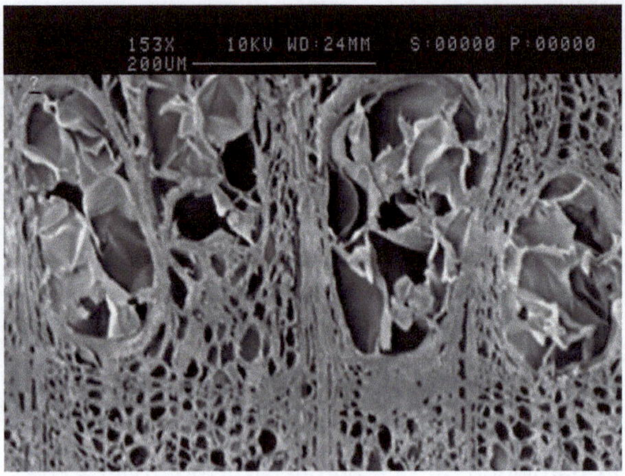

Figure 2.3 Tyloses in heartwood of robinia (cross-section, magnified 153 times) (source: [3], with permission)

The wood resulting from the heartwood formation process has a lower moisture content and a higher density than the wood in the outer zone (sapwood). The less homogeneous structure of wood species with heartwood, e.g. pine, larch, walnut, oak, causes difficult drying in boards that contain both sapwood and heartwood, accompanied by drying stresses and an increased risk of cracking.

The presence of knots is another aspect related to the structure of wood that influences drying quality. Knot wood has a higher density than the surrounding wood. As a result, the knot contracts more strongly during the drying process and tends to crack (Figure 2.4) and sometimes even to become loose from the surrounding wood. The loosening of knots during drying is a common phenomenon, especially when high drying temperatures (>100 °C) are applied.

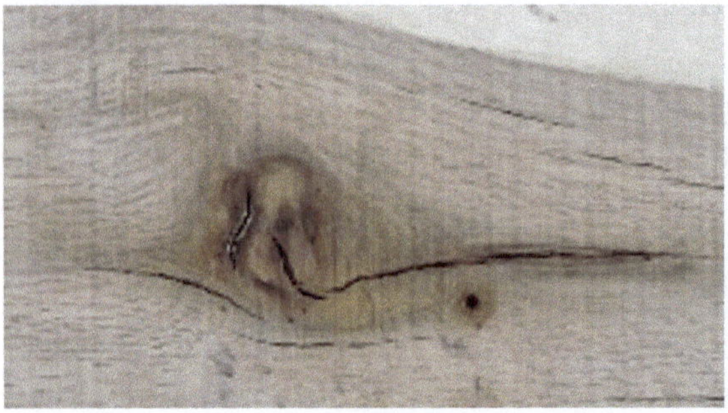

Figure 2.4 Crack formation in the knot and in the surrounding wood

2.1.2 Wood Density

The bulk density (ρ) is, for porous materials such as wood, the ratio between the mass (m) of the wood (including the water contained in the pores) and the external volume (V) of the wood (which, in addition to the cell substance, includes the growth-related cavities), at a defined moisture content [4]:

$$\rho = m/V\,[\text{kg/m}^3] \tag{2.1}$$

Since wood consists of both wood substance and numerous voids that are partially or completely filled with water, water vapor, air, or an impregnation liquid, the wood density can be expressed by various terms.

The following are relevant in drying practice:

- The oven-dry density (ρ_0) is the density of wood in the water-free state (u = 0 %). The oven-dry density is important because it is independent of the wood moisture content and the hygroscopic behavior of the wood. It can therefore be regarded as a material constant and serve as a basis for comparison between wood species. The oven-dry density values of wood range between 100 kg/m^3 and 1300 kg/m^3 [5]. The density varies not only from one wood species to another, but also considerably within a species. Appendix A lists the average oven-dry density values for a few selected European wood species;
- The density of wood at a specific moisture content (ρ_u) is determined using the diagram in Figure 2.5.

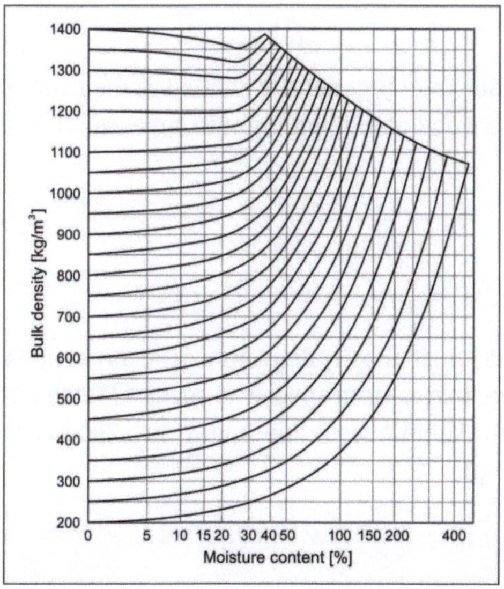

Figure 2.5 Dependence of bulk density on wood moisture content (source: [6])

If one knows that a wood sample at 40% wood moisture content has a density of 500 kg/m³, then one can determine its density at any other wood moisture content (e.g., at 10% it is ρ_{10} = 430 kg/m³).

The density is a particularly important property in wood drying, as it represents the main criterion for the preparation of the drying schedule. Considering that shrinkage coefficients increase with increasing density, it becomes clear that the denser the wood, the more it tends to crack and therefore requires milder drying conditions.

The density affects not only the drying quality but also the drying time. The denser the wood, the more wood substance needs to be dehumidified for a given volume. Furthermore, the diffusion rate decreases with increasing density, and therefore the duration of moisture removal is prolonged. In particular, it is the hardwoods with their thicker cell walls and higher densities that offer greater resistance to water movement in the wood than the loose structure of softwoods [2]. The diffusion rate can be approximated with good accuracy as inversely proportional to the square root of the oven-dry density [7]:

$$\frac{D_2}{D_1} = \left(\frac{\rho_2}{\rho_1}\right)^{1{,}5} \tag{2.2}$$

D_1, D_2 drying times of two different wood species of the same thickness
ρ_1, ρ_2 oven-dry densities of the two wood species

2.1.3 Wood Moisture Content

The wood moisture content (u, or in some books ω) is generally defined as the mass of the total amount of water contained in the wood at a given point in time, relative to the mass of absolutely dry wood:

$$u = \frac{m_u - m_0}{m_0} \cdot 100\,[\%] \tag{2.3}$$

m_u mass of the moist wood [kg]
m_0 mass of the oven-dry wood [kg]

The moisture content of freshly felled wood, which mostly determines the initial moisture content from which drying must begin, varies from species to species (Table 2.1) and can have values between 40 – 60 % (for robinia, ash) and up to 150 – 200 % (for poplar), or even more (600 % for balsa wood).

Table 2.1 Initial Moisture Content of Freshly Felled (Green) Wood of Various Native Wood Species (based on [8], [9])

Class of initial moisture content	Wood species	Wood moisture content in the green state [%]
Moderately wet wood	Heartwood of spruce, pine, larch, Douglas fir, fir (without wet core)	30 - 40
Wet wood	Heartwood of Weymouth pine	85 - 100
	Fir (wet core), ash	40 - 50
	Walnut, robinia	50 - 55
Very wet wood	Beech, oak, birch, maple	80 - 90
	Lime	100
	Alder	115
Extremely wet wood	Elm	90 - 150
	Horse chestnut, Poplar	120 - 180
	Sweet chestnut	120 - 140
	Sapwood of spruce, pine, larch, Douglas fir	130 to over 200

The distribution of water in wood is not uniform. Large moisture differences are sometimes found within a single log, from ground level to the top, or between sapwood and heartwood, particularly in resinous conifers. A large variation in initial moisture content makes drying more difficult and leads to uneven final moisture content.

In order to remove water from wood during a drying process, the forces that bind the water to the wood substance must be overcome and broken. It is already known from wood physics that wood contains free water (in the cell cavities) and bound water (in the cell walls).

The free water bound to the wood substance by weak mechanical forces is removed relatively quickly and easily during drying. In this way, the so-called fiber saturation point is reached, whose value in native wood species lies between 22 % and 35 %, with an average of 28 % (Table 2.2). The fiber saturation point (FSP) describes the moment at which all the free water has been removed from the wood and only bound water remains. It is difficult to determine it as a precise value, and therefore the term fiber saturation range (FSR) is preferred.

The removal of bound water is considerably more difficult due to the much stronger binding forces between the water and the wood substance. In addition, the wood also begins to shrink in this range of moisture reduction, so that the risk of cracking and deformation arises.

After removing the water from the cell lumina, the water bound by capillary condensation is removed during drying. The wood moisture content is thus reduced from the fiber saturation point to approximately 15 %.

This is followed by the removal of water bound by adsorption: the wood moisture content is reduced from 15 % to approximately 6 %. Since much stronger binding forces are involved here, these can only be broken at temperatures above 30 °C. Technical drying below this final moisture value is never required, so that the removal of chemically bound water (below 6 % wood moisture content) is practically never necessary. This situation only occurs during oven drying of test specimens. To break the extremely strong chemical bond, temperatures above 150 °C are required.

Table 2.2 Moisture Content of Various Native Wood Species in the Fiber Saturation Range (based on [4])

Wood species	Fiber saturation moisture content [%]
Diffuse-porous hardwoods without colored heartwood (lime, willow, poplar, alder, beech, birch, hornbeam)	32 - 35
Sapwood of ring- and semi-ring-porous hardwoods with heartwood (robinia, sweet chestnut, oak, ash, walnut, cherry)	
Softwoods without colored heartwood (fir, spruce)	30 - 34
Sapwood of softwoods with colored heartwood (pine, larch)	
Softwoods with colored heartwood and moderate resin content (pine, larch, Douglas fir)	26 - 28
Softwoods with colored heartwood and high resin content (spruce)	22 - 24
Ring-porous and semi-ring-porous hardwoods with mostly pronounced colored heartwood (black locust, sweet chestnut, oak, ash, walnut, cherry)	

Fiber saturation as a physical state is extremely important in wood drying, as it marks the moment at which the air temperature must be increased in order to intensify the heat transfer to the material required to break the bonds between the bound water and the material. Some wood species tolerate the temperature increase even before the fiber saturation range is reached (e.g., spruce, fir, pine), while other wood species must be dried carefully even below fiber saturation in order to minimize the risk of cell collapse and discoloration (e.g., oak). For this reason, a substitute value has been introduced in the control of wood drying, namely the so-called limit moisture content (u_{gr}). This is defined as the wood moisture content below which the drying temperature can be increased without any risk of cracking [14].

The determination of wood moisture content is possible by several methods [11–13]. In wood drying (Table 2.3), electrical measurement methods are most commonly used (resistance method and/or capacitive method) using handheld measuring devices (for measuring initial moisture content, final moisture content) and electrodes (for monitoring the decrease in moisture content during drying). Due to greater effort and delayed results, the oven-dry method is less frequently used in production, but predominantly in research, because it is the most accurate determination method, which can also provide accurate data above the fiber saturation range.

Because the initial moisture content is usually well above the fiber saturation range, the oven-dry method [11] is the only alternative for determining this value with accuracy. To carry out the oven-dry method, a ventilated drying oven is required, the temperature of which can be regulated to 103 ± 3 °C, and a balance with which the mass of the wood sample can be determined to an accuracy of 0.1%. The 20 mm wide wood sample is taken from one of the wettest boards, approximately 0.3 m from the end grain. It is weighed and then oven-dried in the drying oven. Depending on the wood species, oven drying can take up to one week. The oven-dried sample is cooled in a desiccator and then weighed again. The initial moisture content is calculated based on (Equation 2.4).

Table 2.3 Measurement of Wood Moisture Content During Drying

WHEN?	**Before drying**	**During drying**		**After drying**
WHAT?	Determination of initial moisture content	Decrease in moisture content		Final moisture content
		Above fiber saturation	Below fiber saturation	
	Oven-dry method	Oven-dry method (follow-up sample)	Resistance method (electrodes)	Resistance method; capacitive method (handheld measuring devices)
HOW?			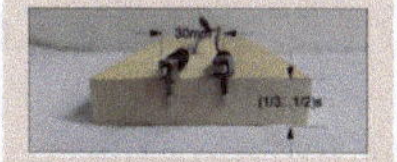	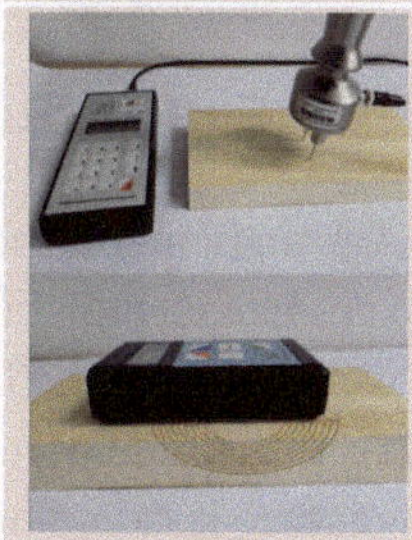

In practice, this step is usually skipped. The moisture measuring electrodes of the automatic control system supply the system with the values of the measured initial moisture contents, so that it can start the drying process even if the values are not entirely accurate. As drying progresses and the wood moisture content approaches the fiber saturation range, the measurement becomes more accurate and reliable.

The continuous measurement/estimation of wood moisture content in drying kilns using electrodes driven in before the start of drying (electrical resistance measurement method [12]) is a component of most control systems in Europe. The measurement based on the capacitive method [13] or using a follow-up sample is less commonly used, mostly only to verify the accuracy of the electrode measurement.

The follow-up sample is a 1.5 – 2 m long piece from one of the wettest boards, with a known initial moisture content. The follow-up sample is weighed and then installed in the stack, using shortened or notched stacking stickers, so that it can be easily removed through the access door during drying for periodic weighing. The current wood moisture content is then determined based on Equation 2.4.

$$u = \frac{m \cdot (u_a + 100)}{m_a} - 100\ [\%] \tag{2.4}$$

m current mass of the follow-up sample [kg]
u_a initial moisture content of the follow-up sample [%]
m_a initial mass of the follow-up sample [kg]

Whether the monitoring of moisture reduction by follow-up samples is representative of the entire kiln charge depends on the number and placement of the samples.

Handheld measuring devices are normally used to verify the final moisture content. Regardless of the measuring method, it is important that the device adjusts the measured value depending on wood species and wood temperature. For capacitive devices, compensation depending on wood thickness is also desirable. For resistive devices, it is important to use insulated pin electrodes so that the measurement is taken at the driven-in depth.

2.1.4 Equilibrium Moisture Content

As already mentioned in Chapter 1, wood is a hygroscopic material that has the ability to change its moisture content by releasing or absorbing moisture from the surrounding environment. This property of wood is attributed to its large internal surface area. The absorption of moisture from the environment is called sorption and the release of moisture is called desorption (Figure 2.6).

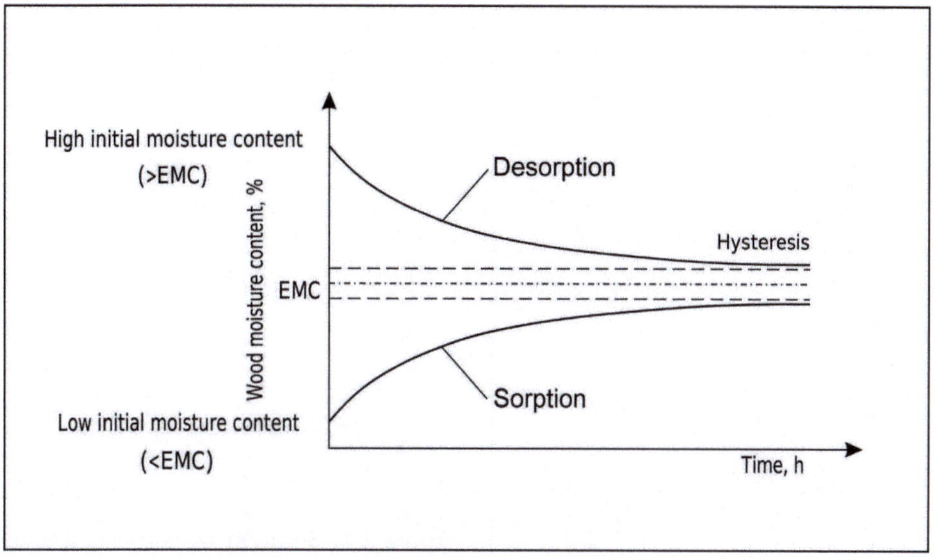

Figure 2.6 Interaction between wood and the environment through sorption and desorption (source: [9], with permission)

Sorption and desorption occur only in the range of bound water. The difference between the moisture content that is established after desorption in an environment with constant parameters and the moisture content that is established after sorption in the same environment is called hysteresis. Its value is approximately 2.5 % for sawn timber and 0.1 – 0.2 % for small pieces (veneers, chips). The hysteresis occurs because the desorption curve (e.g., when wood is dried from 30 % to 0 % moisture content) and the sorption curve (when it is re-moistened from 0 % to 30 %) are not congruent but form a loop (Figure 2.7). The sorption and desorption dynamics also depend additionally on temperature (Figure 2.8).

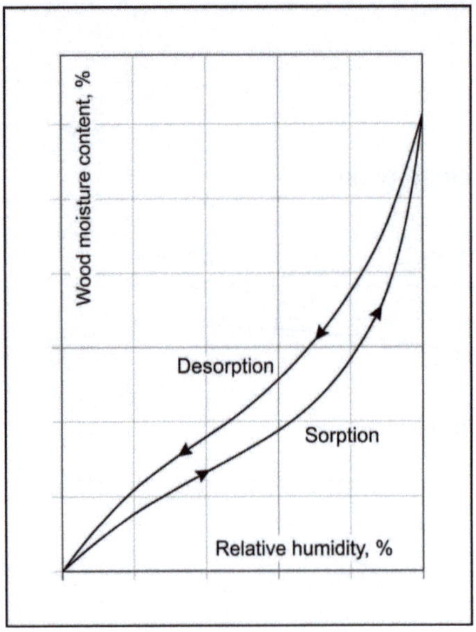

Figure 2.7 Sorption isotherms (based on [14])

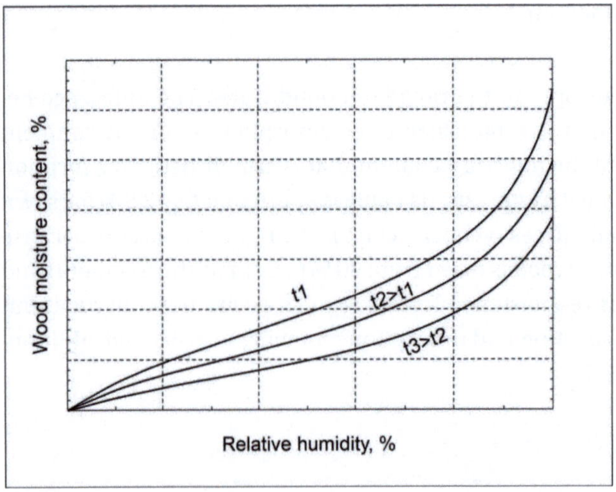

Figure 2.8 Dependence of sorption isotherms on temperature (based on [15])

When wood is kept long enough in an environment with constant parameters (temperature and relative humidity), it tends to change its moisture content to reach equilibrium with the environment. This moisture state of the wood is called the equilibrium moisture content (*EMC* – see Figure 2.6). Therefore, during drying, the air parameters must be selected and controlled such that the resulting equilibrium moisture content is always below the desired final moisture content of the wood, so that the wood is stimulated to release moisture.

The equilibrium moisture content depends mainly on the state parameters of the drying medium, but is also dependent on the wood species (Table 2.4). For practical purposes, the equilibrium moisture content can be determined with good approximation based on the diagram and the table in Appendix B.

Table 2.4 Equilibrium Moisture Contents of Various Wood Species (source: [16])

Initial moisture content class	Wood species	Wood moisture content in the green state [%]
Moderately wet wood	Heartwood of spruce, pine, larch, Douglas fir, fir (without wet core)	30 - 40
Moist wood	Heartwood of eastern white pine	85 - 100
	Fir (wet core), ash	40 - 50
	Walnut, black locust	50 - 55
Wet wood	Beech, oak, birch, maple	80 - 90
	Linden	100
	Alder	115
Very wet wood	Elm	90 - 150
	Horse chestnut, poplar	120 - 180
	Sweet chestnut	120 - 140
	Sapwood of spruce, pine, larch, Douglas fir	130 to over 200

2.1.5 Swelling and Shrinkage

The sorption and desorption of wood are accompanied by changes in the dimensions of the wood: during sorption the wood swells and during desorption it shrinks. The swelling and shrinkage of wood occur only below the fiber saturation point, when the content of bound water changes. However, there are wood species, e.g. beech, in which shrinkage begins at 50 % wood moisture content, probably because the release of moisture from the cell walls starts before the lumina are completely emptied [17].

During drying, shrinkage plays a particularly important role, because by reducing the wood moisture content to final moisture contents of 6 – 20 % it is unavoidable, and the result is a reduction in the dimensions of the wood part – particular attention must be paid to the thickness. The determination of the shrinkage coefficient (β) is described in [18]. The following applies:

$$\beta = \frac{a_{u2} - a_{u1}}{a_{u2}} = 100\ [\%] \qquad (2.5)$$

a_u dimension of the wood part at arbitrary moisture contents ($u_2 > u_1$; $u_2 \leq u_{fs}$) [m]

Therefore, when drying sawn timber, it is necessary to provide a thickness allowance (oversize) in order to maintain the required thickness after drying.

Because wood is anisotropic, it shrinks differently in the three principal anatomical directions. Axial (longitudinal) shrinkage (β_l) is the smallest and ranges between 0.1 – 0.3 %. Shrinkage in the radial direction (β_r) is considerably greater, with values between 2.3 – 7.8 %. Wood shrinks most strongly in the tangential direction to the annual rings. The values of tangential shrinkage (β_t) range between 5.0 – 12.0 %. This results in the wood component deforming to a greater or lesser extent depending on its position across the cross-section of the tree trunk (Figure 2.9).

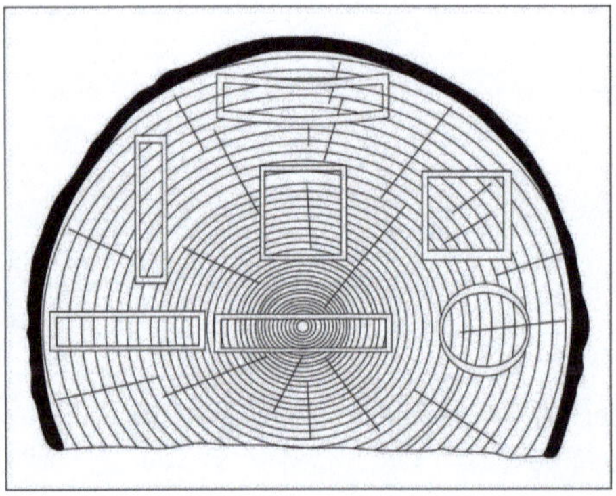

Figure 2.9 Deformation tendency of sawn timber parts depending on their position in the trunk (source: [4])

The volumetric shrinkage is calculated from the shrinkage values of the individual principal directions (β_v):

$$\beta_v = 100 - \frac{(100 - \beta_l) \cdot (100 - \beta_r) \cdot (100 - \beta_t)}{10000}\ [\%] \qquad (2.6)$$

Shrinkage anisotropy (A_β) is defined as the ratio between the tangential and radial drying shrinkage coefficients, where the drying shrinkage coefficients (β_{tN} and β_{rN}) represent the percentage shrinkage during drying from the green state to the standard climate of 20 °C/65 % (wood moisture content of 11 – 13 %). The following applies:

$$A_\beta = \frac{\beta_{tN}}{\beta_{rN}} \qquad (2.7)$$

Shrinkage anisotropy varies among different wood species: for example, poplar, maple, beech, and oak exhibit high shrinkage anisotropy, in contrast to birch, walnut, mahogany, and lime. The higher the shrinkage anisotropy, the more the wood tends to warp (Figure 2.10) and to crack during drying. Proper stacking and the use of press systems are solutions to minimize this tendency.

Figure 2.10 Effects of shrinkage anisotropy: warping of the upper layers of a stack that are free to shrink (source: [10])

Based on the values of shrinkage coefficients and bulk density (see Appendix A), one can already assess what difficulties are to be expected when drying a particular wood species. However, it must be taken into account that all shrinkage values have been measured according to standards on small, defect-free, conditioned samples. In industry, however, one deals with boards that also have knots and grain deviations, are mostly not strictly oriented in the grain direction, so that an accurate specification of the reduction in dimensions is difficult to predict.

2.1.6 Chemical Composition

Dry wood consists of carbon (approximately 50 %), oxygen (43 %) and hydrogen (6.1 %). The remainder consists of nitrogen and mineral components (ash). The elemental composition is practically the same for all wood species [6].

The main chemical components of wood are cellulose, which has the largest share at approximately 40 – 60 %, hemicelluloses and lignin. The secondary components, such as minerals, resins, oils, proteins, tannins and dyes, are present only in smaller quantities [2].

Of the main chemical components of wood, the hemicelluloses are the least stable. They begin to decompose at temperatures above 120 °C. This is the basis for the plasticization of wood during steaming and the stabilization of wood during thermal treatment (e.g., thermally modified wood). Since, however, conventional drying takes place at temperatures below 100 °C, it can be stated that the chemical composition of wood does not change during drying.

Therefore, the influence of the chemical composition of wood during drying is reduced to the reactions of certain secondary chemical substances in the wood, e.g., tannins, starch, sugars, etc., which in the presence of free water in the wood and atmospheric oxygen can cause the occurrence of certain undesirable discolorations during drying.

For example, alder wood tends to oxidative discoloration due to its high starch contentand takes on an orange color as soon as it comes into contact with atmospheric oxygen. Drying immediately after felling at a temperature of 50 – 60 °C and low relative humidity (approximately 40 %) inhibits this tendency by rapidly lowering the wood moisture content below the fiber saturation point. Once the free water is removed from the wood, no further discoloration occurs.

Oak wood tends to darken in the marginal zones, caused by the oxidative reaction of the tannins it contains. The reaction takes place above the fiber saturation range when the temperature is raised above 30 °C. To avoid discoloration, the temperature must be kept low, but this extends the drying time.

Chemically speaking, wood is an acidic substance, with a pH value of 5 – 6 or even lower (e.g., oak has pH = 4). This results in a considerable risk of corrosion for various metals. For this reason, only corrosion-resistant materials such as aluminum and stainless steel are used for the entire interior fittings of drying chambers.

2.1.7 Thermal Properties

The thermal properties of wood are:

- the thermal conductivity λ [W/mK];
- the specific heat capacity c [J/kgK];
- the thermal diffusivity a [m^2/s].

The thermal conductivity (λ) is the quantity of heat that flows through a cube with a side length of 1 m per unit time when a temperature difference of 1 K exists between the two opposite faces [4]. Thermal conductivity characterizes the thermal insulating ability of a material. The smaller it is, the better the material insulates, and wood is a good insulating material thanks to its high pore fraction.

The thermal conductivity of wood depends on the bulk density, the wood moisture content and the wood temperature. It increases with increasing bulk density, decreases with decreasing moisture content (Figure 2.11) and increases with increasing temperature. The influence of temperature is greater the higher the pore fraction. Furthermore, due to the anisotropy of wood, this property of wood is also influenced by the grain direction. The thermal conductivity perpendicular to the grain ($\lambda_\perp$) is smaller than that parallel to the grain ($\lambda_\parallel$) (Equation 2.8 and Equation 2.9), and in the tangential direction the λ value is about 10% lower than in the radial direction.

For the thermal conductivity of wood with a moisture content of 12%, the following applies [4]:

$$\lambda_\perp = 0{,}026 + 0{,}195 \cdot \rho \cdot 10^{-3} \text{ [W/mK]} \tag{2.8}$$

and

$$\lambda_{\parallel} = 0{,}026 + 0{,}46 \cdot \rho \cdot 10^{-3}\ [\text{W/mK}] \tag{2.9}$$

ρ bulk density of wood [kg/m³]

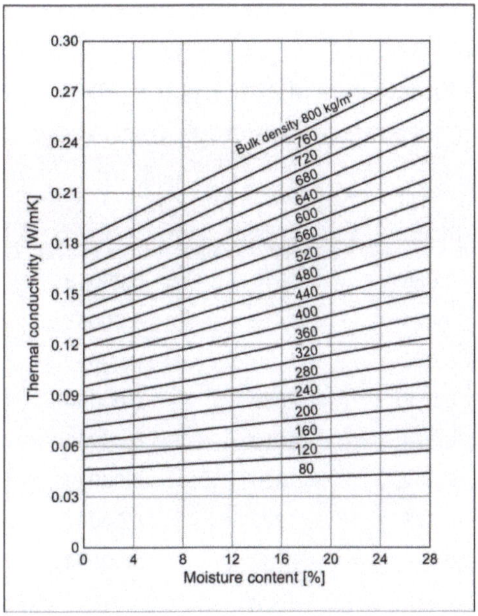

Figure 2.11 Thermal conductivity of wood perpendicular to the grain direction as a function of bulk density and moisture content (source: [16])

The specific heat capacity (c) is the amount of heat required to raise the temperature of 1 kg of a material by 1 K. It depends on the wood temperature and the wood moisture content and can be determined using the following equation [9]:

$$c = 1{,}172 \cdot \left[u\left(1 + \frac{T}{100}\right)\right]^{0{,}2}\ [\text{J/kgK}] \tag{2.10}$$

u wood moisture content [%]
T wood temperature [K]

The thermal diffusivity (a) plays an important role in drying, as it determines the rate of heat conduction in wood. The following applies:

$$a = \frac{\lambda}{c \cdot \rho}\ [\text{m}^2/\text{s}] \tag{2.11}$$

λ thermal conductivity [W/mK]
c specific heat capacity [J/kgK]
ρ bulk density [kg/m³]

The thermal diffusivity therefore depends on the bulk density of the wood and, through the thermal conductivity, also on the wood moisture content, the wood temperature, and the grain direction.

The thermal properties of wood as a function of density and wood moisture content are presented in Table 2.5.

Table 2.5 Thermal Properties of Wood at a Temperature of 27 °C (source: [6])

Oven-dry density [kg/m³]	Moisture content [%]	Density [kg/m³]	Thermal conductivity [W/mK]	Specific heat capacity [J/kgK]	Thermal diffusivity [m²/s]
200	10	217	0.066	1610	0.00068
	20	233	0.074	1810	0.00063
	30	252	0.082	1990	0.00059
	50	287	0.098	2280	0.00054
600	10	627	0.144	1610	0.00051
	20	657	0.162	1810	0.00049
	30	690	0.178	1990	0.00047
	50	776	0.216	2280	0.00044

2.1.8 Electrical Properties

The electrical properties of wood relevant to wood drying are:

- the electrical resistance R [Ω];
- the dielectric constant ε [F/m];
- the loss factor tgδ.

They play an important role in wood moisture measurement using resistive or capacitive instruments and in the context of electrical drying methods, e.g. in high-frequency drying.

The electrical resistance of wood (R) is the resistance that wood opposes to the passage of electric current. Wood moisture content has a decisive influence on the electrical resistance. From 0 % to the fiber saturation range, an almost linear relationship exists between wood moisture content and electrical resistance (Figure 2.12). Therefore, in this range the resistive measurement method is precise. Above the fiber saturation range, the influence of wood moisture content is significantly lower and therefore also more difficult to measure accurately.

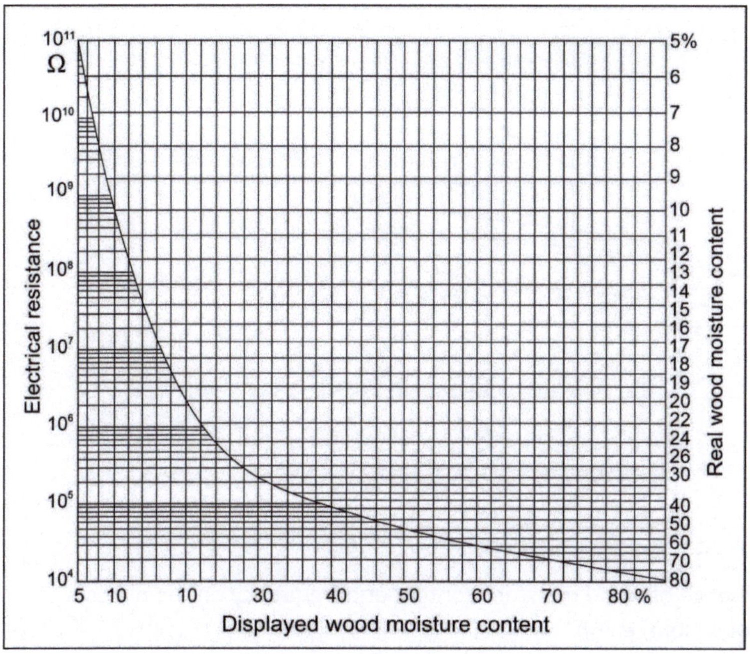

Figure 2.12 Influence of wood moisture content on the electrical resistance of wood at a wood temperature of 20 °C (based on [19])

The dielectric constant (ε) indicates by what multiple the dielectric conductivity of a material (e.g. wood) is greater than that of a vacuum. It depends on the wood moisture content [6], and the capacitive method for measuring wood moisture content is based on this dependency function.

The loss factor (tgδ): When wood is placed between the plates of an alternating current capacitor, it heats up due to the so-called dielectric loss. The loss factor (tgδ) characterizes the intensity of the heating. Therefore, it is desirable to have it as high as possible during drying.

Both the dielectric constant and the loss factor increase with increasing wood moisture content and decrease with increasing frequency (Figure 2.13). They have different values in the three principal anatomical directions and also depend on the wood species and wood density (Table 2.6).

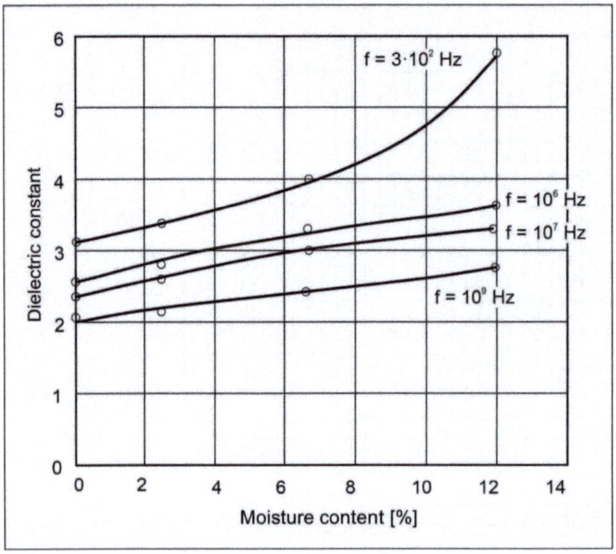

Figure 2.13 Influence of wood moisture content and frequency on the dielectric constant of beech wood (source: [6])

Table 2.6 Dependence of the Dielectric Properties of Various Wood Species on the Fiber Direction and on the Wood Species (Values for Oven-Dry Wood, at a Frequency of 1 kHz) (Source: [6])

Wood species	Dielectric constant ε			Loss factor tg$\delta \cdot 10^{-4}$		
	along	radial	tangential	along	radial	tangential
Fir	3.06	1.98	1,91	625	345	310
Book	3,18	2,20	2,40	585	298	319
Oak	2.86	2,30	2,46	465	285	325

2.2 Important Properties of the Drying Medium

The liquid or gaseous medium that surrounds the material (wood) during the drying process is referred to as the drying medium referred to. Its task is to supply heat to the material and to absorb the moisture extracted from it and remove it from the wood surface. In drying practice, different drying media can be used, which influence both the duration and the quality of drying through their parameters (temperature, humidity, pressure).

One can use the following agents as drying media:

- moist air;
- superheated steam;
- flue gases;
- hot oils.

Moist air is the most commonly used drying medium, characterized by a temperature below 100 °C, a relative humidity in the range of 20 – 95 % and atmospheric pressure (1 bar). Its most important properties and their influence on the drying process are discussed in Section 2.2.2 explained in detail.

In contrast to moist air, all other drying media mentioned above are characterized by temperatures above 100 °C and thus shorter drying times.

2.2.1 Drying Media for High-Temperature Drying

2.2.1.1 Superheated Steam

Superheated steam (also called superheated vapor) is steam whose temperature is higher than the saturation temperature at a given pressure, or whose pressure is lower than the saturation pressure at a given temperature.

Superheated steam is used as a drying medium in the following processes:

- In high-temperature drying, when the temperature inside the plant is raised to 105 – 130 °C at atmospheric pressure (1 bar). In this process, the dry air contained in the mixture is expelled from the plant and only the steam remains in the chamber, which at this temperature transforms into superheated steam (Figure 2.14). It is an intensive drying method that can only be used for softwoods (fir, spruce, pine). The main parameters of superheated steam at normal pressure (1 bar) as a function of temperature are listed in Table 2.7.
- In the combined V/VS process (drying under vacuum and superheated steam), when the pressure is reduced below atmospheric level, which lowers the boiling point of water to below 100 °C (e.g., at 0.2 bar the boiling temperature of water is 60 °C).

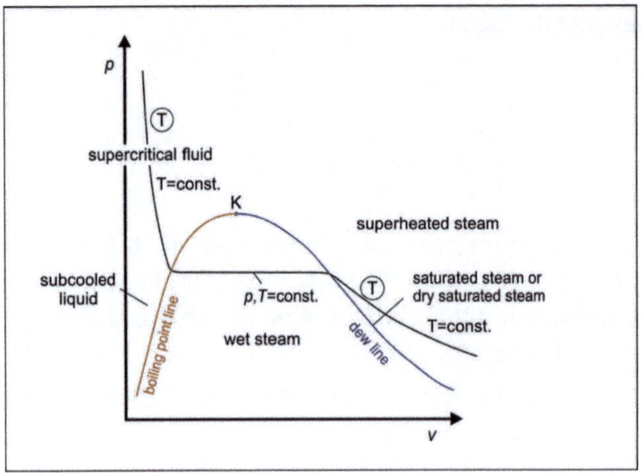

Figure 2.14 *p-v*-diagram (based on [20])

Table 2.7 Parameters of Superheated Steam at Normal Pressure (1 bar) [9]

Temperature [°C]	Specific volume [m^3/kg]	Density [kg/m^3]	Relative humidity [%]	Enthalpy [kJ/kg]	Specific heat [kJ/kg°C]
105	1.750	0.570	83	2685.8	2.035
110	1.780	0.562	70	2697.1	2.031
120	1.830	0.546	50	2717.2	2.018
130	1.880	0.533	37	2738.2	2.010
140	1.926	0.520	27	2756.6	1.997
160	2.023	0.495	16	2795.9	1.985
180	2.119	0.475	10	2814.4	1.976
200	2.214	0.452	7	2874.7	1.980

The use of superheated steam as a drying agent is limited by the nature of the material to be dried; it must be possible to heat it to temperatures above the boiling point of water without damage.

It should also be noted that at constant pressure (p = 1 bar) the relative humidity decreases rapidly with increasing temperature. This considerably increases the tendency to crack. This disadvantage is eliminated in vacuum drying by the reduction of pressure, since the reduction of the boiling point of water allows lower temperatures and thus mild drying conditions. In addition, the reduction in pressure also influences a reduction in the resistance that the wood tissue offers to the passage of water vapor (diffusion) (Figure 2.15), which also contributes to high-quality drying, even for hard hardwoods.

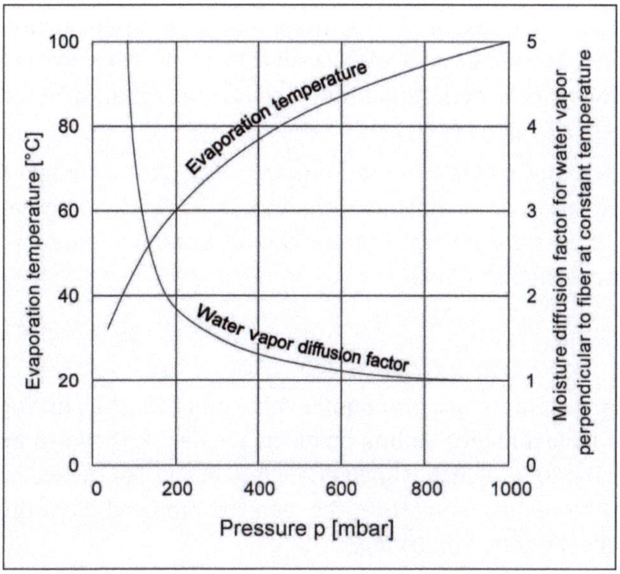

Figure 2.15 Influence of negative pressure on the evaporation temperature and on the mobility factor for water vapor inside the wood during vacuum drying (based on [10])

2.2.1.2 Flue Gases

Flue gases was the first drying agent used; primarily for the artificial drying of building materials, including wood. Their advantage lies in their very high thermal efficiency and in the fact that they are obtained by burning waste, thus enabling better utilization of wood as a raw material.

The combustion gases used in wood drying technology are produced in the furnaces of steam boilers (exhaust gases) or in special furnaces. For their use as a drying agent, prior cooling is necessary, since they have a temperature of approximately 600 °C when leaving the furnace.

The wood fuel used to produce combustion gases is characterized by a low sulfur content, a low ash content (1 – 2 %), a high proportion of volatile substances (85 %), and generally a relative humidity of 30 %. Its chemical composition in mass percent is represented as follows:

$$C\ (35.3\ \%) + H\ (4.2\%) + O\ (29.4\%) + N\ (0.4\%) + A\ (0.7\%) + X\ (30\%) = 100\%$$

Ash (A) and water (X) represent the non-combustible part or ballast of the fuel, and the sum of the other elements represents the combustible mass.

In the case of complete combustion the combustion gases contain nitrogen (N), oxygen (O), carbon dioxide (CO_2) and water vapor (H_2O). In the case of incomplete combustion – not recommended, but easier to achieve – combustion gases also contain carbon, ash, soot, and gases such as carbon monoxide (CO), hydrogen (N), and various hydrocarbons. The use of incompletely burned gases leads to a superficial blackening of the wood – the layer that is completely removed by planing – and an impregnation of the wood with an unpleasant smell of smoke.

The combustion gases used as drying agents must not contain any smoke, so complete combustion with a significant excess of air must be achieved. In addition, the gases must not contain sparks and, if possible, no ash. The presence of a certain amount of CO_2 in the drying system reduces the risk of fire.

2.2.1.3 Hot Oils

Although it may seem paradoxical, there are also liquids that can be used as drying agents. Certain vegetable oils have a higher boiling point than water, so they can be heated to a temperature of 120–140 °C, which is high enough to evaporate the water from the wood immersed in the hot oil bath. Thus, the water is removed from the wood, which corresponds to the definition of drying.

Fundamentally, this is also a superheated steam drying process, with the oil acting as a heat transfer medium.

Due to the pressure difference that arises between the cell cavities inside the wood emptied of water and the wood surface, the oil penetrates the cell cavities of the wood and impregnates it. Thus, a combined drying and impregnation process takes place, with the aim of increasing the durability of wood used outdoors (e.g., railway sleepers).

2.2.2 Moist Air and Its Properties As a Drying Medium

Moist air is a mixture of dry air (clean air) and water vapor. As long as it is unsaturated, it can be used as a drying medium. The parameters that characterize the respective state of moist air are:

- temperature (t, or in some books also symbolized by ϑ);
- partial pressure of clean air (p_L), partial pressure of water vapor (p_D) and saturation pressure (p_{DS});
- absolute humidity (also called moisture content) (x);
- relative humidity (φ);
- density (ρ) and specific volume (v);
- enthalpy (also called heat content) (h).

2.2.2.1 Temperature

The temperature (t or ϑ) is the main accelerating factor of the drying process, as it favorably influences both the evaporation rate of water from the wood surface and the diffusion rate of water and water vapor inside the wood.

At atmospheric pressure, the following applies: the higher the temperature of the drying medium, the higher its moisture absorption capacity (Figure 2.16). In this case, less air needs to be exhausted from or supplied to the system and heated, which shortens the drying time.

The risk of cracking and deformation increases with increasing temperature. Therefore, the maximum value to which the temperature can be raised during the drying of a charge is limited. It depends mainly on the wood species (density) and wood thickness (e.g., for oak boards 25 – 40 mm thick, t_{max} = 70 °C should under no circumstances be exceeded; for thicker boards, t_{max} = 60 °C should apply).

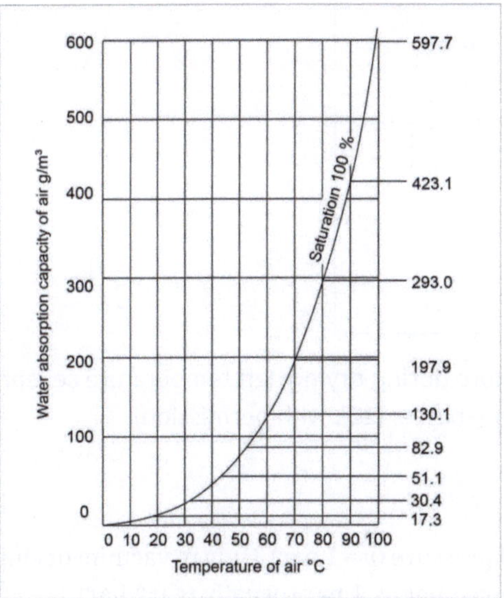

Figure 2.16 Water absorption capacity of air, as a function of temperature (based on [21])

The drying temperature also depends on the required quality (permissible proportion of cracks and discoloration). Criteria that additionally limit the temperature are:

- color preservation for light-colored wood species (maple, ash, alder, beech): $t_{max} \leq$ 45 °C;
- avoidance of excessive resin exudation (Douglas fir): $t_{max} \leq$ 60 °C;
- avoidance of reddish edge discoloration in oak: t_{max} = 30 °C at the beginning of drying.

During drying, not only the ambient temperature is of particular importance, but also the temperature at the wood surface, which in certain situations can deviate from the air temperature, e.g. when the heating pipes are located near the wood stacks or in the case of direct sunlight during open-air drying. Overheating of the wood surface leads to rapid removal of free water from the surface layers and thus to case hardening of the surface and to cracks.

Usually, the air temperature in the drying chamber is measured and controlled. Only rarely is the wood temperature also measured, although this would provide a much more accurate indication for the correct management of the drying process.

For measuring the air temperature during drying, resistance thermometers are commonly used (Figure 2.17).

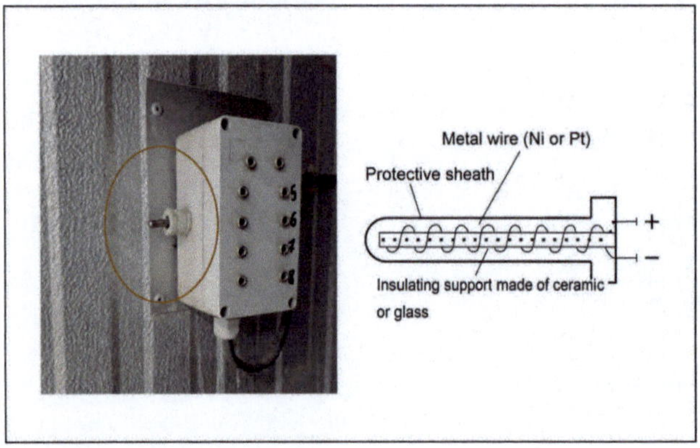

Figure 2.17 Measurement of air temperature during drying: left temperature sensor; right structure of a resistance thermometer (source: [22], with permission)

2.2.2.2 Pressure

Most drying processes take place at normal pressure ($p = 1$ bar). Only in vacuum drying is the pressure in the drying chamber reduced below 1 bar (usually to 0.2 bar).

The total pressure (p) of moist air as a water vapor–air mixture can be expressed using Dalton's law as the sum of the partial pressures of pure air (p_L) and water vapor (p_D):

$$p = p_L + p_D \text{ [N/mm}^2\text{]} \tag{2.12}$$

In the temperature range from –30 °C to +150 °C, pure air, water vapor and thus their mixture (moist air) can be regarded as ideal gases and calculated from Clapeyron's equation:

$$p_L \cdot V = m_L \cdot R_L \cdot (t + 273) \tag{2.13}$$

and

$$p_D \cdot V = m_D \cdot R_D \cdot (t + 273) \tag{2.14}$$

V volume of the mixture [m³]
m_L mass of pure air [kg]
m_D mass of water vapor [kg]
$R_L R_D$ gas constants (R_L =287.1 J/kg·K; R_D = 461.5 J/kg·K)
t air temperature [°C]

The partial pressure of water vapor can take all values between zero (absolutely dry air) and the saturation vapor pressure. This depends only on temperature (Figure 2.18).

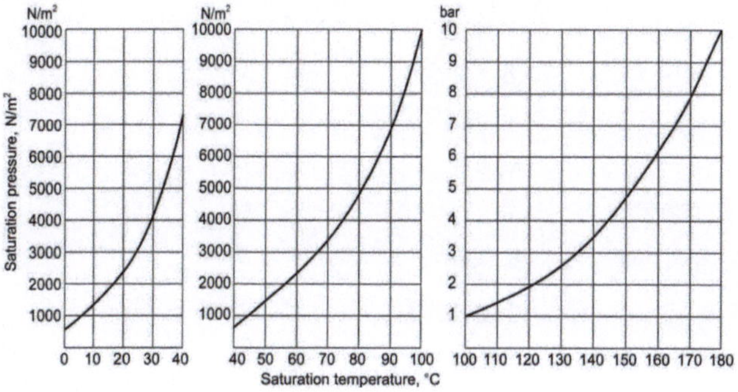

Figure 2.18 Saturation pressure of water vapor as a function of temperature (source: [9])

2.2.2.3 Absolute Humidity

The absolute humidity, also called moisture content (x), is the ratio of the water vapor mass (m_D) to the dry air mass (m_L). From the division of Equation 2.14 and Equation 2.13 the following is obtained:

$$x = \frac{m_D}{m_L} = \frac{p_D \cdot R_L}{p_L \cdot R_D} = 0{,}622 \cdot \frac{p_D}{p - p_D} \text{ [kg/kg]} \tag{2.15}$$

A special feature of moist air is that it can only absorb water vapor until the partial pressure (p_D) equals the saturation pressure (p_{DS}) (see Figure 2.18).

For numerical calculations in which precise values of the saturation pressure are required, one may use Table 2.8 prove more practical.

Table 2.8 Saturation Pressure of Water Vapor As a Function of Temperature (source: [9])

Temperature [°C]	Saturation pressure [N/m²]
40	7,375
45	9,584
50	12,335
55	15,740
60	19,917
65	25,010
70	31,170
75	38,550
80	47,360
85	57,810
90	70,110
95	84,570
100	101,325

In the saturation state, the following applies:

$$x_s = 0{,}622 \cdot \frac{p_{DS}}{p - p_{DS}} \text{ [kg/kg]} \tag{2.16}$$

Thus x_s is the maximum amount of moisture that can be present in the air in vapor form at a given pressure (p) and a given temperature (t).

If $x < x_s$ one speaks of unsaturated moist air. This can be used as a drying medium because it has the ability to absorb the moisture removed from the wood. As soon as x reaches the value x_s, however, the air is saturated and can no longer serve as a drying medium, but must first be dehumidified.

2.2.2.4 Relative Humidity

The relative humidity (φ) is defined as the ratio between the partial pressure (p_D) of the water vapor present in the air and the saturation pressure (p_{DS}):

$$\varphi = \frac{p_D}{p_{DS}} \cdot 100\,[\%] \tag{2.17}$$

The relative humidity thus represents the degree of saturation of pure air with water vapor. It takes values between 0 % (pure air) and 100 % (air saturated with water vapor).

In contrast to temperature, relative humidity is an inhibiting factor for drying: the higher the humidity, the longer the drying takes. Therefore, the relative humidity of the drying medium should be kept as low as possible in order to promote the reduction of wood moisture content. However, at the beginning of the process the humidity must be kept at high values (85 – 90 %) to prevent a sudden decrease in wood moisture from the surface layers and to avoid drying defects. After the entire amount of free water has been removed and the moisture content of the wood falls below the fiber saturation range, the relative humidity can be gradually reduced (toward a value of 30 – 40 %).

The measurement of relative humidity during drying is carried out using the so-called psychrometer (Figure 2.19). This consists of two identical thermometers: one is called the dry-bulb thermometer and measures the air temperature in the drying plant; the second is called the wet-bulb thermometer (or wet thermometer) and is covered with an absorbent fabric ("sock") that is continuously moistened. In the airstream, heat is extracted from the wet-bulb thermometer as a result of the evaporation that sets in, and thus its reading (the so-called wet-bulb temperature) is lower than the dry-bulb temperature.

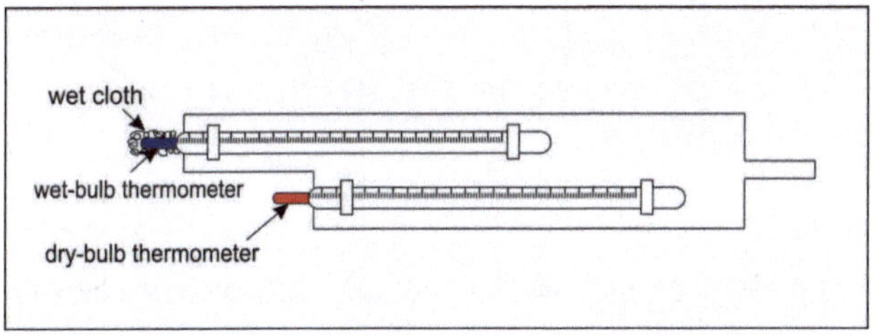

Figure 2.19 Psychrometer

The difference ($\Delta t = t - t_{wet}$) is called the psychrometric difference and it is a measure of the relative humidity:

$$\varphi = \frac{p_{DS\_wet} - C \cdot \Delta t}{p_{DS\_dry}} \; [\%] \tag{2.18}$$

p_{DS_wet} saturation pressure of water vapor at the wet-bulb temperature [N/m^2]
p_{DS_dry} saturation pressure of water vapor at the dry-bulb temperature [N/m^2]

Both p_{DS_wet} and p_{DS_dry} are taken from Table 2.8, but for different temperatures (p_{DS_wet} for t_{wet} and p_{DS_dry} for t).

C is the psychrometric coefficient, which depends on the air velocity (w):

$$C = \left(65 + \frac{6{,}75}{w}\right) \tag{2.19}$$

The relationship between the three quantities (t, Δt and φ) is also represented in the so-called psychrometric table (Table 2.9).

Table 2.9 Psychrometric Table for Determining the Relative Humidity When Using a Psychrometer As a Measuring Instrument (source: [9])

Dry-bulb temperature [°C]	Relative humidity at a psychrometric difference °C [%]																
	1	**2**	**3**	**4**	**5**	**6**	**7**	**8**	**9**	**10**	**11**	**12**	**13**	**14**	**16**	**18**	**20**
30	92	85	78	71	65	58	52	46	41	–	–	–	–	–	–	–	–
32	93	86	80	74	68	62	57	51	46	41	–	–	–	–	–	–	–
35	93	87	81	75	69	64	59	54	49	44	39	–	–	–	–	–	–
38	94	88	82	76	70	65	60	55	51	46	41	37	–	–	–	–	–
40	94	89	82	77	72	67	62	58	53	49	45	41	37	34	27	22	15
41	94	88	82	77	72	67	63	58	54	50	46	42	38	34	28	22	16
42	94	89	83	78	73	68	63	59	54	50	46	42	39	35	29	23	17
43	94	89	83	78	73	68	64	59	55	51	47	43	40	36	30	23	17
44	94	89	83	78	73	69	64	60	56	52	48	44	41	37	31	24	18
45	94	89	84	79	74	69	64	60	56	52	48	45	41	38	31	25	19
46	94	89	84	79	74	69	65	61	57	53	49	45	42	38	32	26	20
47	95	89	84	79	74	70	66	61	57	53	49	46	42	39	33	27	21
48	95	89	84	79	75	70	66	62	58	54	50	46	43	40	34	28	22
49	95	90	84	80	75	70	66	62	58	54	50	46	43	40	34	28	22
50	95	90	84	80	75	70	66	62	58	54	51	47	44	40	34	29	23

Dry-bulb temperature [°C]	Relative humidity at a psychrometric difference °C [%]																
53	95	90	85	80	75	71	67	63	59	56	52	49	45	42	36	31	25
55	95	90	85	81	76	72	68	64	60	56	53	50	47	44	38	32	26
58	95	90	85	81	77	73	69	65	61	58	54	51	48	45	39	33	28
60	95	90	85	82	77	73	69	66	62	58	55	52	49	46	40	34	29
63	96	91	86	82	78	74	70	66	63	60	56	53	51	47	41	36	30
65	96	91	86	82	78	74	71	67	64	60	57	54	51	48	42	37	31
68	96	91	87	83	79	75	71	68	64	61	58	55	52	49	43	38	33
70	96	91	87	83	79	75	72	68	65	62	59	56	53	50	44	39	33
73	96	91	87	83	80	76	72	69	66	62	59	57	54	51	45	40	35
75	96	91	88	84	80	76	73	69	66	63	60	57	54	52	46	41	36
78	96	91	88	84	80	77	73	70	67	64	61	58	55	53	47	42	37
80	96	91	88	84	81	77	74	70	67	65	62	59	56	53	48	43	38
85	96	91	88	85	81	78	75	71	68	65	63	60	56	55	49	44	40
90	96	91	89	85	82	78	75	72	69	66	64	61	59	56	51	46	41
95	96	92	89	86	82	79	75	73	70	68	65	62	60	57	52	48	43
100	96	93	90	86	83	80	77	74	71	68	66	63	61	58	54	49	44

The psychrometer of a drying kiln shows at a certain point in time during drying: t = 70 °C and t_{wet} = 65 °C. The relative humidity corresponding to this condition is φ = 79 %.

Psychrometers should be used for measuring air temperature and humidity during drying when careful maintenance can be ensured and when the air velocity in the drying chamber is at least 2 m/s.

However, the preferred alternative today is to measure the equilibrium moisture content (*EMC*) of the air instead of the relative humidity. For measuring the equilibrium moisture content during the drying process, special sensors are used that have as their actual sensing element a thin plate made of a highly hygroscopic material (Limba veneer or cellulose) (Figure 2.20). This plate has the property of coming almost instantaneously into equilibrium with the surrounding environment. It is positioned closely between two stainless steel electrodes, with the entire assembly mounted beneath a protective plate on the wall of the drying plant to prevent the ingress of condensation water. The measuring principle is resistive, i.e. it is based on the change in electrical resistance of the hygroscopic plate as a function of its moisture content.

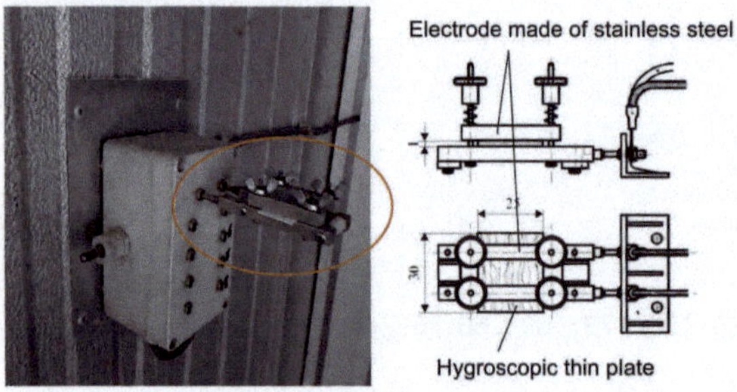

Figure 2.20 Measurement of air humidity during drying: left *EMC*-sensor; right assembly of a *EMC*-sensor (source: [22], with permission)

The transition from the data of the psychrometric measurement (t, φ) to those with temperature and *EMC*-sensors is carried out using the graph in Figure 1.1 or the table in Appendix B.

2.2.2.5 Density and Specific Volume

As in the case of total pressure, the density (ρ) of air is calculated as the sum of the dry air density (ρ_L) and the water vapor density (ρ_D):

$$\rho = \rho_L + \rho_D \; [\text{kg/m}^3] \tag{2.20}$$

The partial densities ρ_L and ρ_D can be expressed from Clapeyron's equations (Equation 2.13) and (Equation 2.14):

$$\rho_L = \frac{m_L}{V} = \frac{p}{R_L \cdot (t + 273)} \text{ [kg/m}^3\text{]} \tag{2.21}$$

and

$$\rho_D = \frac{m_D}{V} = \frac{p}{R_D \cdot (t + 273)} \text{ [kg/m}^3\text{]} \tag{2.22}$$

By summation one obtains:

$$\rho = \frac{p}{R_L \cdot (t + 273)} + \frac{p}{R_D \cdot (t + 273)} = \frac{0{,}00342p - 0{,}00129p_D}{t + 273} \text{ [kg/m}^3\text{]} \tag{2.23}$$

The specific volume (v) of air is the inverse of density:

$$v = \frac{1}{\rho} \text{ [m}^3\text{/kg]} \tag{2.24}$$

2.2.2.6 Enthalpy

The enthalpy (h) represents the total heat content of dry air (h_L) and water vapor (h_D), referred to 1 kg of mixture:

$$h = h_L + 0{,}01 \cdot x \cdot h_D \text{ [kJ/kg]} \tag{2.25}$$

The enthalpy of dry air at temperature t is:

$$h_L = c_L \cdot t \text{ [kJ/kg]} \tag{2.26}$$

c_L specific heat of dry air at constant pressure (c_L = 1 kJ/kg grd)

The enthalpy of water vapor at temperature t is:

$$h_D = r_0 + c_D \cdot t \text{ [kJ/kg]} \tag{2.27}$$

r_0 specific heat of vaporization at 0 °C (r_0 = 2,490 kJ/kg)
c_D specific heat of water vapor at constant pressure (c_D =1.93 kJ/kg grd)

Therefore, the enthalpy of moist air is:

$$h = c_L \cdot t + (r_0 + c_D \cdot t) \cdot \frac{x}{1000} \text{ [kJ/kg]} \tag{2.28}$$

As can be seen from the above equation, the enthalpy is directly proportional to the temperature and the moisture content. Conventionally, the enthalpy of absolutely dry air at a temperature of t = 0 °C is considered equal to zero.

Using the given equations, the numerical values of the moist air parameters can be calculated if two of them are known. Appendix C lists the state data of moist air at atmospheric pressure as a function of temperature and relative humidity.

2.2.2.7 The *h-x*-diagram and Its Applications

The relationship between the physical state parameters of air was represented in chart tables to avoid calculations. Mollier's *h-x*-diagram, for example, uses the moisture content (*x*) for the abscissa and the enthalpy (*h*) for the ordinate. A simplified form of this diagram is shown in Figure 2.21. Each point in the chart field represents an air state. The red lines correspond to the temperature (*t*). The blue curves correspond to the relative humidity (φ). The green lines correspond to the specific volume (*v*).

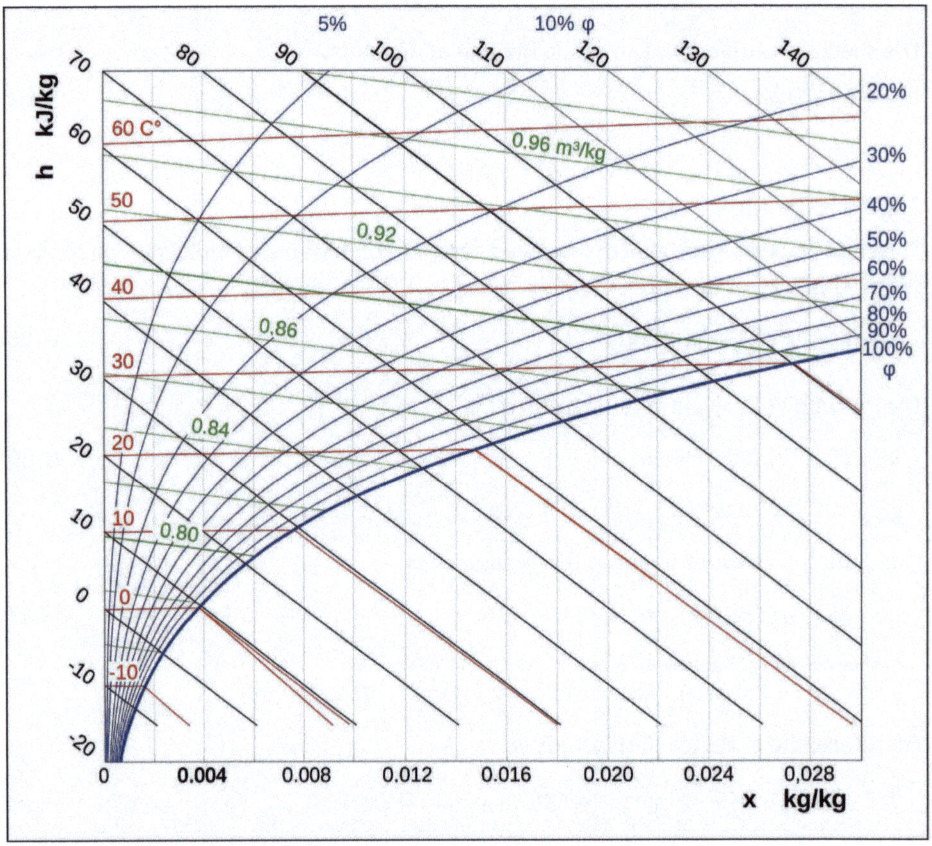

Figure 2.21 Simplified form of the *h*-*x*diagram (based on [23])

Using the *h*-*x*-diagram, the rapid determination of the numerical values of all moist air parameters is possible, depending on two parameters that are known.

The air parameters in a drying chamber at a certain point in time are: t = 40 °C and φ = 60 %. Using the h-x-diagram in Figure 2.21, the corresponding state point is determined and then the values of the other parameters are read off: x = 0.019 kg/kg$_{\text{dry air}}$, h = 90 kJ/kg, v = 0.91 m^3/kg.

In addition, the diagram offers the possibility of representing state change processes of the air and drying-specific processes, as well as establishing mass and energy balances.

The **air heating** is a process characterized by an increase in air temperature without a change in moisture content (x = constant). Therefore, it is represented in the h-x-diagram by a vertical line directed upward (segment 4–1 in Figure 2.22). It can be seen that heating causes the relative humidity to decrease ($\varphi_1 < \varphi_4$). This effect is significant for drying control. When the temperature rises, an automatic decrease in air humidity must be expected. Therefore, the temperature increase should not be too steep, so that the air humidity can follow. In addition, the air enthalpy increases with rising temperature ($h_1 > h_4$).

The **air cooling** is the reverse process of heating, also represented in the diagram by a vertical line, but directed downward (segment 2–3 in Figure 2.22). If the cooling process is continued until the intersection with the curve φ = 100 %, the so-called dew point (point 3) is reached. This corresponds to the state in which the air is saturated with water vapor. The temperature corresponding to this state is called the dew point temperature.

The **condensation**. If cooling is continued, part of the water vapor in the air condenses and the air changes its water content. This process no longer takes place at constant moisture content, but at the lower end of the curve φ = 100 % (segment 3–4 in Figure 2.22).

The **evaporation of water** from the wood surface occurs when the partial pressure of the air vapor is lower than the saturation pressure at the given temperature. It is assumed that when water evaporates into the air, the heat content of the air remains constant (adiabatic process).

Therefore, the evaporation of water is regarded as a process with constant enthalpy, and as such is represented in the h-x-diagram as a straight inclined segment, parallel to the line h = constant, represented, which is oriented in the direction of decreasing temperature (segment 1–2 in Figure 2.22).

It can be noted that in this case the process is accompanied by an increase in moisture content ($x_2 > x_1$). If the evaporation process lasts long enough, the air can reach the saturation state (point 5 in Figure 2.22), at which the phase equilibrium between the gaseous and liquid state of water in the air sets in and evaporation ceases. The temperature corresponding to this state is called the wet-bulb temperature, and corresponds to the wet-bulb temperature measured by the psychrometer.

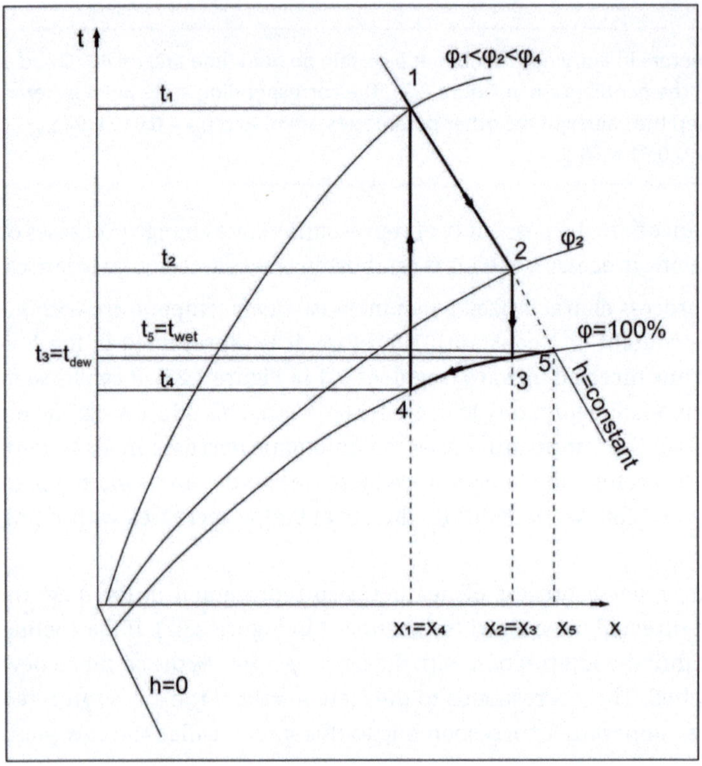

Figure 2.22 Representation of heating, cooling, condensation and evaporation on the *h*-*x*-diagram (source: [9], with permission)

Mixing Process of Air with Different States (e.g. Fresh Air with Recirculated Air)

Most drying systems operate on the principle of repeated air recirculation. When the humidity inside the chamber rises too much, the exhaust air dampers open and a small portion of the air is discharged into the atmosphere; however, the largest portion is mixed with fresh air and, after heating, reintroduced into the timber stacks. The mixing process of air with two different states is shown in Figure 2.23, where point 0 represents the fresh air and point 2 represents the recirculated air (exhaust air). The two are present in different quantities in the mixture, where n is referred to as the mixing ratio coefficient:

$$n = \frac{m_{\text{exhaust air}}}{m_{\text{fresh air}}} \tag{2.29}$$

The quantity of air is further calculated based on the mass of its dry component. This means that for a quantity of n kg of recirculated air with state 2 (h_2, x_2), 1 kg of fresh air with state 0 (h_0, x_0) is added, whereby the quantity of the mixture in this case is (n+1) kg.

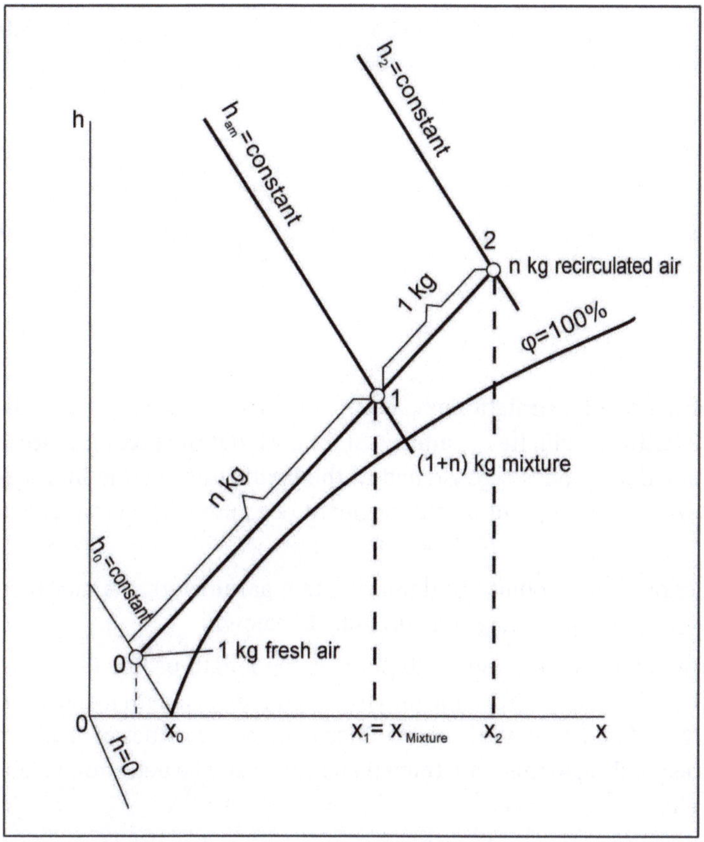

Figure 2.23 Mixing of air in two different states, representation on the *h*-*x*-diagram (source: [9], with permission)

To determine the mixing parameters, the heat and moisture balance equations must be established. The heat contained in the fresh air and in the recirculated air before mixing must equal the heat contained in the mixture. The same condition applies to the moisture content. Therefore, the balance relationships can be written as:

$$1 \cdot h_0 + n \cdot h_2 = (1 + n) \cdot h_M \tag{2.30}$$

$$1 \cdot x_0 + n \cdot d_2 = (1 + n) \cdot x_M \tag{2.31}$$

From these balance relationships, the enthalpy (h_M) and the moisture content (x_M) of the mixture are derived:

$$h_M = \frac{h_0 + nh_2}{1 + n} \tag{2.32}$$

$$x_M = \frac{x_0 + nx_2}{1+n} \tag{2.33}$$

and furthermore:

$$h_M - h_0 = n \cdot (h_2 - h_M) \tag{2.34}$$

$$x_M - x_0 = n \cdot (x_2 - x_M) \tag{2.35}$$

or

$$\frac{h_M - h_0}{x_M - x_0} = \frac{h_2 - h_M}{x_2 - x_M} \tag{2.36}$$

This represents the equation of a straight line passing through points 0 (with coordinates h_0 and x_0) and 2 (with coordinates h_2 and x_2). It follows that the mixing process between fresh air and exhaust air is represented in the *h-x*diagram by the line segment 0–2 (Figure 2.24), and every point on this segment (e.g., point 1) characterizes the mixing state.

To determine the exact position of point 1 and thus its state parameters, the moisture content of the mixture (x_M) or the mixing ratio (n) must be known.

To determine the parameters corresponding to point 1, the length of line 0–2 is set equal to 1+n assumed. Point 1 divides this line into two segments of length n and 1, respectively. If the position of point 1 in the *h-x*diagram is known, the values of all other state parameters corresponding to this state (mixed air) can easily be determined: enthalpy, temperature, etc.

Above, it was taken into account that the 0–2 mixture line lies above the curve φ = 100 %. Sometimes, however, it may occur that this line intersects the curve φ = 100 % and the point 1′ characterizing the mixture state lies below this curve (Figure 2.24). This situation is common when low drying temperatures are used (e.g., for oak).

In this case, the mixing process is accompanied by condensation. This means that point 1′ does not characterize moist air, but air that is saturated with vapor and contains water droplets in suspension, which is why point 1′ is regarded as unreal (fictitious).

To determine the real point 1, the constant enthalpy line (h_M = constant) is drawn through the fictitious point (Figure 2.24). The intersection of this line with the saturation curve (point 1) represents the real state point of the mixture.

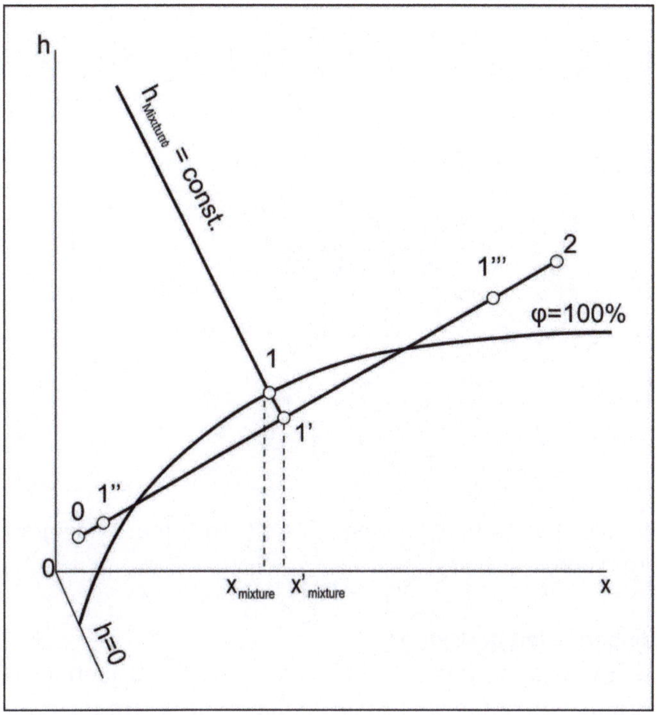

Figure 2.24 Determination of the properties of the air mixture when the state point characterizing the mixing state lies below the saturation curve (source: [9], with permission)

Illustration of Conventional Drying

Another application of the *h*-*x*-diagram is the representation of the drying process in various drying systems. In Figure 2.25 on the left, the principle of fresh air/exhaust air drying is illustrated, and in Figure 2.25 on the right, its representation in the *h*-*x*-diagram. Point 0 represents the state of the fresh air, segment 0–2 is the mixture of fresh air with the recirculated air, point 3 characterizes the mixture, segment 3–1 represents the heating of the mixture, and segment 1–2 is the passage of the heated air mixture through the stack. The drying process is assumed to be an isenthalpic process.

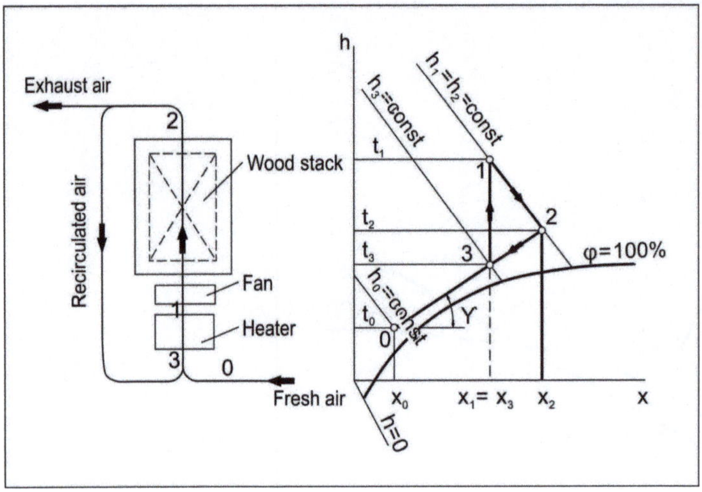

Figure 2.25 Representation of a conventional drying system (left) in the *h*-*x*-diagram (right) (source: [9], with permission)

Determination of Heat Consumption in Drying

Another application of the *h*-*x*-diagram, this time in energy balance calculations of drying systems, is the determination of the heat consumption required for the evaporation of water from wood. It is assumed that the state of the fresh air (point 0) is characterized by the enthalpy h_0 and the absolute humidity x_0 and the state of the exhaust air (point 2) is characterized by h_2 and x_2 (Figure 2.24). When passing through the material stack, 1 kg of air (the mass of its dry portion) absorbs (x_2–x_0) g of water.

Therefore, the air consumption required to remove 1 kg of water from the drying system is:

$$m_0 = \frac{1000}{x_2 - x_0}\ [\mathrm{kg_{dry\ air}/kg_{evap\ water}}] \tag{2.37}$$

$$m = \frac{1000}{x_2 - x_1}\ [\mathrm{kg_{dry\ air}/kg_{evap\ water}}] \tag{2.38}$$

$$q_{\mathrm{evaporation}} = m_0 \cdot (h_2 - h_0) = 1{,}000 \cdot \frac{h_2 - h_0}{x_2 - x_0} \tag{2.39}$$

From Equation 2.39, the following facts can be derived:

- The heat consumption for the evaporation of 1 kg of water depends exclusively on the initial state (h_0, x_0) and final state (h_2, x_2) of the air, and not on the intermediate changes of the air in the drying system.
- The heat consumption for water evaporation includes the consumption required for heating the supplied fresh air (from t_0 to t_2).

- The larger the angle γ is (see Figure 2.25 right), so the steeper the segment 0–2, the greater the heat consumption.

2.2.2.8 Air Velocity and Airflow

In addition to the state parameters of moist air as the drying medium, the air velocity and the air flow also have a decisive influence on wood drying.

The Air Current through the drying material must fulfill the following tasks:

- the heat exchange between drying medium and wood;
- the removal of the moisture extracted from the wood from its surface;
- the destruction of the "boundary layer" forming in the immediate vicinity of the wood surface"; when this becomes saturated with water vapour, further moisture extraction from the wood is impeded;
- the uniform reduction of the wood moisture.

To ensure the first two functions, the magnitude of the air velocity is crucial. The third condition can only be met if a turbulent airflow is generated through the drying material. To fulfil the fourth function, a uniform distribution of air velocity in the timber stacks must be ensured. Each of these aspects will be discussed in further detail.

The air speed decisively influences the drying dynamics, particularly during the removal of free water from the wood. The higher the air velocity, the faster the drying process. At the same time, however, with increasing velocity the risk of end checks and also the electricity consumption increase, causing the drying costs to rise. The optimal air velocity in the lumber stack is on average w = 2 m/s, but can be up to w = 5 m/s are achieved in the drying of softwoods.

At higher flow velocities, the power increases only slightly, but the costs rise exponentially (Figure 2.26).

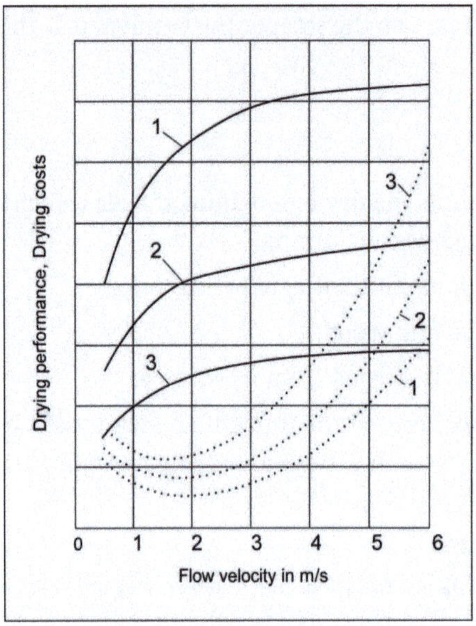

Figure 2.26 Dependence of the drying performance (continuous lines) and the drying costs (dotted lines) on the air velocity, for 25 mm thick sawn timber: 1-pine, 2-birch; 3-beech (based on [24])

The Dimming of the passive channels in the dryer (Figure 2.27) can lead to an increase of up to 74% in air velocity through the wood stacks (from 2.7 m/s to 4.7 m/s).

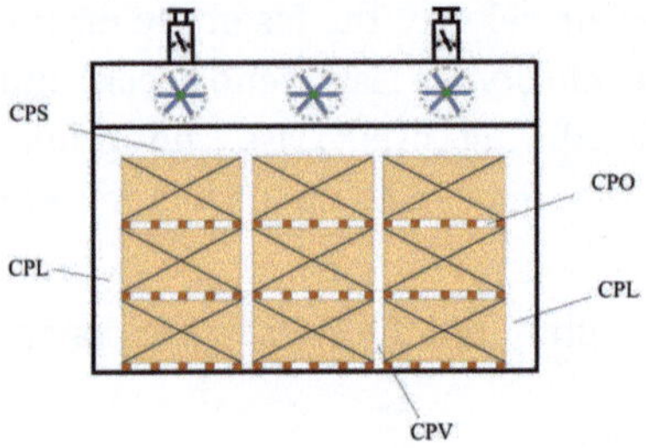

Figure 2.27 Passive air ducts in a drying chamber: CPO - horizontal air channels; CPL - lateral air channels; CPV - vertical air channels; CPS - upper air channel [25]

The type of flow can be determined by the value of the Reynolds number (*Re*) can be distinguished, where:

$$Re = \frac{d \cdot w}{\nu} \tag{2.40}$$

d characteristic length [m]
w velocity [m/s]
ν kinematic viscosity [m^2/s]

In laminar flow, in which the air particles move in well-ordered layers sliding alongside one another (Figure 2.28 left), $Re < 2{,}300$. In turbulent flow, additional movements are superimposed at several points in the flow, so that the resulting flows run in all directions and form small vortices (Figure 2.28 right). In this case, $Re > 2{,}300$.

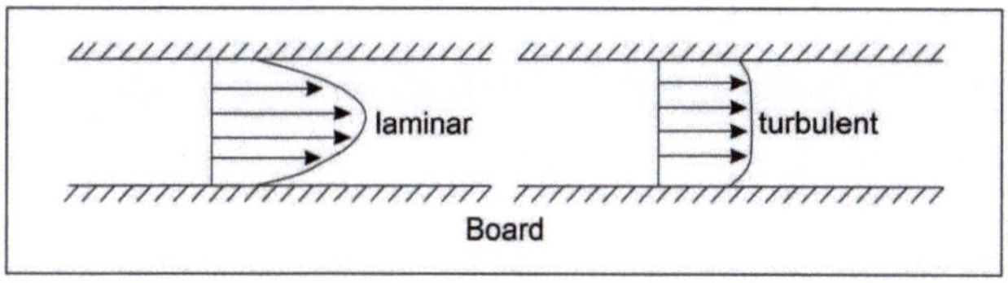

Figure 2.28 Velocity profile in laminar (left) and turbulent (right) flow (based on [10])

A turbulent flow that drives and removes water particles from the immediate vicinity of the wood surface is desirable during drying. However, this does not occur on its own in a dryer. It can be generated by the following measures:

- alternating direction of rotation of the fans, optimally every 2.5 hours;
- impulse ventilation, by repeatedly switching the fans on and off at short time intervals. The drying time can thereby be reduced by 15 – 20 % [26];
- the use of air deflection devices in the drying system, which are controlled by the wood moisture measuring points so that they direct the air to the part of the stack where the wood moisture is highest [27];
- drying with rapidly alternating air movement through the stack [28].

A sufficient air velocity is important during drying, but even more important is the uniformity of the airflow. In order to achieve a uniform reduction in wood moisture, the stacking rules should be observed (in particular the arrangement of the stacking stickers, exactly vertically above one another in the stack).

Measuring air velocity during drying is not common, but as a maintenance measure it should be carried out annually. The periodic examination of the flow conditions in the drying system should include a measurement (M_1) across the stack height on the air outlet side, a measurement in the inlet duct (M_2), a further measurement at the fan edge (M_3) and another measurement in the duct (M_4) (Figure 2.29).

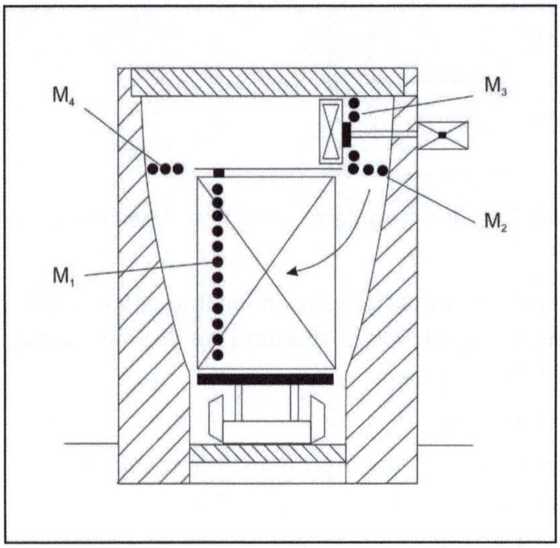

Figure 2.29 Recommended flow measurements in a drying system, as a maintenance measure (based on [24])

The air velocity is measured using anemometers (Figure 2.30). The operating principle of the vane anemometer (Figure 2.30 left) is based on measuring the rotational speed of a vane wheel placed perpendicular to the direction of the airflow. Hot-wire anemometers (Figure 2.30 right) measure air velocity according to a thermal principle: a heated wire is cooled by the passing airflow, causing its electrical resistance to change. This enables precise determination of the flow velocity, even at very low airflows. High measurement accuracy requires high-quality manufacturing and periodic calibration.

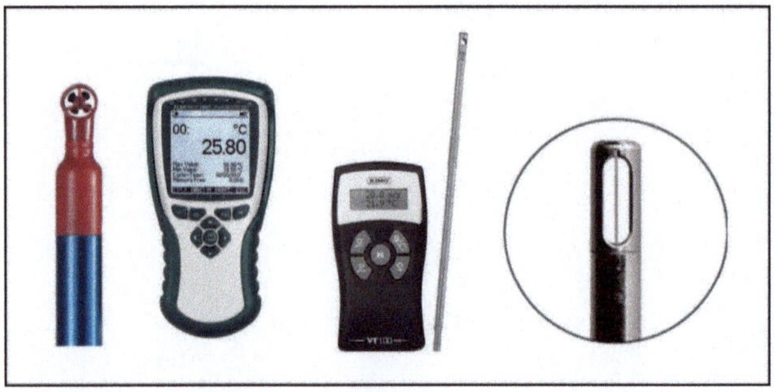

Figure 2.30 Measuring instruments for determining air velocity in a drying system: left vane anemometer; right hot-wire anemometer (source: [29])

The uniformity of the airflow in the dryer can be checked using smoke or fog devices.

The modeling of the airflow in the drying system is an advanced method for simulating the velocity profiles at various key positions in the dryer and offers solutions for optimizing the air guidance in order to achieve better, more uniform drying.

2.2.2.9 Modeling of the Airflow in a Drying System

The airflow modeling can be carried out throughout the entire system and/or only locally (e.g., in the fan, heat exchanger, distribution duct, fresh air/exhaust air shaft). It is performed using numerical methods [30, 31].

The main phases of airflow modeling are listed in Figure 2.31. The preprocessing phase consists mainly of creating the 3D model, its discretization, and the definition of the boundary conditions at the boundaries of the analysis domain.

Since the air velocity in a drying chamber is below 150 m/s, the flow is considered incompressible [32]. During the processing phase, the mass and momentum conservation equations are solved iteratively using a dedicated program (e.g., Ansys Fluent).

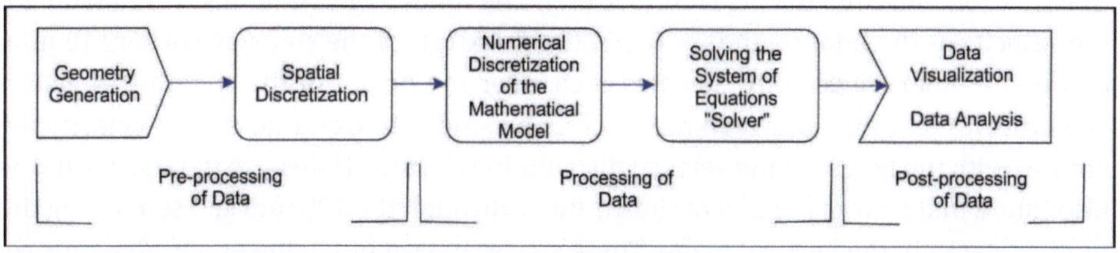

Figure 2.31 Stages of numerical simulation (source: [33], with permission)

The simulation of the airflow in a drying system involves solving the Reynolds-averaged Navier-Stokes equations (RANS), which describe the fluid motion [32]. These equations are solved using fluid dynamics simulation software (CFD).

$$\frac{\partial \overline{v}_i}{\partial x_i} = \frac{\partial \overline{v}_1}{\partial x_1} + \frac{\partial \overline{v}_2}{\partial x_2} + \frac{\partial \overline{v}_3}{\partial x_3} = 0 \tag{2.41}$$

$$\rho\left(\frac{\partial \overline{v}_i}{\partial t} + \overline{v}_j \frac{\partial \overline{v}_i}{\partial x_j}\right) = \rho gi - \frac{\partial \overline{p}}{\partial x_i} + \frac{\partial}{\partial x_j}\left[\mu\left(\frac{\partial \overline{v}_j}{\partial x_i} + \frac{\partial \overline{v}_i}{\partial x_j}\right) - \rho\overline{v_i' v_j'}\right] \tag{2.42}$$

$$\rho\left(\frac{\partial k}{\partial t} + v_j \frac{\partial k}{\partial x_j}\right) = \frac{\partial}{\partial x_j}\left[\left(\mu + \frac{\mu_t}{Pr_k}\right)\frac{\partial_k}{\partial x_j}\right] + \mu_t \frac{\partial v_i}{\partial x_j}\left(\frac{\partial v_i}{\partial x_j} + \frac{\partial v_j}{\partial x_i}\right) - \rho\varepsilon \tag{2.43}$$

$$\rho\left(\frac{\partial \varepsilon}{\partial t}+v_j\frac{\partial \varepsilon}{\partial x_j}\right)=\frac{\partial}{\partial x_j}\left[\left(\mu+\frac{\mu_t}{Pr_\varepsilon}\right)\frac{\partial_\varepsilon}{\partial x_j}\right]+C_1\frac{\varepsilon}{k}\mu_t\frac{\partial v_i}{\partial x_j}\left(\frac{\partial v_i}{\partial x_j}+\frac{\partial v_j}{\partial x_i}\right)-C_2\rho\frac{\varepsilon^2}{k} \qquad (2.44)$$

$\mu_t = \rho C_\mu k^2/\varepsilon$

Pr_k and Pr_ε Prandtl numbers
C_μ, C_1, C_2 constants

The degree of air turbulence in a drying room is estimated at 10%. The highest turbulence value is recorded in the material stack. Turbulence modeling is performed using the standard *k*-*ε*- or RNG *k*-*ε*-model [34, 35].

The post-processing phase enables a qualitative analysis in the form of a color map as well as a quantitative analysis based on diagrams and/or by determining the values of the monitored parameters (flow rate, velocity, and uniformity).

A) Modeling of Airflow in the Entire System

The airflow simulation can be performed without simplifying assumptions in a 2D or 3D domain. Figure 2.32 shows a computational domain used to investigate air circulation in a conventional drying chamber [31]. The following conditions were assumed: the velocity at the inlet boundary 6 m/s, the thickness of the stacking stickers 19 mm, and two lumber stacks with 30 rows each, over the height of the chamber. A gap of 100 mm was assumed between the stacks. The study showed that the air velocity decreases with the height of the vertical distribution channel (lateral to the stacks). It was also found that a large portion of the air flows through the 100 mm gap separating the lumber stacks in the vertical plane. These spaces have a much lower aerodynamic resistance than the spaces between the boards (active channels), and therefore air flows through them. The inclined ceilingshape promotes an increase in air velocity at the entry into the vertical distribution channel lateral to the stacks. Due to this phenomenon, the geometry of the false ceiling and the upper baffle creates a turbulence zone (vortex) that reduces the air velocity in the upper part of the distribution channel. Therefore, the air velocity over the height of the vertical stack group is non-uniform (Figure 2.32 c).

The recirculation zone (circulation zone) that forms at the entry of air into the lateral distribution channel (Figure 2.33) is highlighted by Bedelean (2009) [25]. In this study, a 2D domain with a velocity of 3 m/s at the inlet face was considered. Three stacks at room height and four stacks across the chamber width were taken into account.

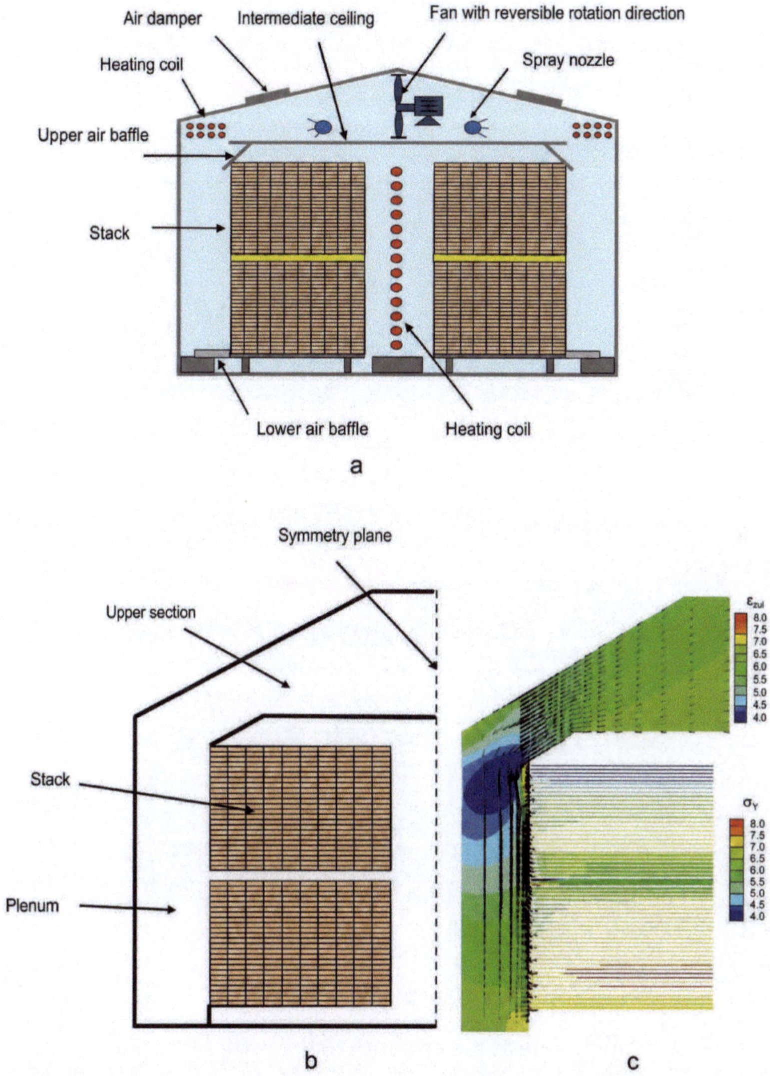

Figure 2.32 Simulation of airflow in a drying chamber: a – geometry of the chamber; b – computational domain and boundary conditions; c – air velocity profiles (adapted from [30])

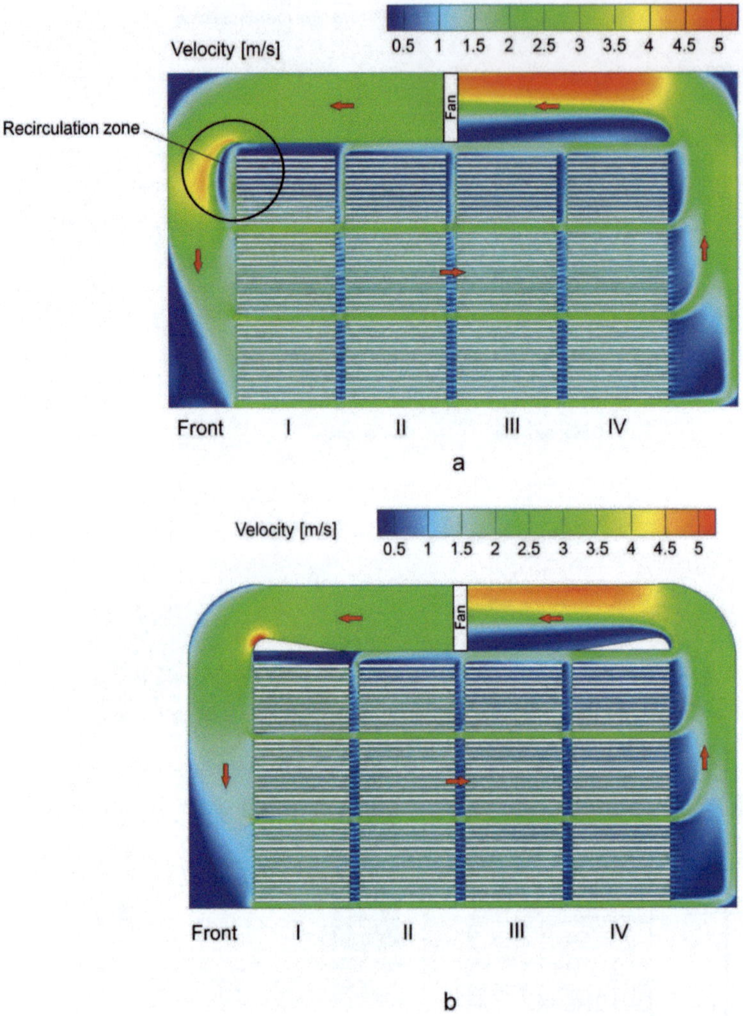

Figure 2.33 Recirculation zone at the entrance to the vertical distribution channel: a – ceiling with right-angle corners; b – ceiling with rounded corners (source: [25])

By rounding the ceiling corners (Figure 2.33 b), the negative influence of the recirculation zone on the value of the air velocity and its uniformity can be reduced (Figure 2.34 b). The air velocity increases at the top, since the forming vortex reduces its effective area, causing a higher proportion of air to flow through the upper stacks (Figure 2.34 a).

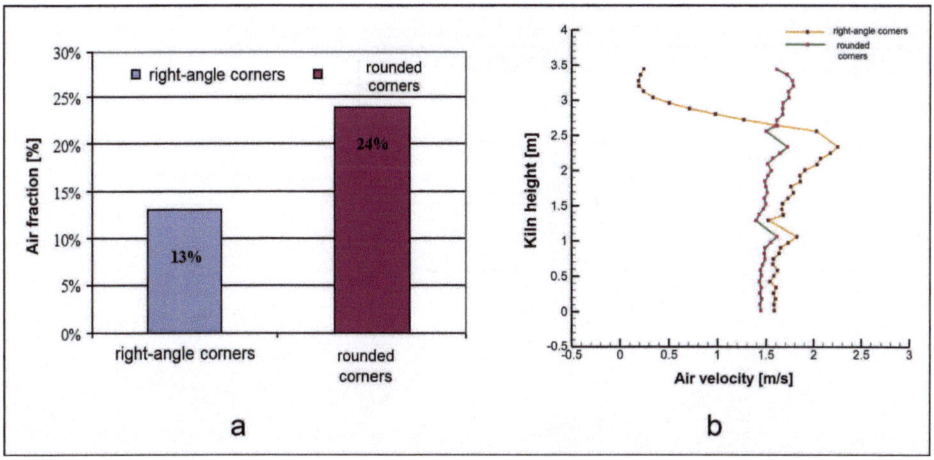

Figure 2.34 Influence of the geometry of the drying chamber on the air proportion and velocity: a – percentage of air flowing through the vertical air channel to the side of the stacks; b – air velocity profile over the room height (source: [25])

An increase in the uniformity of air distribution in the drying chamber can also be achieved by the correct use of baffles. The way in which the baffles are positioned plays an important role for the airflow [25]. If only one upper baffle is used at the air inlet in the stack group, the percentage of air flowing through the first one increases slightly. However, the influence is minimal compared to the absence of baffles. If the upper baffle were installed in the middle part of the upper channel, an increase in air velocity at all four fronts would result, according to the graph in Figure 2.35. Two upper baffles ensure a sudden increase in air in the last stack group.

In industry, upper baffles are provided for each stack group. In this situation, the air velocity increases in all four stack groups. When all passive channels are blocked off, both a significant increase in air velocity through all four stack groups and their uniformity are ensured.

An incomplete loading of the chamber affects the airflow [25]. Figure 2.36 shows the situation when the top stack in the first stack front is missing. In this case, a large portion of the airflow passes through the upper stacks of fronts II, III, and IV. Therefore, the value of the air velocity increases in the subsequent stack groups (Figure 2.37). If the stack is missing from the second stack row (Figure 2.38), a sudden drop in air occurs in the second stack group (Figure 2.37). Therefore, the placement option in Figure 2.36 provides better air circulation compared to the option in Figure 2.38. From a theoretical point of view, the possibility of using a "dummy stack" instead of the missing one was also analyzed. This variant ensures adequate air movement in all four stack groups (Figure 2.37).

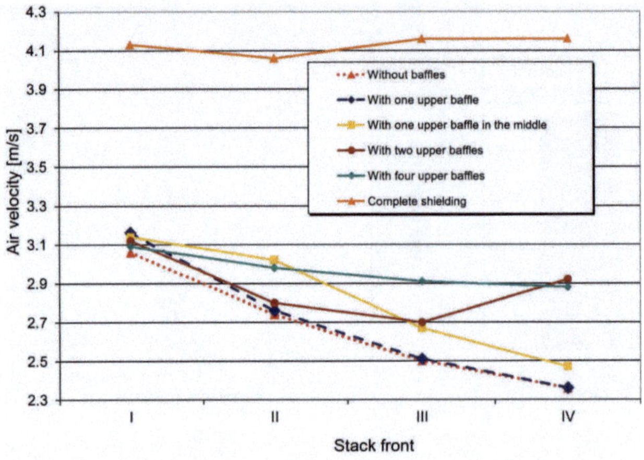

Figure 2.35 Influence of the baffling strategy on the average velocity through the four stack fronts (I – IV) arranged in the flow direction (source: [25])

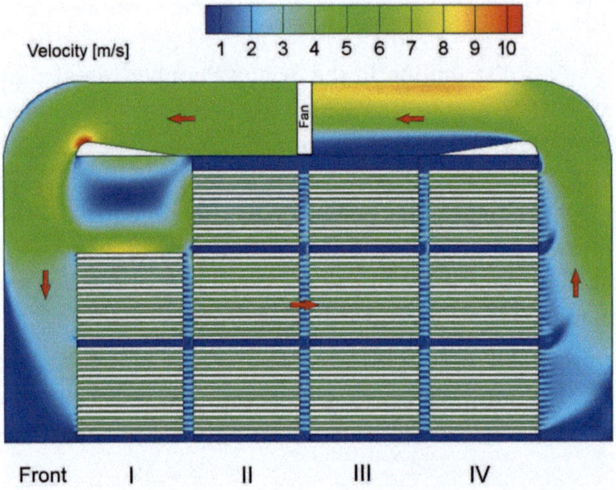

Figure 2.36 Modeled air flow velocity through the drying chamber in the case where the topmost stack in the first stack group is missing (source: [25])

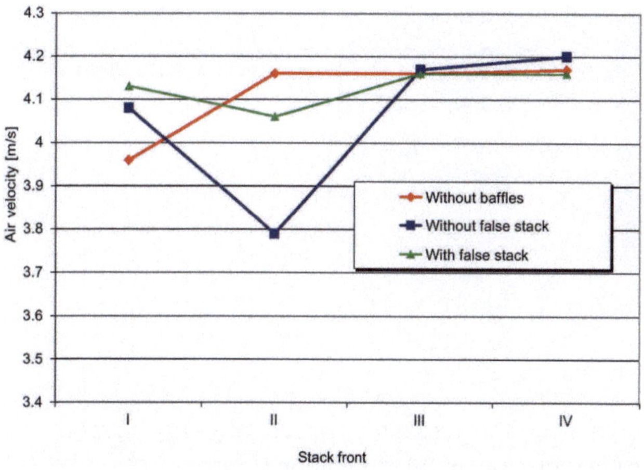

Figure 2.37 Influence of the use of a baffle instead of the missing stack (source: [25])

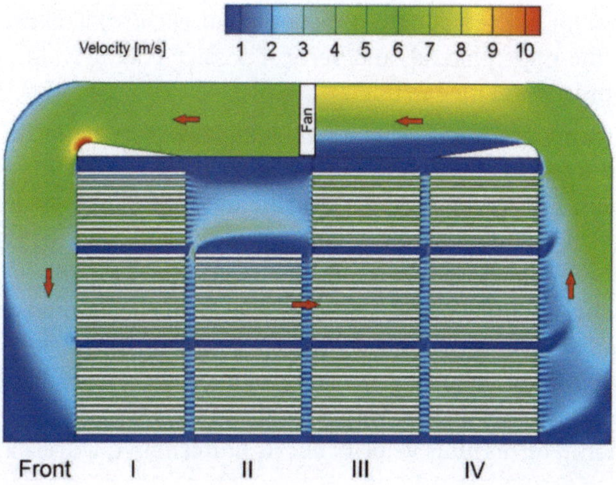

Figure 2.38 Modeled air flow velocity through the drying chamber for the case where a top stack is missing in the second stack group (source: [25])

The aerodynamics of the drying chamber can also be analyzed using a 3D model in the case of reversible fan operation [34]. Figure 2.39 shows the computational domain used in the simulation. Based on the numerical results, it was found that the energy loss coefficient differs at most of the analyzed points depending on the direction of rotation of the fan (Figure 2.40 a, b), with the exception of the air inlet area in the vertical air channel to the side of the stacks (S3/S3B).

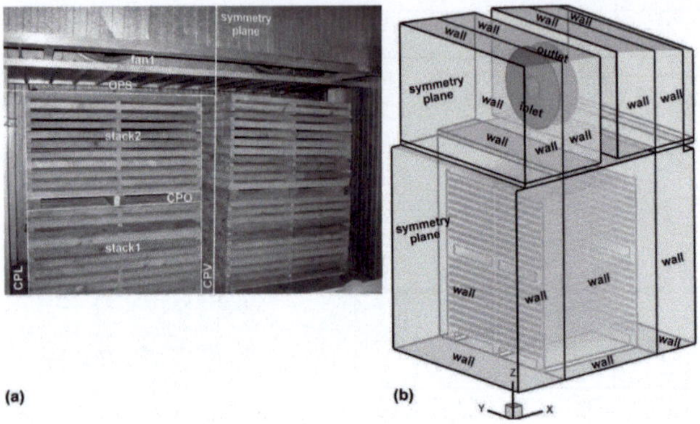

Figure 2.39 Computational domain for the numerical simulation: a – drying room considered in the study; b – the 3Ddomain used in the simulation (source: [34])

The air non-uniformity coefficient is much higher in the timber stack when the fan runs in the reverse direction than in the standard direction (Figure 2.40 c). Furthermore, the air losses through the upper, lateral, and vertical passive channels differ depending on the operating direction of the fan (Figure 2.40 d).

The geometry of the lateral distribution channel has a significant influence on the distribution of airflow in this channel and at the entry of air into the stack face [35]. Four variants were analyzed (Figure 2.41). The air velocity distribution is shown in Figure 2.41 a, b, c. First in the normal distribution channel with constant width (Figure 2.41 a), then in a conical distribution channel (Figure 2.41 b), then in a channel with double the width compared to the initial situation (Figure 2.41 c), which leads to a reduction in air velocity at the lower end of the channel (Figure 2.41 d). Due to the fact that the channel resistance decreases, an improvement in air uniformity is also achieved. In the case of the conical distribution channel, no improvement in airflow was found in the analyzed base case, neither in terms of its velocity nor its uniformity (Figure 2.41 b and d).

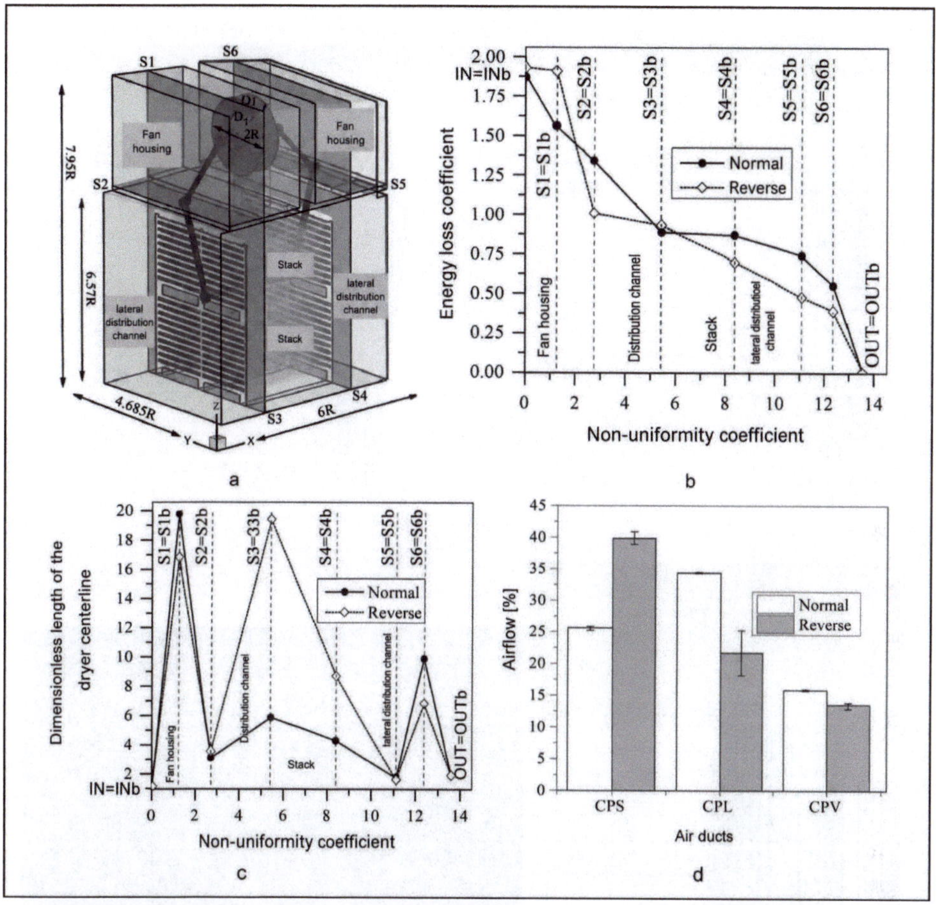

Figure 2.40 Variation of the energy loss coefficient, the non-uniformity coefficient, and the distribution of air volume in certain defined sections for both operating directions of the fan: a – the sections defined within the analysis; b – energy loss coefficient; c – non-uniformity coefficient; d – distribution of airflow through the upper, lateral, and vertical passive channels (source: [34])

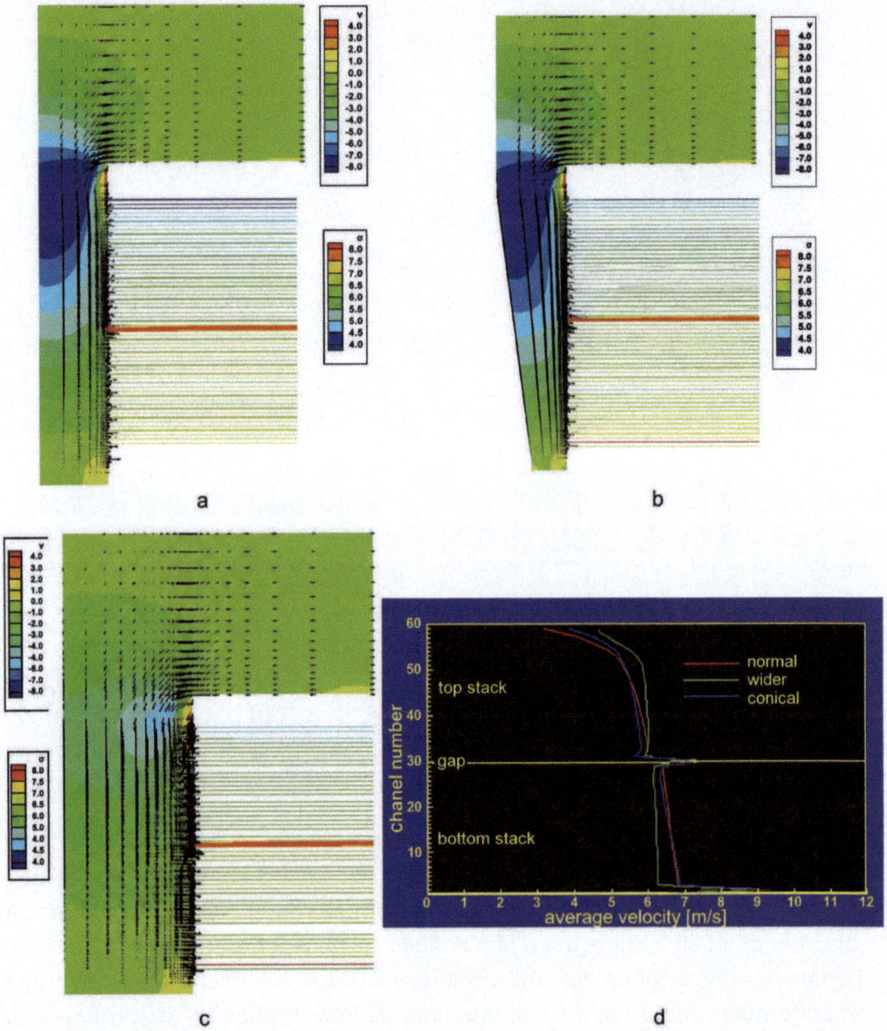

Figure 2.41 Average air velocity at the exit from the timber stacks for the analyzed variants: a – normal distribution channel (reference); b – conical distribution channel; c – wider distribution channel; d – average velocity (source: [36], with permission)

By introducing a vertical plate in the distribution channel (Figure 2.42 b), a subdivision of the channel was sought in order to achieve greater air uniformity. Results from Hua et al. (2001) show that this solution deteriorates the airflow compared to the base case (Figure 2.42 a and c) [35].

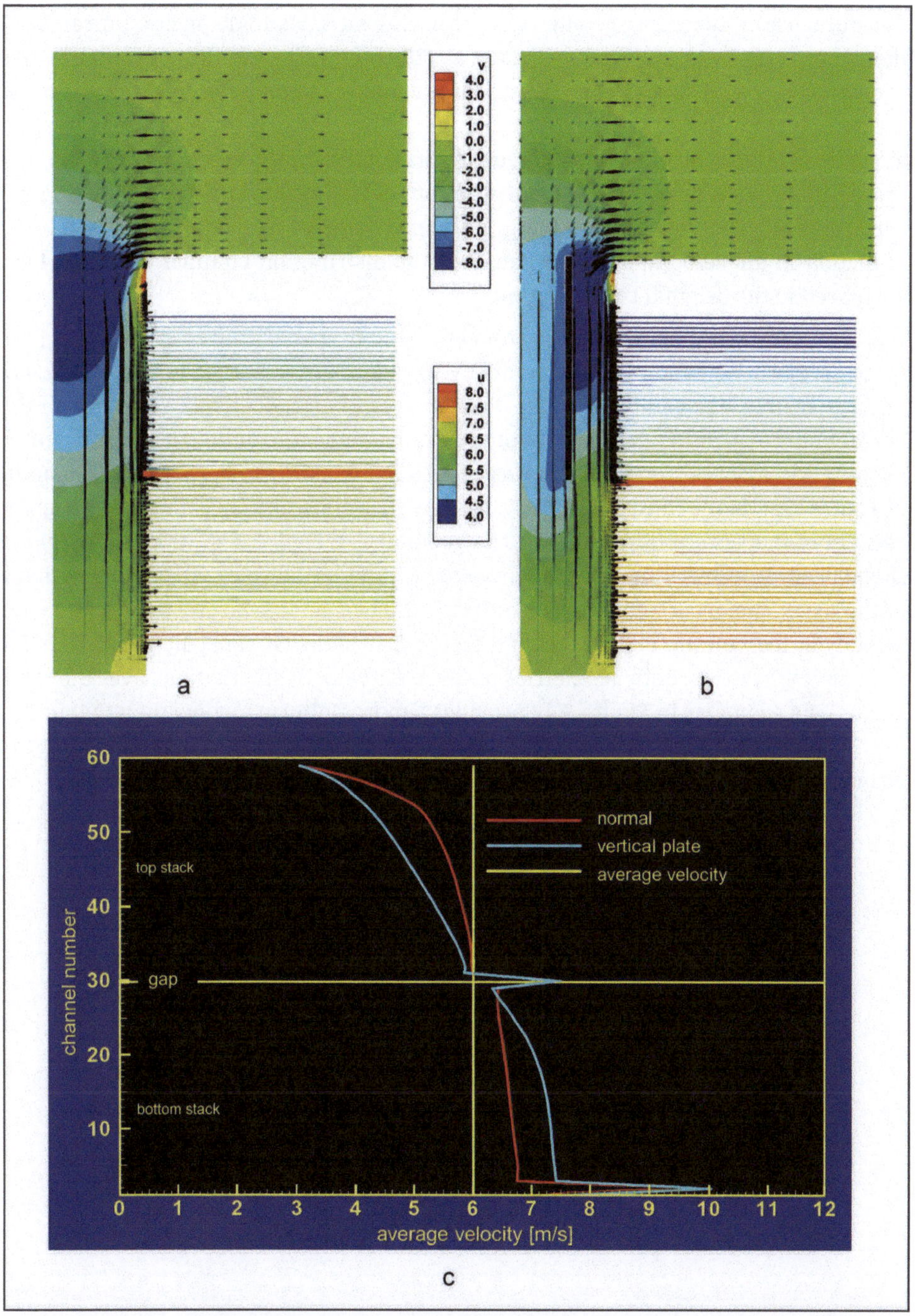

Figure 2.42 Average air velocity at the exit from the timber stacks for the analyzed variants: a – normal distribution channel (reference); b – insertion of a vertical plate into the distribution channel; c – velocity profile (source: [36], with permission)

The influence of the ceiling geometry of the drying facility and the inclination angle of the upper baffle was analyzed by Hua et al. (2001) [35]. The resulting geometries are shown in Figure 2.43 a, b. The inclined ceiling enables an acceleration of the airflow at the entry into the vertical distribution channel on the side of the stacks. An inclination angle of the baffle position of 30° leads to a reduction in air velocity at the top, but favors an increase in air velocity at the bottom of the stack compared to the original configuration. The variant of placing the upper baffle at an angle of 45° leads to a reduction in air velocity over the height of the distribution channel (Figure 2.43 c) compared to the original configuration.

The numerical analysis was also carried out by Rodriguez-Ramirez et al. (2014) [37], using six specific geometries of timber drying chambers (Figure 2.44 to Figure 2.49). In these figures, one can observe the development of air velocity as a function of the height of the distribution channel both at the inlet and at the outlet of the air channel. The numbering of the wood pieces was carried out from top to bottom. The geometry in Figure 2.44 can lead to an improvement in air uniformity by rounding the upper corners of the system (Figure 2.45). The geometry in Figure 2.46 (triangular ceiling shape) leads to a high variability of air velocity in the wood stack. The rounded shape of the dryer shown in Figure 2.47 provides a higher velocity at the top and a lower velocity at the bottom. Figure 2.48 and Figure 2.49 show two specific geometries for solar drying systems. Both geometries produce air velocity fluctuations over the stack height. The geometry in Figure 2.49 provides a more uniform airflow than that in Figure 2.49. However, the recommended variant for conventional drying systems is the geometry shown in Figure 2.45.

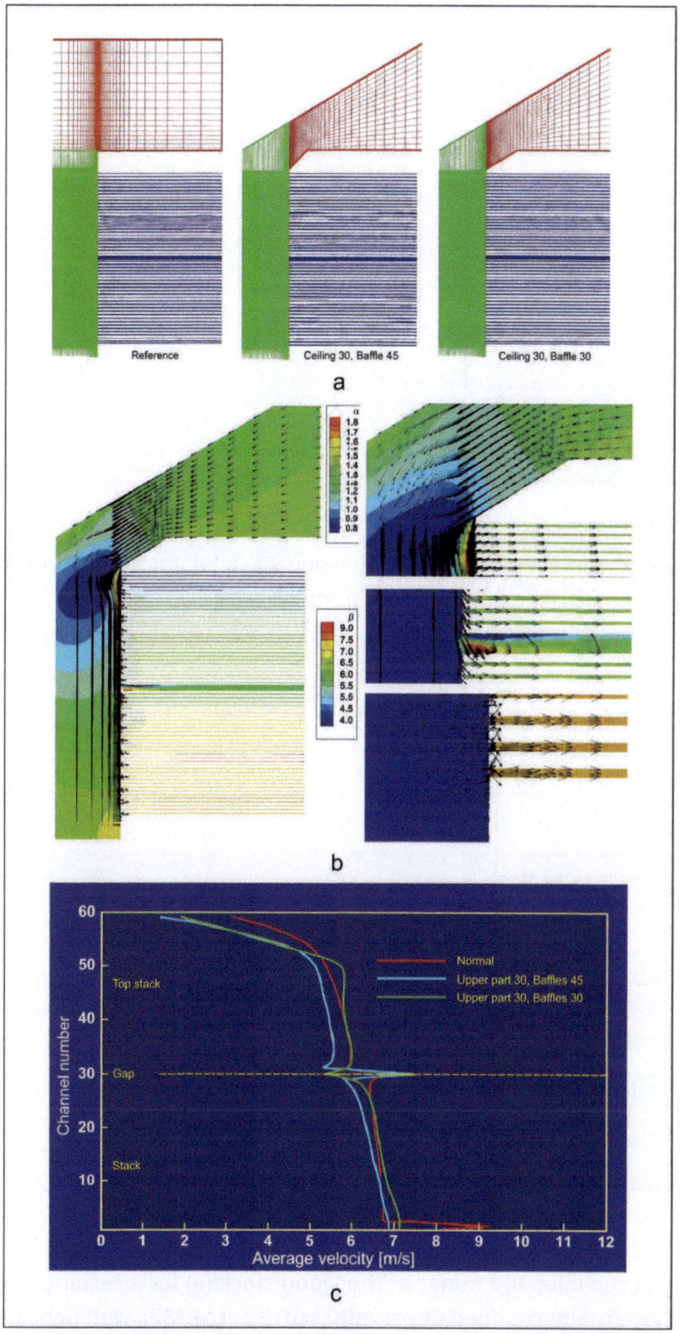

Figure 2.43 Calculation domain (a and b) and velocity distribution (c) as a function of ceiling geometry and the degree of inclination of the baffle (source: [36], with permission)

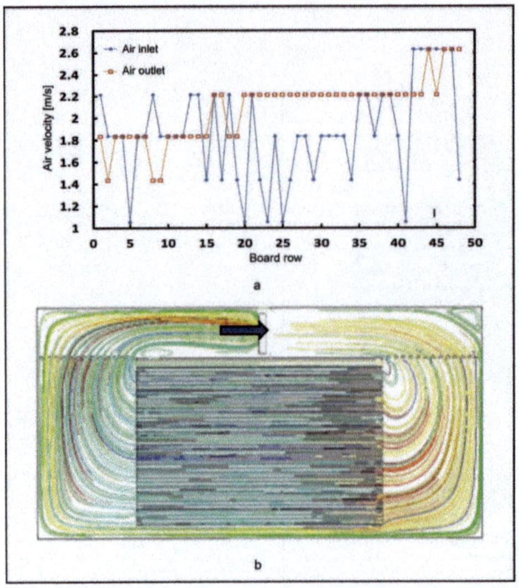

Figure 2.44 Air velocity at the inlet and outlet of the wood stack (a), for a rectangular geometry of the drying system (b) (source: [37], with permission)

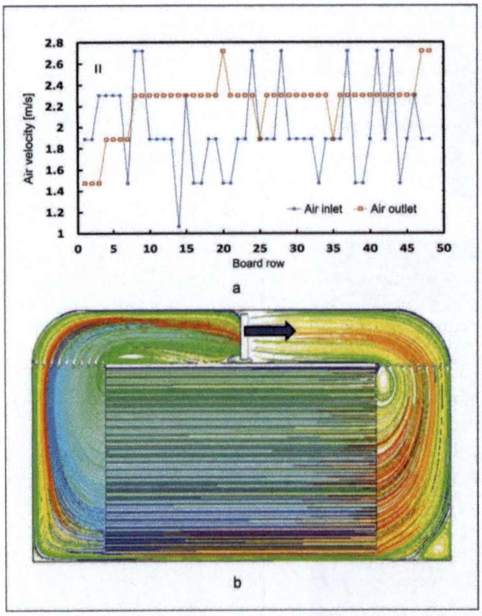

Figure 2.45 Air velocity at the inlet and outlet of the wood stack (a) for a rectangular geometry of the drying system with rounded upper edges (b) (source: [37], with permission)

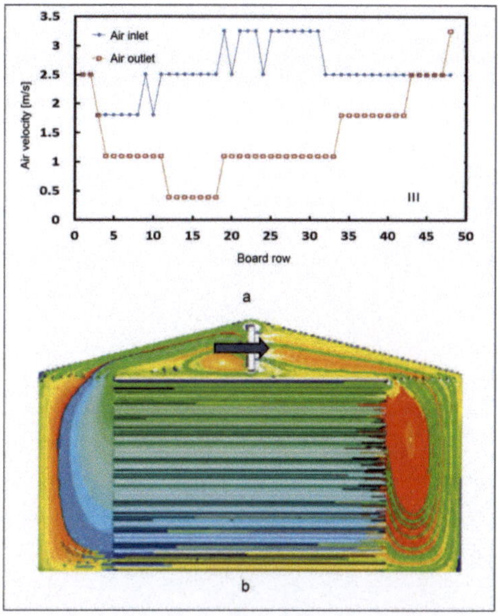

Figure 2.46 Air velocity at the inlet and outlet of the wood stack (a) for a triangular geometry of the drying system (b) (source: [37], with permission)

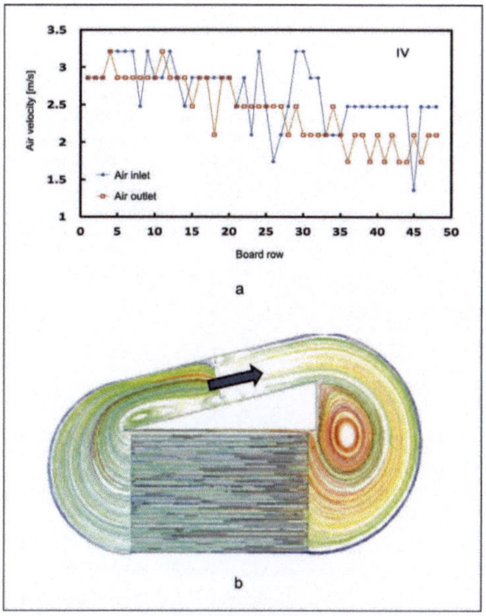

Figure 2.47 Air velocity at the inlet and outlet of the wood stack (a) for a rounded geometry of the drying system (b) (source: [37], with permission)

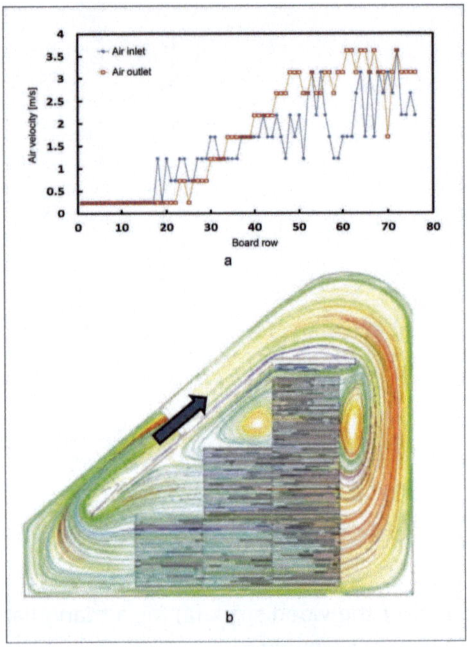

Figure 2.48 Air velocity at the inlet and outlet of the wood stack (a) for a rounded geometry of a solar drying system (b) (source: [37], with permission)

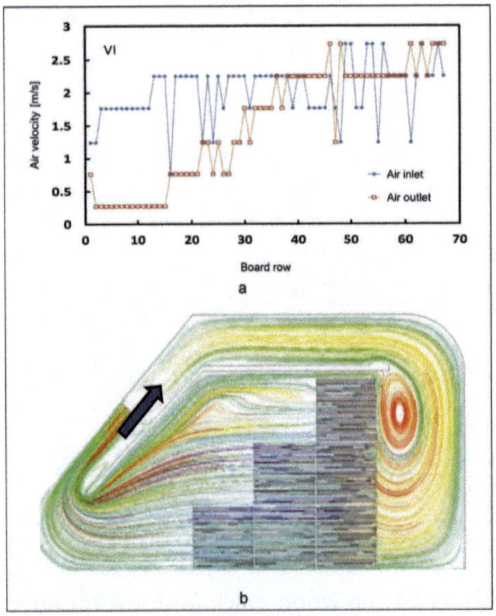

Figure 2.49 Air velocity at the inlet and outlet of the wood stack (a) for a rounded geometry of a second type of solar drying system (b) (source: [37], with permission)

B) Modeling of Airflow in the Wood Stack

An example of local airflow modeling in the air channel formed by the stacking stickers between two rows of boards placed on top of each other is shown in Figure 2.50.

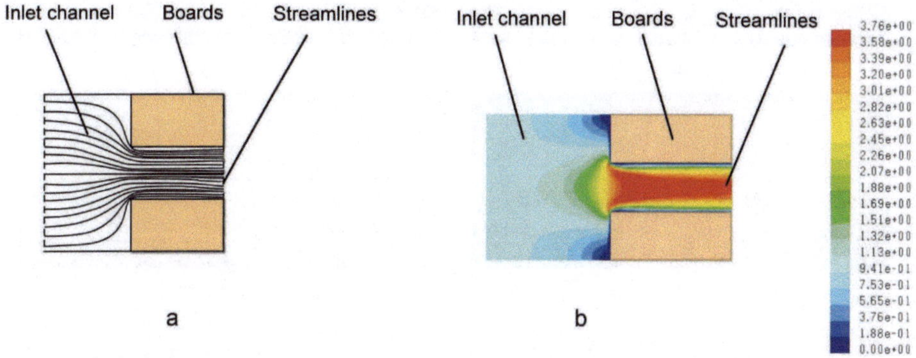

Figure 2.50 Air streamlines (a) and air velocity (b) at the inlet of the airflow between the boards (source: [25])

The airflow profile in the active channels (between the boards) is shown in Figure 2.51. The highest velocity prevails in the center of the channel and the lowest near the channel walls. Therefore, a velocity gradient develops in the channel.

When the air enters the stack, the flow separates according to Figure 2.51 b. Due to this phenomenon, a recirculation zone forms at both the upper and lower ends of the channel, leading to a non-uniform flow over a certain section of the channel. This non-uniformity is reduced by periodically changing the air direction in the drying plant.

The gaps that form between the wood pieces arranged in a row generate a turbulent airflow and thus an intensification of mass transfer next to each wood piece (Figure 2.51 c). These gaps can lead to non-uniform airflow if they are of different sizes and/or not evenly distributed in the wood stack [35].

Stacking wood pieces of different thicknesses in the same row (Figure 2.51 d) can lead to an increase in the degree of air turbulence and thereby contribute to an intensification of mass transfer in the respective areas.

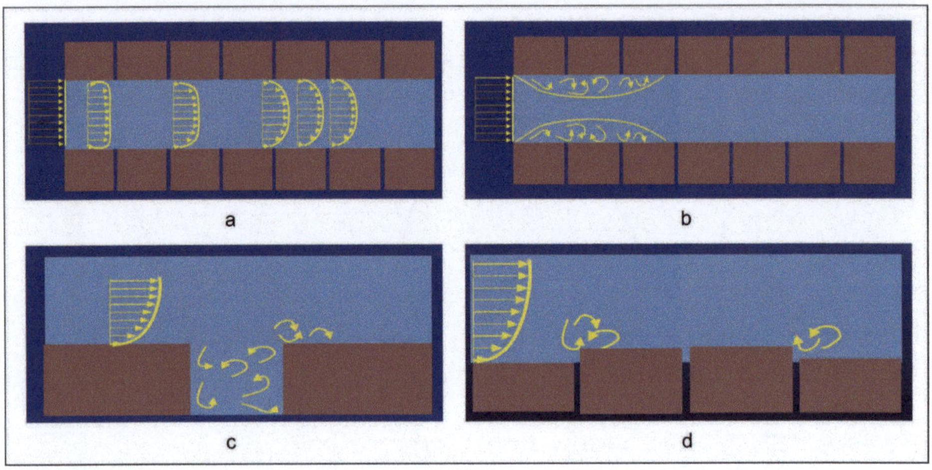

Figure 2.51 Phenomena occurring when air flows through the wood stack: a – air velocity profile in a channel formed by the stacked wood pieces; b, c, d – separation of the airflow (source: [36], with permission)

Another situation that can occur during wood drying is the absence of a board within a row (Figure 2.51 c). This circumstance leads to a turbulent airflow in the stack and thus to an intensification of mass transfer [35].

The thickness of the stacking stickers influences both the air movement in the stack (Figure 2.52) and the capacity of the drying chamber.

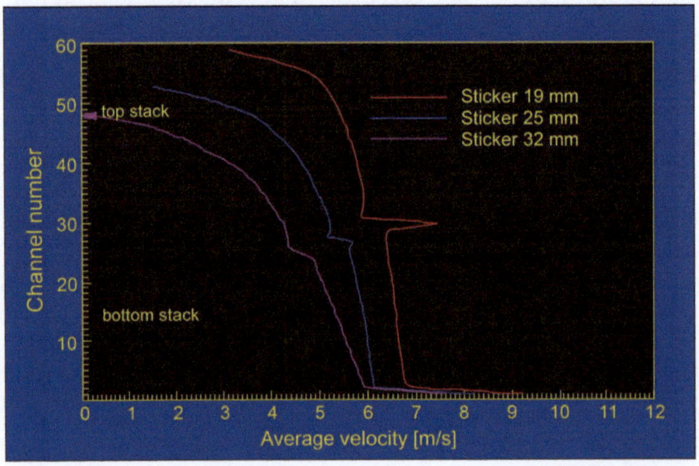

Figure 2.52 Air velocity profile over the height of the wood stack as a function of the sticker thickness: 19, 25 and 32 mm (source: [36], with permission)

The numerical results show that the air velocity distribution over the height of the stack does not change significantly with a change in the thickness of the stacking stickers (see Figure 2.52). The use of stickers with a thickness of 32 mm leads to a reduction in air velocity in the wood stack compared to the sticker thickness of 19 mm. Greater sticker thicknesses cause a reduction in the local resistance generated by the stacks and an increase in the amount of air flowing through the active channels between the wood pieces. A greater thickness of the stacking stickers also affects the size of the recirculation zone that forms when the air enters the distribution channel and blocks part of the topmost stack [35].

When using stacking stickers with a smaller thickness, it was found that non-uniform airflow as well as a longer drying time result.

A simulation of the airflow at the inlet and outlet of air from the stacks was carried out by [25] using the analysis domains shown in Figure 2.53. The analysis domain consists of three parts, namely the distribution channel at the inlet of air into the stacks, the active channel, and the distribution channel at the outlet of air from the stacks. Figure 2.53 b and d show the analysis domain in connection with a theoretical situation in which aerodynamic profiles are attached at the end of the stack [25]. The airflow in both variants is illustrated in Figure 2.54.

It is evident that by attaching some aerodynamic profiles, a gradual transition from the surface of the lateral distribution channel to the surface of the active channels and vice versa is ensured (Figure 2.54 to Figure 2.58), as well as the elimination of the apparently conditionally induced recirculation zones up to the edges of the wood pieces. Furthermore, the pressure loss decreases by approximately 30%. Although attaching the aerodynamic profiles achieves an improvement in airflow, the drying performance of the plant decreases by approximately 25% [38], which proves to be uneconomical.

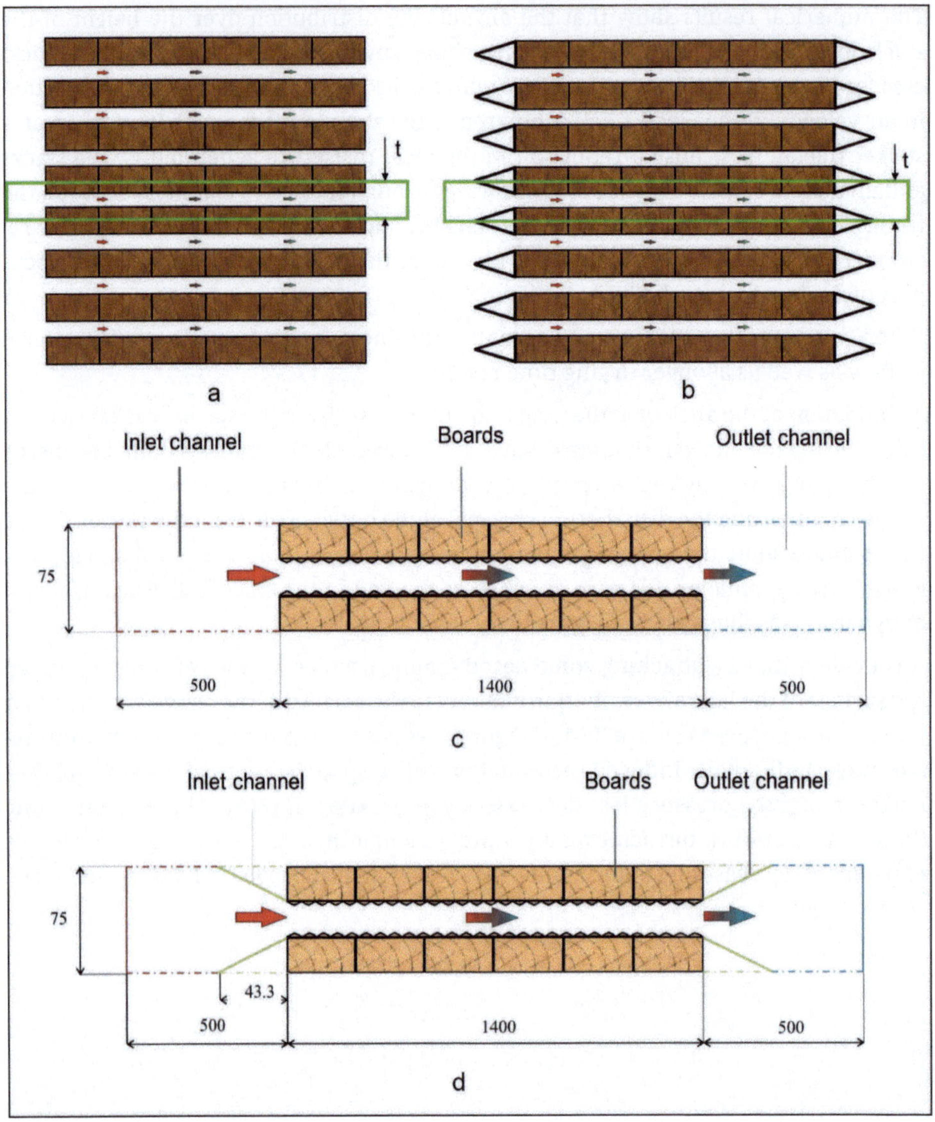

Figure 2.53 Definition of the geometry of the active channel (source: [25]): a – the normal appearance of the stack; b – attachment of some aerodynamic profiles on the inlet and outlet side of the air in the stack; c – calculation domain for the current channel shape; d – calculation domain for the proposed channel shape; *t* = network pitch

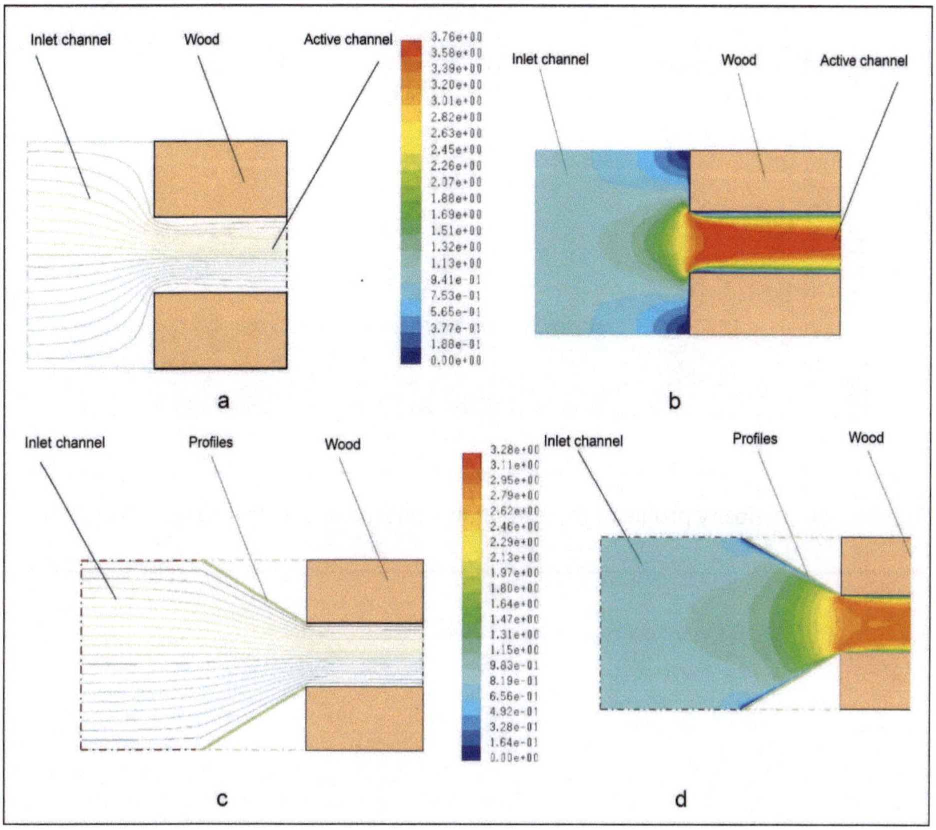

Figure 2.54 Streamlines (a, c) and air velocity (b, d) at the inlet into the active channels (source: [25])

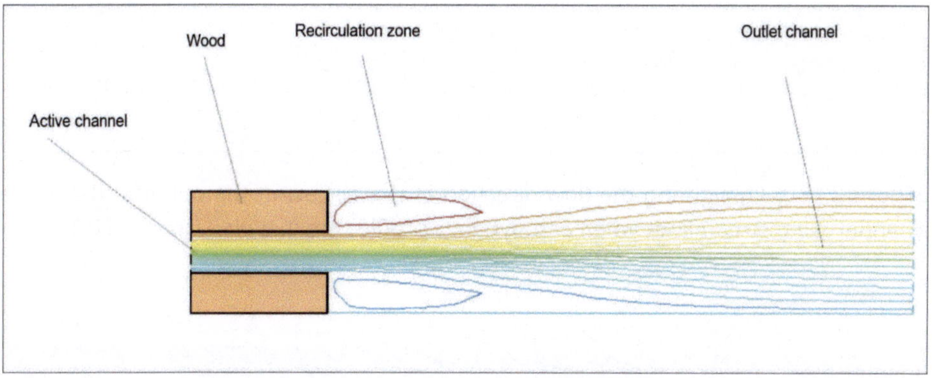

Figure 2.55 Streamlines at the exit of air from the active channel: 1 – active channel; 2 – wood; 3 – recirculation zone; 4 – outlet channel (source: [25])

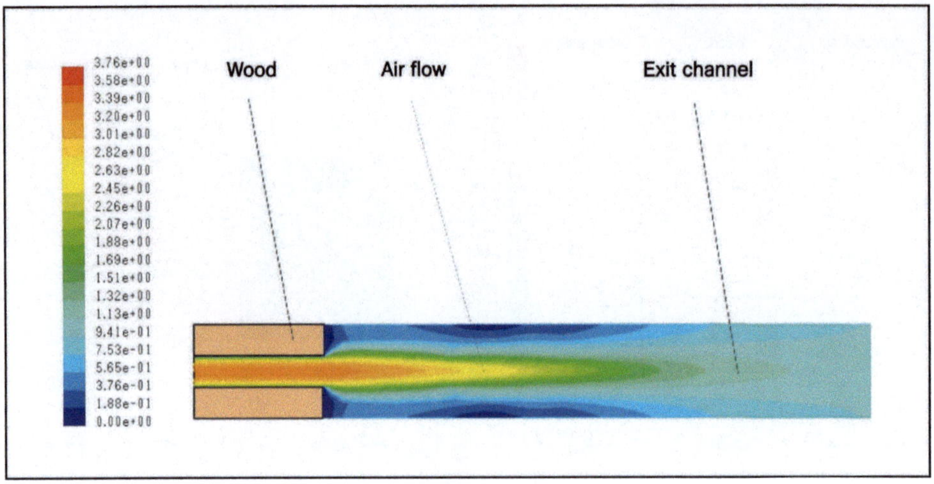

Figure 2.56 Velocity profile at the exit of air from the active channel (source: [25])

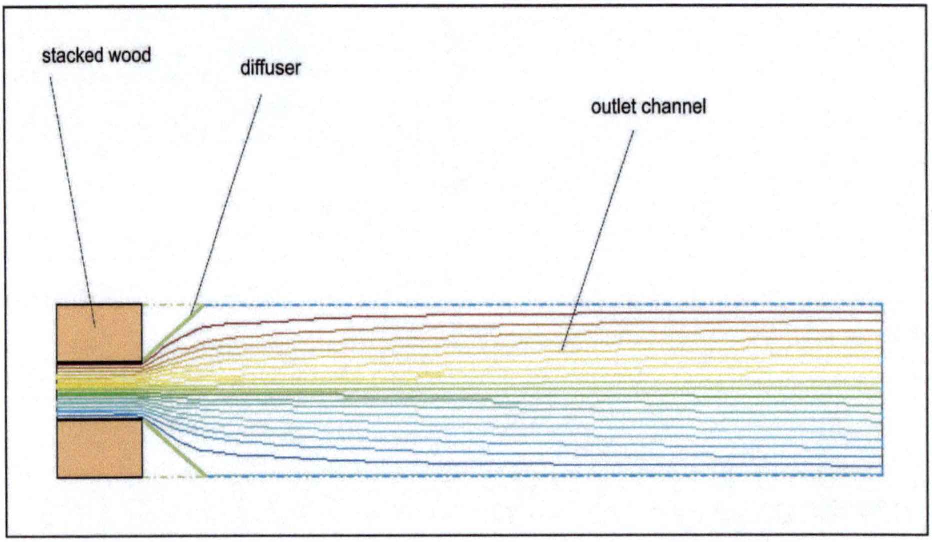

Figure 2.57 Highlighting the flow direction of the airflow at the outlet of the active channel using streamlines when aerodynamic profiles are installed (source: [25])

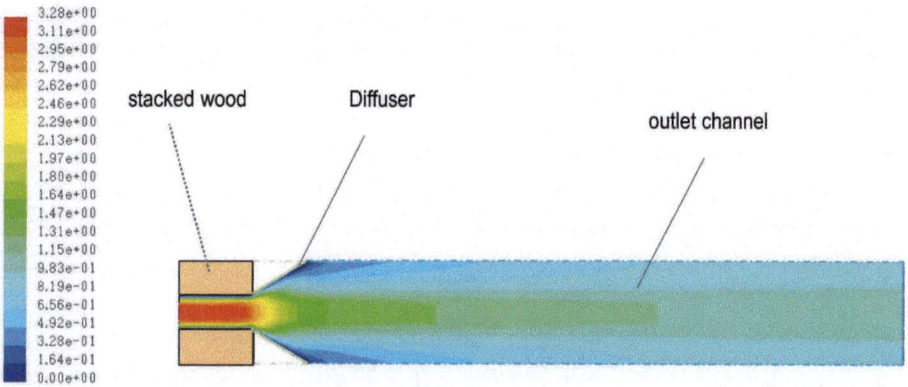

Figure 2.58 Highlighting the airflow at the outlet of the active channel using the color map (source: [25])

References

[1] Timar, M. C. (2009): *Elektronischer Holzarten-Katalog*. Transilvania University, Brasov, Romania

[2] Janik, W. (1965): *Handbuch der Holztrocknung*. Fachbuchverlag, Leipzig

[3] Porojan, M. (2007): *Contributions to the investigation of the physical, mechanical and technological properties of acacia wood* (in Romanian). Doctoral thesis, Transilvania University, Brasov, Romania

[4] Niemz, P.; Sonderegger, W. (2021): *Holzphysik. Eigenschaften, Prüfung und Kennwerte*. 2nd ed., Carl Hanser Verlag, Munich, Germany

[5] Wagenführ, A. (2008): *Holzatlas*. Carl Hanser Verlag, Munich, Germany

[6] Kollmann, F. F. P. (1951): *Technologie des Holzes und der Holzwerkstoffe. Band 1*. 2nd ed., Springer Verlag, Berlin/Göttingen/Heidelberg

[7] Kollmann, F. (1933): *Die Abhängigkeit der Trocknungsgeschwindigkeit des Holzes vom Raumgewicht und der Struktur*. In: Forsch.-Ber. Holz H. 1, Berlin, Germany

[8] Niemz, P.; Sandberg, D. (2025): *General Data Related to Wood*. Chapter 1 in *Wood Material and Process Data. The Most Relevant Data, Tables and Figures*. Springer Nature, Switzerland

[9] Marinescu, I. (1979): *Uscarea lemnului*. Editura Tehnică, Bucharest, Romania

[10] Trübswetter, T. (2009): *Holztrocknung. Verfahren zur Trocknung von Schnittholz – Planung von Trocknungsanlagen*. 2nd ed., Carl Hanser Verlag, Leipzig, Germany

[11] DIN EN 13183-1 (2002): Feuchtegehalt eines Stückes Schnittholz – Teil 1. Bestimmung durch Darrverfahren

[12] DIN EN 13183-2 (2002): Feuchtegehalt eines Stückes Schnittholz – Teil 2. Schätzung durch elektrisches Widerstands-Messverfahren

[13] DIN EN 13183-3 (2005): Feuchtegehalt eines Stückes Schnittholz – Teil 3. Schätzung durch kapazitives Messverfahren

[14] Informationsdienst Holz (2008): *Holz als konstruktiver Baustoff. Holzbau Handbuch, Reihe 4, Teil 1, Folge 1.* Holzabsatzfonds Absatzförderungsfonds der deutschen Forst- und Holzwirtschaft, Bonn, Germany

[15] Perre, P.; Keey, R. B. (2006): *Drying of Wood: Principles and Practices.* Chapter 36 in *Handbook of Industrial Drying.* 3rd ed., Taylor and Francis Group, CLC Press

[16] Keylwerth, R. (1969): *Praktische Untersuchungen zum Holzfeuchtigkeits-Gleichgewicht.* In: Holz als Roh- und Werkstoff, 27 (8), p. 285 – 290

[17] Frühwald, A.; Ressel, J. (1982): *Holztrocknung. Teil 1. Grundlagen der Holztrocknung.* Universität Hamburg, Germany

[18] DIN 52184 (1979): Prüfung von Holz – Bestimmung der Quellung und Schwindung

[19] Keylwerth, R.; Noack, D. (1968): *Über den Einfluss höherer Temperaturen auf die elektrische Holzfeuchtigkeitsmessung nach dem Widerstandsprinzip.* In: Holz als Roh- und Werkstoff, 14, p. 162 – 172

[20] Pitsch, H. (o. J.): Thermodynamik I. Institut für Technische Verbrennung, Universität Aachen. Online: *https://www.itv.rwth-aachen.de/fileadmin/LehreSeminar/Thermodynamik_I/Vorlesungen_SS14/Thermo_I_Kap_2.pdf*, retrieved 27. 04. 2025

[21] Lohmann, U. (1998): *Holz Handbuch.* 5th ed., DRW Verlag, Leinfelden-Echterdingen, Germany

[22] Popa, V. (2012): *Automatizari in industria lemnului, Vol. 1.* Transilvania University, Brasov, Romania

[23] Krischer, O.; Kröll, K. (1963): *Die wirtschaftlichen Grundlagen der Trocknungstechnik.* 2nd ed., Springer Verlag, Berlin/Göttingen/Heidelberg

[24] Langendorf, G.; Eichler, H. (1982): *Holzvergütung.* 3rd ed., VEB Fachbuchverlag, Leipzig, Germany

[25] Bedelean, B. (2009): *Contributii cu privire la studiul aerodinamic al camerelor de uscare a lemnului.* Doctoral thesis, Transilvania University, Brasov, Romania

[26] Klinkmüller, H. (1987): *Versuche zur Bestimmung des Einflusses von wechselnden Luftrichtungen auf die Trocknung.* Bericht auf DGfH-Sitzung

[27] Brunner-Hildebrand (Hrsg.) (1987): *Die Schnittholztrocknung.* 5th ed., Hannover, Germany

[28] Câmpean, M.; Marinescu, I.; Ispas, M. (2003): *Evaluation of a Kiln Incorporating an Oscillating Plate Airflow System.* In: Maderas Ciencia y Tecnología, 5(2), p. 137 – 143. *https://doi.org/10.4067/S0718-221X2003000200005*

[29] Bedelean, B.; Campean, M. (2013): *Tratamente termice ale lemnului. Îndrumar pentru lucrări practice.* Transilvania University, Brasov, Romania

[30] Bian, Z. (2001): *Airflow and Wood Drying Models for Woodkilns.* Master Thesis, University of British Columbia. *https://doi.org/10.14288/10080969*

[31] Ledig, S. F.; Paarhuis, B.; Riepen, M. (2007): *Airflow within Kilns.* In: Perre, P. (Hrsg.), Fundamentals of Wood Drying, European Cost A.R.BO.LOR.

[32] Idelcik, I. E. (1984): *Îndrumător pentru calculul rezistențelor hidraulice.* Editura Tehnică, Bucharest, Romania

[33] Muntean, S. (2008): *Analiza numerică a curgerii în turbinele hidraulice*. Editura Orizonturi Universitare, Timisoara, Romania

[34] Bedelean, B.; Muntean, S.; Campean, M. (2016): *Analysis of Drying Kiln Aerodynamics Based on a Full Three-Dimensional Turbulent Numerical Computation*. In: Drvna Industrija, 67(1), p. 53 – 64. *https://doi.org/10.5552/drind.2016.1515*

[35] Hua, L.; Bibeau, E.; He, P.; Gartshore, I.; Salcudean, M.; Bian, Z.; Chow, S. (2001): *Modeling of Airflow in Wood Kilns*. In: Forest Product Journal, 51, p. 74 – 81

[36] Bibeau, E. (2000): *Modelling of Airflow in Wood Kilns*. Kiln Drying Course, University of British Columbia, Canada

[37] Rodriguez-Ramirez, J.; Ignacio-Caballero, F.; Mendez-Lagunas, L. L.; Sandoval-Torres, S. (2014): *Air Flow Simulation in Timber (Lumber) Kilns: Shape and Geometry Effects*. In: 19th International Drying Symposium (IDS), Lyon, France

[38] Bedelean, I. B.; Sova, D. (2010): *Investigations into the Possibility of Minimising Aerodynamic Resistance of Kiln Drying Stacks*. In: International Wood Products Journal, 1(2), p. 62 – 69. *https://doi.org/10.1179/2042645310Y.0000000004*

3 Heat and Moisture Transport in Wood Drying

The drying of wood is accompanied by a continuous exchange of heat and moisture between the material and its surroundings. These processes belong to the category of transfer phenomena and are generally referred to as heatand mass transport (moisture transport).

Quantitatively, all transfer phenomena are characterized by similar mathematical formulas based on one and the same property: the amount of matter, whether in the form of energy or substance, that moves per unit time is proportional to the potential gradient and inversely proportional to the length of movement. The direction of transport always proceeds from higher to lower potential.

The transfer potential is represented by the intensity of a physical field (e.g., temperature is the transfer potential for heat transport processes, moisture is the transfer potential for moisture transport, etc.).

The gradient of a physical quantity arises when it varies within the material. It is calculated as the difference between the maximum and minimum values of that quantity within the material. Once a potential gradient exists between two points of the field, it acts as the driving force of the transfer phenomenon. This takes place as long as a potential difference exists, and stops as soon as it is zero (e.g., heat transfer from the warm air to the wood surface only occurs as long as the temperature of the wood surface is lower than the air temperature).

3.1 Heat Transfer in Wood Drying

3.1.1 Types of Heat Transfer

There are three types of heat transfer:

- by convection – this is heat transfer by gas or liquid flow;

- by conduction (or in common usage, thermal conduction) – this is heat transfer by direct contact;
- by radiation – this is heat transfer by electromagnetic oscillations.

In a drying process, two or even all three forms of heat transfer can occur, e.g., in fresh air–exhaust air drying: the heating of the wood surface occurs by convection, the inner layers of the wood are heated by conduction, and the lateral parts of the stacks are additionally heated by radiation when the heating registers are installed laterally in the immediate vicinity of the stacks. In high-frequency drying (HF drying), on the other hand, the heating of the inner layers of the wood is first ensured by radiation, after which heat transfer by conduction occurs throughout the entire wood mass.

Depending on the dominant form of heat transfer, the wood drying methods are classified according to Table 3.1.

Table 3.1 Classification of Drying Methods According to the Form of Heat Transfer

Drying by convection		Drying by conduction	Drying by radiation
in gases	in liquids		
▪ Air drying ▪ Fresh air–exhaust air drying ▪ Condensation drying ▪ High-temperature drying ▪ Vacuum drying ▪ Solar drying ▪ Drying in vapors of organic substances ▪ Centrifugal drying ▪ Freeze-sublimation drying	▪ Drying in hot oils ▪ Drying in organic solvents	▪ Press drying	▪ IR drying ▪ High-frequency drying ▪ Microwave drying

Table 3.1 shows that most methods are convective and use a gaseous drying agent (in most cases air). For these methods, the scheme in Figure 3.1.

It is noted that the heat and moisture exchange takes place both in the boundary layer between the wood surface and the drying medium and inside the wood.

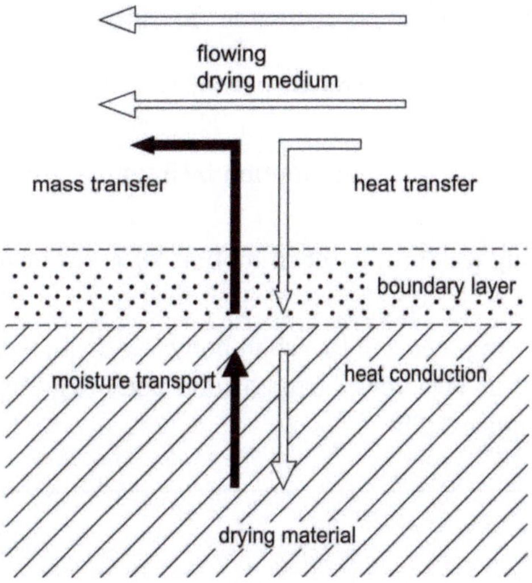

Figure 3.1 Heat and moisture transport during convective drying (based on [1])

In the boundary layer between the wood surface and the drying environment, the surface processes take place:

- the heat transfer by convection, oriented from the environment to the wood surface (this ensures the heating of the wood surface)

  and

- the mass transfer (water evaporation), oriented from the wood surface to the drying environment – this ensures the removal of the escaping water.

The two transport phenomena depend significantly on the condition of the wood surface (temperature, surface moisture, roughness, etc.) and on the parameters of the drying medium (air temperature, air velocity, type of flow and degree of turbulence, etc.).

The transport phenomena that take place inside the wood are:

- the heat transfer by conduction, oriented from the surface into the interior of the wood piece; this ensures the heating of the entire volume of the wood piece

  and

- the movement of water in the wood (by capillary movement and diffusion), oriented from the interior of the wood to the surface.

These two transport phenomena are determined mainly by the wood properties (wood species, density, structure, hygroscopicity, porosity, permeability, temperature, thermal conductivity, moisture, diffusion resistance, etc.).

3.1.2 Heat Transfer by Convection

In this type of transfer, the energy (heat) moves with the molar particles of the drying medium. A distinction is made between:

- free convection – in which the flowing medium is set in motion by a temperature or density difference;
- forced convection – in which the flowing medium is set in motion by a fan or a pump.

The heat transfer can take place from a flowing medium to a solid body (e.g. from the air flow to the wood surface) (Figure 3.2 a) or vice versa, from a solid body (e.g. the heating register) to the flowing medium (Figure 3.2 b).

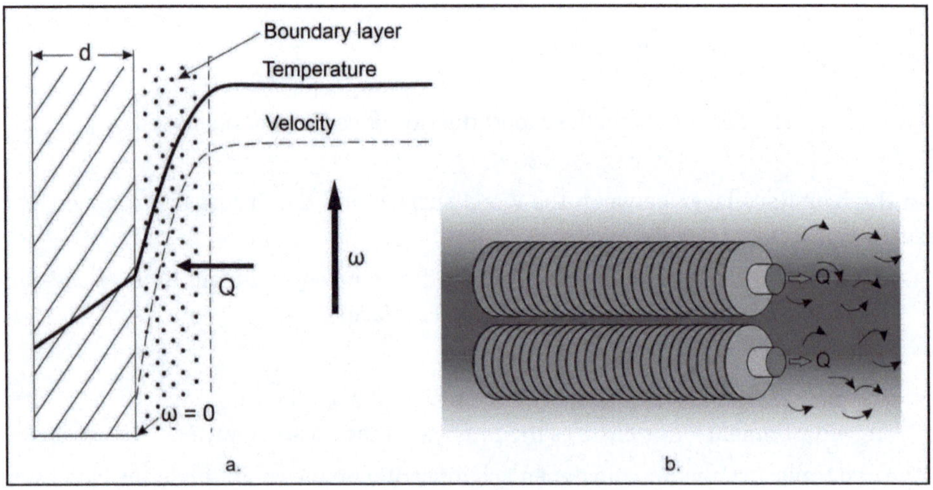

Figure 3.2 Schematic representation of heat transfer by convection: a – from a flowing medium to a solid body (based on [2]); b – from a solid body to the flowing medium

In the first case, when heat is transferred from the flowing medium to the material being dried (Figure 3.2 a), the amount of heat transferred per unit time ($\dot{Q}$) is calculated from the following equation:

$$\dot{Q} = \alpha \cdot A \cdot (T_1 - T_2)\ [\mathrm{W}] \tag{3.1}$$

α heat transfer coefficient [W/m^{2}K]
A surface through which heat transfer takes place [m^2]
T_1 temperature of the flowing medium [K]
T_2 temperature of the wood surface [K]

The heat transfer coefficient (α) thus indicates the amount of heat transferred per unit time through 1 m^2, when the temperature difference between the flowing medium and the wood surface is 1 K.

The process of heat transfer between solids and gases is particularly complex and depends on numerous factors: type of flowing medium, temperature, pressure, flow velocity, orientation of the flow to the surface, surface roughness, thickness of the boundary layer, etc. All these influences are summarized in the heat transfer coefficient (α). The αvalue can only be determined accurately by empirical means. For example, for airflow along a plate, an approximation of the αvalue is possible using the following equations [3]:

for $w < 5$ m/s:

$$\alpha = 5{,}8 + 4 \cdot w \text{ [W/m}^2\text{K]} \tag{3.2}$$

for $w > 5$ m/s:

$$\alpha = 7{,}15 \cdot w^{0{,}78} \text{ [W/m}^2\text{K]} \tag{3.3}$$

w velocity of the airflow [m/s]

When heat is transferred from a solid body to the flowing medium (Figure 3.2 b) the amount of heat transferred per unit time ($\dot{Q}$) is calculated using the following equation:

$$\dot{Q} = U \cdot A \cdot (T_1 - T_2) \text{ [W]} \tag{3.4}$$

U overall heat transfer coefficient [W/m^2 K]
A surface through which heat exchange takes place [m^2]
T_1 temperature of the solid body [K]
T_2 temperature of the flowing medium [K]

The overall heat transfer coefficient (U) characterizes the material of which the solid body is made. In the case of bimetallic finned tubes, which are nowadays used in the heating coils of most drying chambers, U = 34 – 47 W/m^{2}K.

3.1.3 Heat Transfer by Conduction

In contrast to convection, this type of transfer does not involve the transport of matter, but only of energy. Energy transfer occurs through collisions between neighboring elementary particles. Heat transfer by conduction (Figure 3.3) is specific to solid bodies (such as in the case of contact drying, where the wood surface is heated by contact with a hot plate).

According to Fourier's law, the amount of heat moving per unit time between two isothermal surfaces within the body is proportional to the surface area (A), to the temperature gradient (ΔT) and is inversely proportional to the distance (d) between the two surfaces (Figure 3.3):

$$\dot{Q} = \frac{\lambda \cdot A}{d} \cdot (T_1 - T_2) \tag{3.5}$$

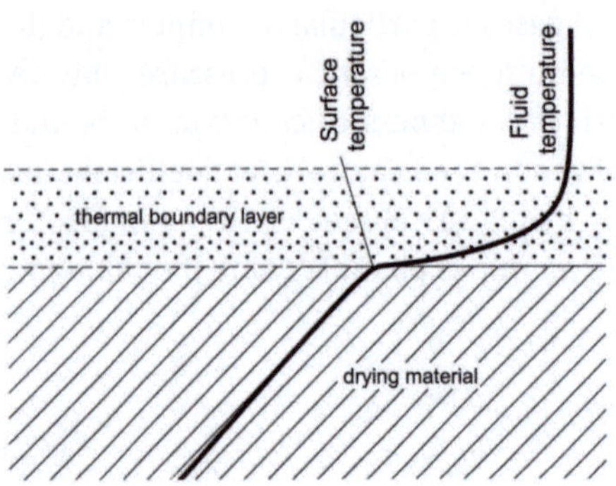

Figure 3.3 Schematic representation of heat transfer by conduction (adapted from [2])

The thermal conductivity (λ) is a material characteristic. In the case of wood, it depends on the wood species, density, moisture content, temperature, and the orientation of the heat flow relative to the grain direction (see Section 2.1.7).

3.1.4 Heat Transfer by Radiation

Radiation is the process of heat transfer from one body to another through electromagnetic waves (oscillations). When these waves strike a body that is receptive to them, the radiant energy is converted into thermal energy. The wavelength range of thermal radiation extends from 0.02 to 800 µm.

Heat transfer by radiation is encountered in IR drying and in HF drying.

Heat transfer by infrared radiation. The particularly effective wavelength range of infrared radiation extends from 0.8 to 400 µm.

The total radiant energy (E), which a heated body can emit, is proportional to the fourth power of its absolute temperature (T) according to the Stefan-Boltzmann law:

$$E = C \cdot \left(\frac{T}{100}\right)^4 \ [\mathrm{W/m^2}] \tag{3.6}$$

C radiation constant [$\mathrm{W/m^2K^4}$]
T absolute temperature [K]

This law applies to the perfectly black body, which absorbs all radiation. This is practically accurate enough for all technical surfaces [2].

The radiation constant (C) for the perfectly black body is $C = C_s = 5.77\ \mathrm{W/m^2K^4}$.

The radiated energy is absorbed by the wood surface and penetrates to a limited depth into the wood. The penetration depth depends on the wavelength of the radiation, but also on the type of wood: fir and spruce – 6 mm; poplar, pine – 4 mm; beech – 3 mm; oak – 2 mm; walnut – 0.5 mm. Therefore, drying with IR rays is only suitable for thin wood assortments (lamellae, friezes, veneers, chips). To achieve uniform drying, all points of the irradiated surface must be equidistant from the radiation source.

Heat transfer in an alternating electric field. The wavelength range of high-frequency radiation (HF) is considerably higher than that of IR radiation. Waves from two different frequency ranges are used in wood drying: radio frequency waves (with $f < 100$ MHz) and microwaves (with $f > 500$ MHz) (Table 3.2).

Table 3.2 High-Frequency Bands for Industry (source: [4])

	Frequency Hz	Band Hz	Wavelength m
Radio frequencies	13.56	13.530 - 13.567	22.11
	27.12	26.975 - 27.283	11.05
	40.68	40.660- 40.700	7.37
Microwaves	915	902 - 928	0.328
	2450	2,400 - 2,500	0.122
	5,800	5,725 - 5,875	0.052

Wet wood naturally contains various types of magnetizable particles: positive and negative ions, electrons, dipoles (e.g. water molecules) (Figure 3.4 a). As soon as the wood is placed as a dielectric between the plates (electrodes) of an HF generator, the water molecules contained in the wood begin to oscillate and tend to orient themselves toward the pole with the opposite charge (Figure 3.4 b). This requires a certain amount of time, as they mutually impede one another. The polarization cannot follow the extremely rapid reversals of the electric field (Figure 3.4 c), resulting in friction between the molecules. To overcome this, a precise amount of energy is required. This is provided by the alternating electric field, and thus the electrical energy is converted into thermal energy ("frictional heat").

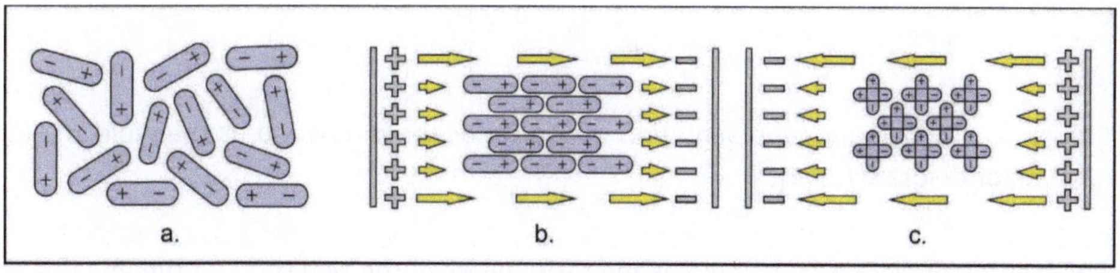

Figure 3.4 Polarization of water molecules in an alternating electric field (source: [5])

The wettest spots inside the wood absorb more energy and heat up the most (Figure 3.5).

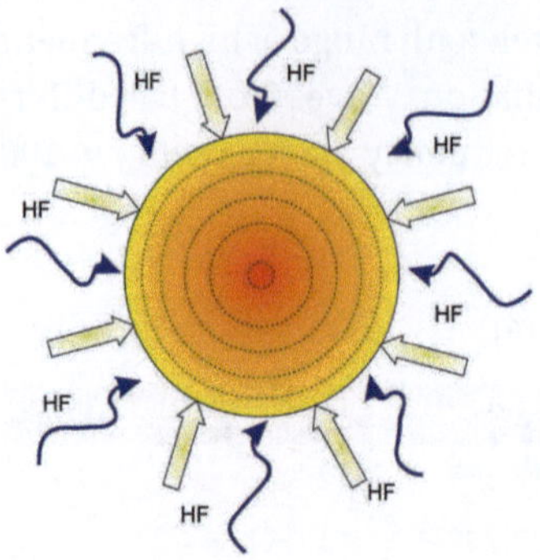

Figure 3.5 Principle of wood heating in an alternating electric field (source: [5])

The specific power density (P), which the material absorbs, can be determined using the following equation [4]:

$$P = 2\,\pi \cdot f \cdot \varepsilon_0 \cdot \varepsilon' \cdot \mathrm{tg}\delta \cdot E^2\ [\mathrm{W/m^3}] \tag{3.7}$$

f frequency [s^{-1}]
ε_0 permittivity of free space ($\varepsilon_0 = 8.8 \cdot 10^{-12}$ F/m)
ε' relative permittivity of moist wood [F/m] (see Section 2.1.8)
tgδ loss angle (Figure 3.6)
E electric field strength, $E = U/d$ [V/m], where U voltage [V], d distance between the two electrodes [m]

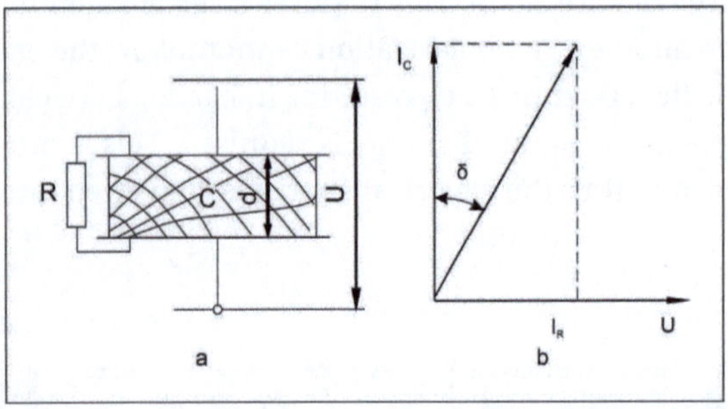

Figure 3.6 Schematic representation of the capacitor with wood as the dielectric (a) and vector diagram (b) [4]

The loss factor (tgδ) is calculated as the ratio between the real (I_R) and the imaginary component (I_C) of the dielectric impedance (Figure 3.6 b).

Most applications of volumetric radio-frequency heating require a uniform heat distribution and complete energetic penetration of the wood part. Therefore, knowledge of the penetration depth is of great importance. Penetration into a dielectric material attenuates the RF wave.

The attenuation factor (α) of a low-loss material (e.g. wood), where $\varepsilon''_{\mathrm{eff}}/\varepsilon' \ll 1$, depends on its dielectric properties and on the free-space wavelength (λ'_0) [4]:

$$\alpha = \frac{\lambda_0' \cdot (\varepsilon')^{0,5}}{\pi \, \varepsilon_{\text{eff}}''} \tag{3.8}$$

$\varepsilon_{\text{eff}}''$ effective loss factor

The wavelength (and frequency) of the applied radiation is decisive with respect to the penetration depth, but the size and shape of the wood part also play an important role. The longer the wavelength of the radiation, the deeper it can penetrate. Since radio frequencies have wavelengths on the order of meters, insufficient penetration is rarely a problem. When using microwaves, however, this is a problem due to the much smaller wavelengths (see Table 3.2), and therefore microwave heating and drying of wood is limited to thin materials and small objects.

Table 3.3 shows the approximate penetration depths of 2.45 GHz microwaves for wood with two densities. The values show the increase in penetration depth with decreasing wood moisture content and with decreasing temperature.

Table 3.3 Penetration Depths of Microwaves with *f* = 2.45 GHz in Wood Perpendicular to the Grain (source: [4])

Density [kg/m³]	Temperature [°C]	Wood moisture content [%]	Penetration depth [mm]
400	20	10	164
		30	58
		100	46
	50	10	160
		30	61
		100	50
	90	10	153
		30	70
		100	54
800	20	10	67
		30	22
		100	15
	50	10	160
		30	24
		100	17
	90	10	62
		30	27
		100	20

3.2 Moisture Transport During Wood Drying

3.2.1 Evaporation of Water from the Wood Surface

The evaporation of water from the wood surface is comparable to the evaporation of water from an uncovered surface. This involves a one-sided permeable interface, because only one of the substances involved (the water vapor escaping from the wood) can pass through it. The air diffusing in the opposite direction cannot be absorbed by the moist surface.

In this case, the amount of water evaporating per unit area per unit time is determined using the following equation [7]:

$$\dot{m}_W = \frac{\beta_H}{R_D \cdot T} \cdot (p_{D0} - p_{DL}) \text{ [kg/m}^2\text{s]} \quad (3.9)$$

β_H mass transfer coefficient for the one-sided permeable wall [m/s]
R_D gas constant for water vapor: R_D = 461.5 [J/kg·K]
T absolute temperature [K]
p_{D0} vapor pressure at the surface [N/m^2]
p_{DL} partial vapor pressure at the boundary layer edge [N/m^2]

At low partial vapor pressures, i.e., when $p_{DL} \rightarrow 0$, $\beta_H \cong \beta$.

Equation 3.9 applies at constant temperature, velocity, and humidity of the airflow and an airflow parallel to the surface.

With an airflow perpendicular to the water surface, the amount of evaporated water is almost twice as high.

From the end grain, free water evaporates approximately 25% more intensively than from the side faces (edges), which is probably due to the higher roughness that increases the effective evaporation area compared to the geometric area [2].

The diffusion of water vapor extracted from the wood into the air, i.e., the moisture exchange between wood and air at the contact boundary, is achieved by the pressure difference between the two environments. A prerequisite for drying is that the vapor pressure at the wood surface is greater than the partial pressure of water vapor in the air layer immediately adjacent to the wood surface ($p_{D0} > p_{DL}$).

The vapor pressure at the wood surface depends on the temperature and on the equilibrium moisture content of the wood. The partial pressure of water vapor in the air depends on the temperature and the relative humidity. This shows that evaporation is a heat-requiring process that occurs more intensively the higher the temperature of the ambient air, the lower the relative humidity, and the higher the air velocity at the surface to be dried.

If $p_{D0} < p_{DL}$ wetting of the material occurs (sorption).

If $p_{D0} = p_{DL}$, the wood is in equilibrium with the environment and no evaporation takes place.

3.2.2 Moisture Transport Inside the Wood

The goal of technical drying is to dehumidify the wood to a certain moisture content in the shortest possible time. In doing so, cracks and deformations should be avoided so that a large percentage of the dried goods meets the specified requirements.

To achieve this purpose, it is important in wood drying to understand how the movement of water and water vapor from the interior of the wood to the wood surface proceeds. If the movement is too rapid, stresses occur. If the movement is too slow, drying is uneconomical due to the high energy expenditure.

It is therefore particularly important to know the fundamentals of moisture transport during wood drying in order to control the movement of water such that an optimal ratio between duration and quality is achieved.

As already mentioned in Section 2.1.3, wood contains both free water in the cell cavities and bound water in the cell walls. The free water is liquid; the bound water is present both in a liquid state and as water vapor. Accordingly, the movement of water differs depending on the binding forces with the wood and the state of aggregation. As a result, moisture movement in capillary-porous bodies has a very complex character.

Above the fiber saturation range, the movement of free water is caused by capillary forces. Below the fiber saturation range, the high capillary forces of the fine capillaries are initially still involved in the movement of the liquid portion of the bound water, but the predominant transport process is diffusion.

The proportions of the liquid quantities moved by capillary forces and/or diffusion change under the influence of temperature. The higher the temperature, the smaller the proportion of capillary water movement becomes, so that in the hygroscopic range the moisture movement can be regarded as pure diffusion.

As long as drying proceeds in the range above fiber saturation, the wood properties do not affect the movement of water. The determining factors are the external conditions (temperature, relative humidity, air velocity, degree of turbulence of the airflow, etc.). If these conditions are kept constant, the drying rate (amount of water released per unit of time) remains constant. Only the distribution of water in the wood changes. This section of drying is represented by the straight portion in Figure 3.7.

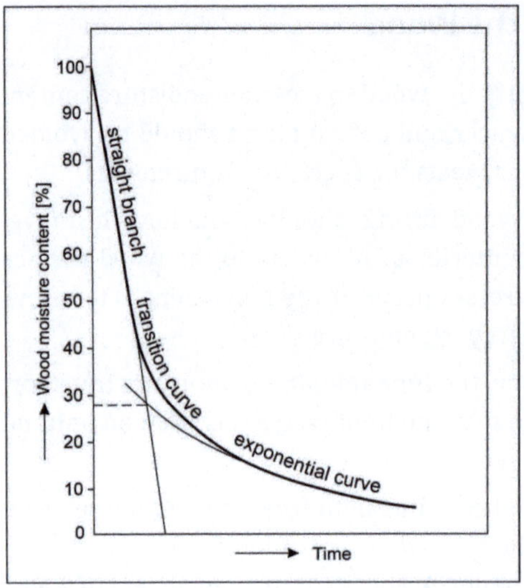

Figure 3.7 Course of the drying curve (based on [8])

Below the fiber saturation range, the wood properties, in particular the shrinkage forces, have an inhibiting effect, so that the drying rate decreases according to an exponential curve (Figure 3.7). However, since capillary condensation is not abruptly replaced by diffusion, but both processes run partially in parallel, a transition curve forms that connects the straight branch and the exponential curve (see Figure 3.7). Therefore, without significant error, the entire drying process from 100% to 0% wood moisture content can be assumed to follow an exponential curve.

Capillary Movement of Free Water in Wood

Above the fiber saturation point, water is present in the larger cell spaces in the form of droplets. These are not entirely "free", as they are connected to the wood by capillary forces. The movement of water in the water-filled capillaries is determined by the surface forces in the liquid menisci. The surface tension T, in a meniscus formed in a capillary tube with radius R (Figure 3.8), is:

$$T = \frac{2 \cdot \beta}{R} \quad (3.10)$$

β surface tension coefficient of water [kg/m]

This is calculated using the Weinberg formula, as a function of temperature t (in °C):

$$\beta = 0.00769 \cdot (1 - 0.00225 \cdot t) \, [\text{kg/m}] \quad (3.11)$$

Equation 3.10 shows that the capillary pressure is inversely proportional to the radius of the liquid meniscus. If a capillary system consists of tubes of different diameters (as is the case with wood), different tensile forces occur in their liquid menisci, which cause water to migrate from the wide capillaries into the narrow ones during moisture absorption, and during moisture release (drying) the water is drawn from the narrowest capillaries into the wider capillaries. As a result, the capillaries with the smallest radii are filled first. In other words, in this case the movement of water in wood is driven by the capillary pressure gradient.

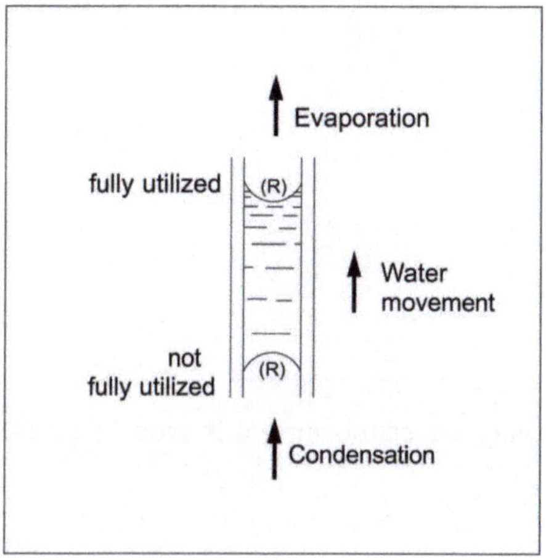

Figure 3.8 Diagram of water movement in capillaries (based on [8])

Figure 3.9 illustrates capillary moisture movement during wood drying using a simplified configuration [10]. At the beginning of drying (Figure 3.9 a) all cell lumens are filled with water, with the exception of cell lumens f and h, which contain air bubbles (b_1 and b_2). It should be noted that in capillary a, which is narrower, the water level is higher than in similar capillaries (b and c), which are located at the evaporation surface ($m - n$). When evaporation begins at the wood surface, the water level in the exposed cell lumens (a, b and c) first drops until it is completely removed (Figure 3.9 b). At the same time, drying of the cell walls begins. The dashed curve $x - y$ indicates the progression of the fiber saturation moisture content in the cell walls. Since the meniscus radius in capillary a is smaller than the radius of the air bubbles in capillaries f and h, the bubbles expand and continuously push the water to the surface until these capillaries are also completely empty (Figure 3.9 c). Below the fiber saturation range, the proportion of water movement by diffusion increases significantly.

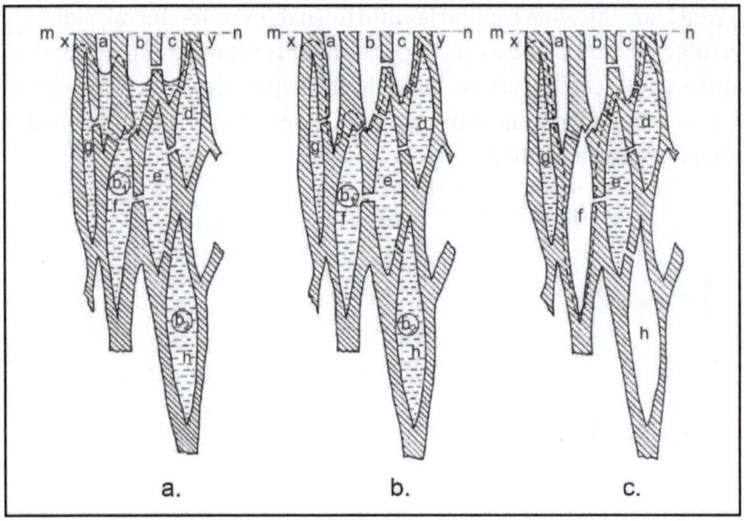

Figure 3.9 Schematic representation of capillary water movement in wood (source: [9])

Diffusion

By diffusion one understands all molecular movement processes in a fluid (gas or liquid) due to a partial pressure- or concentration difference. During wood drying, the movement of bound water (as water vapor) takes place by diffusion.

The diffusion rate (g) is determined using Fick's law:

$$g = -D \cdot \frac{\partial c}{\partial x} \tag{3.12}$$

D diffusion coefficient [m^2/s]
$\partial c/\partial x$ concentration gradient [kg/m^4], where c concentration [kg/m^3] and x length [m]

This equation applies under steady-state conditions and at constant temperature.

The diffusion coefficient (D) indicates the amount of liquid that, per unit time, at a moisture concentration difference of 1 kg/m^3 passes through a cross-section of 1 m^2 flows. It depends on the wood species, temperature, moisture content of the wood, and the direction of water movement relative to the wood grain (Table 3.4). The Dvalues are maximum for movement along the grain and 2 – 4 times smaller in the tangential direction (perpendicular to the grain).

The diffusion rate can analogously also be expressed by means of other gradients (e.g., the moisture gradient, the temperature gradient, or the pressure gradient) [10].

Under the influence of the moisture gradient ($\partial u/\partial x$) water tends to move within the wood from the most moist area to the area with the lowest moisture content.

Table 3.4 Diffusion Coefficients of Selected Wood Species (source: [9])

Wood species	Cutting direction	Temperature [°C]	Diffusion coefficient [$m^2/s \cdot 10^{-10}$], at a moisture content of:		
			10 %	15 %	20 %
Spruce	Radial	20	0.4	0.8	1.6
		40	0.4	0.6	0.8
		60	1.1	1.5	2.1
		75	1.9	2.9	4.1
		100	6.2	8.7	12
	Tangential	20	0.3	0.7	1.7
	Longitudinal	20	5.8	5.6	
		60	28	17	14
Pine	Radial	20	0.6	1.4	2.4
	Tangential	20	0.6	1.7	4.0
Beech	Radial	20	0.5	0.8	1.3
	Tangential	20	0.2	0.4	0.7
	Longitudinal	20	2.2	4.1	-
Oak	Radial	20	0.4	0.7	1.1
	Tangential	20	0.3	0.5	0.6

Under the effect of the temperature gradient ($\partial t/\partial x$) the water inside the wood moves from the warmest point to the point with the lowest temperature.

Under the effect of the pressure gradient ($\partial p/\partial x$), which is created by an increase in pressure inside the wood compared to the external environment (e.g., during vacuum drying), the water tends to move in the direction of the zone with lower pressure.

In fresh air/exhaust air drying (Figure 3.10 a), which is most commonly taken as a reference because it is the most widespread drying method, the diffusion of bound water takes place solely due to the moisture gradient. Since in high-frequency drying (Figure 3.10 b) the interior of the wood is heated first, a temperature gradient is created in addition to the moisture gradient, also oriented from the interior of the wood to the surface. This reduces the drying time compared to conventional fresh air/exhaust air drying. Another situation in which two of the three gradients contribute to drying is vacuum drying (Figure 3.10 c). In this case, the pressure gradient is added and transports the bound water to the wood surface. High-frequency vacuum drying (HFV) is the only method in which all three gradients act and are also oriented in the same direction, from the interior of the wood to the surface (Figure 3.10 d). For this reason, HFV drying is also the fastest drying method developed to date.

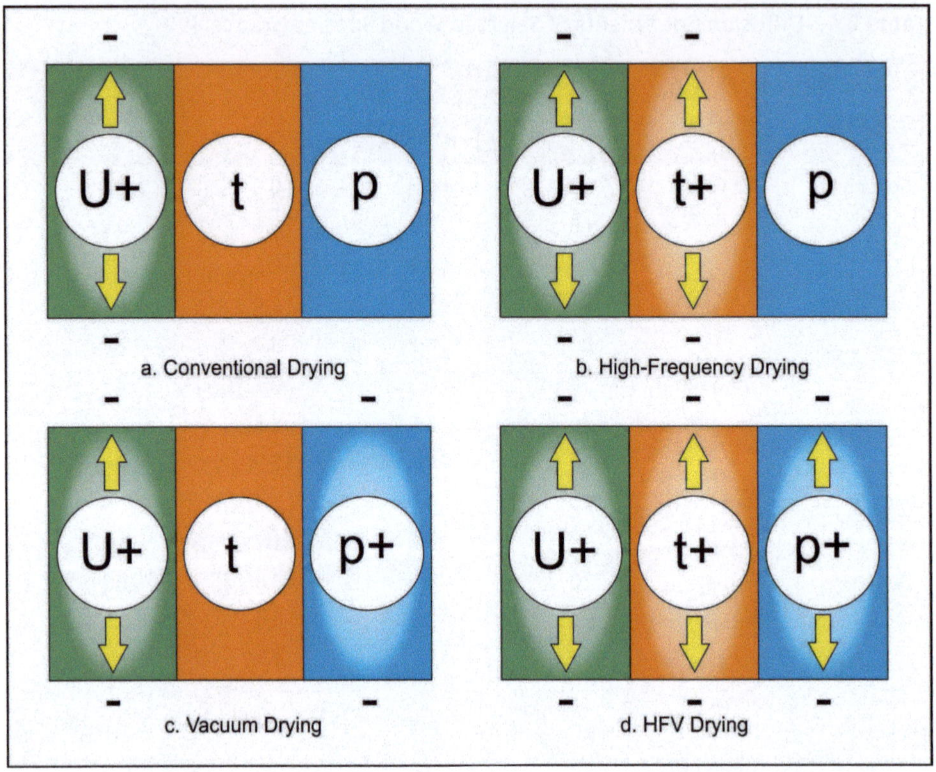

Figure 3.10 Orientation of the moisture, temperature, and pressure gradients in various drying methods

The removal of water from wood during drying is the result of a complex coupling between the heat and mass transfer processes mentioned above. Based on the fundamental equations describing these processes, and assuming various simplifications and boundary conditions, models have been developed that simulate the decrease in wood moisture content and allow an approximate calculation of the drying time for a specific wood assortment (see Chapter 5).

References

[1] Ressel, J. (1986): *Untersuchungen über den Energieverbrauch bei der technischen Schnittholztrocknung und Möglichkeiten zu dessen Verringerung*. Dissertation, Universität Hamburg, Germany

[2] Frühwald, A.; Ressel, J. (1982): *Holztrocknung. Teil 1. Grundlagen der Holztrocknung*. Universität Hamburg, Germany

[3] Recknagel, H.; Sprenger, E. (1981): *Taschenbuch für Heizung und Klimatechnik*. Oldenbourg Verlag, Munich, Germany

[4] Resch, H. (2009): *Drying Wood with High-Frequency Electric Current*. University of Natural Resources and Applied Life Sciences, Vienna, Austria und Society of Wood Science and Technology, Wisconsin, USA

[5] Câmpean, M. (2005): *Heat Treatments of Wood*. Transilvania University, Brasov, Romania

[6] Marinescu, I. (1980): *Uscarea și tratarea termică a lemnului. Vol. 2*. Editura Tehnică, Bucharest, Romania

[7] Kneule, F. (1959): *Das Trocknen*. Verlag H. R. Sauerländer & Co., Aarau und Frankfurt am Main, Germany

[8] Janik, W. (1965): *Handbuch der Holztrocknung*. Fachbuchverlag, Leipzig, Germany

[9] Niemz, P.; Sonderegger, W. (2021): *Holzphysik. Eigenschaften, Prüfung und Kennwerte*. 2nd ed., Carl Hanser Verlag, Munich, Germany

[10] Siau, J. F. (1995): *Wood: Influence of Moisture on Physical Properties*. Department of Wood Science and Forest Products, Virginia Polytechnic Institute and State University, Keene, NY

4 The Conventional Kiln Drying

The conventional kiln drying takes place in an air-steam mixture (e.g., moist air), which is first heated and then transfers the thermal energy necessary for drying to the wood. At the same time, the air absorbs the moisture escaping from the wood and carries it away.

4.1 Kinematics and Mechanics of Conventional Drying

The kinematics of wood drying investigates the movement of water in the wood during drying. For illustration purposes, the moisture distribution curves across the thickness of the piece of wood are shown at successive important stages of the process.

The mechanics of wood drying investigates the causes of drying stresses. It is already known that wood shrinks as soon as the removal of bound water begins. This inevitably leads to stresses in the wood during drying. The anisotropy of the wood and the uneven distribution of moisture within the wood further intensify these undesirable stresses.

To highlight the occurrence of stresses, a piece of wood can be used, which is cut in half along its thickness. If stresses exist in the wood, the two halves deform toward the side that is subjected to tensile stress.

To also quantify the intensity of the stresses, several cuts (teeth) are made across the thickness of the piece of wood. Once the connections are severed, each tooth will behave freely, i.e., it will shrink if its moisture content is below the fiber saturation range, and the lower its moisture content, the more it will shrink. The greater the difference in shrinkage between different teeth within the same piece of wood, the greater the resulting internal stresses will be.

The kinematic and mechanical behavior of wood during drying are closely intertwined. The temporal progression of the drying process is shown in Figure 4.1. Figure 4.2 illustrates the most important stages of water movement in the wood during this process, based on the moisture distribution curves across the wood thickness. Figure 4.3 illustrates the stresses that occur during these stages. To understand the behavior of wood during drying, the three figures must be analyzed in parallel.

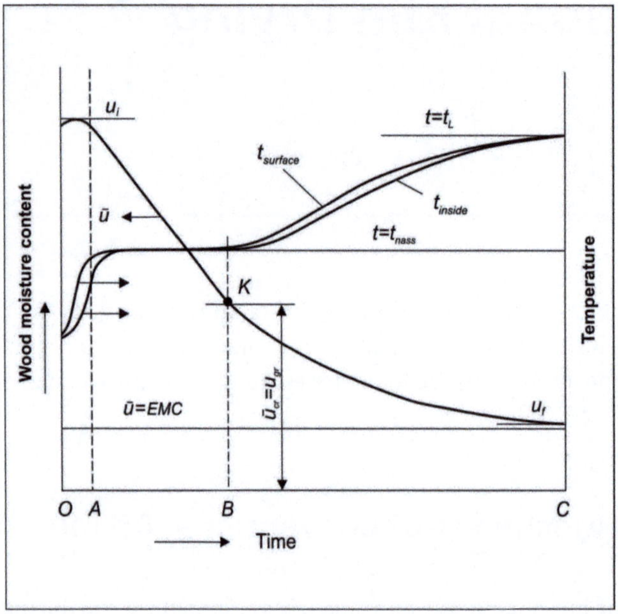

Figure 4.1 Temporal progression of convective drying at temperatures < 100 °C (source: [1], with permission)

A piece of wood with thickness d and initial moisture content u_i ($u_i > u_{fs}$) is therefore considered (stage 0 in Figure 4.1 and Figure 4.2), which is placed in a gaseous environment (warm and moist air) at a temperature below 100 °C for drying.

In the first phase of the process (OA in Figure 4.1), the material is heated, with its moisture content remaining approximately constant at the initial value (u_i) remains. The surface temperature of the material (t_{Ob}) rises rapidly and after a relatively short time a superficial evaporation begins, which initially leads to a decrease in moisture content of the outer layers (stage 0 to 1 in Figure 4.2). In this phase there are theoretically no stresses in the wood (Figure 4.3 a), because the wood moisture content throughout the entire wood volume is still above the fiber saturation range and therefore no shrinkage occurs.

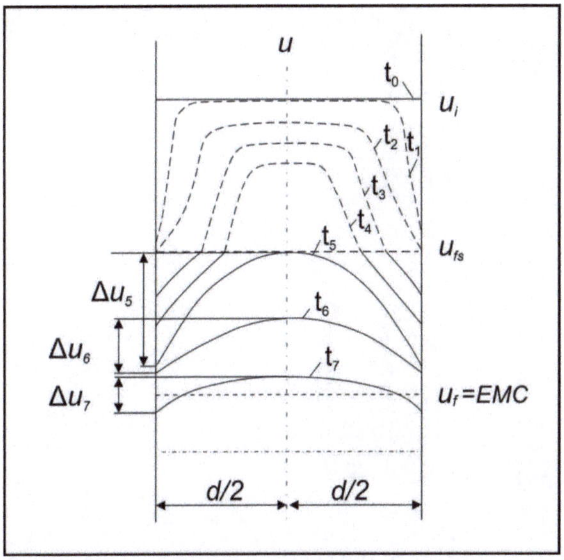

Figure 4.2 Distribution curves of the moisture content over the wood thickness (source: [1], with permission)

After the free water has been completely removed from the surface (point B in Figure 4.1 and stage 1 in Figure 4.2), a capillary pressure gradient develops between the inner layers (where the macrocapillaries are still filled with water) and the surface (where the macrocapillaries are free and water is only present in the microcapillaries of the cell walls), which ensures the movement of free water to the surface.

Initially, the movement of free water occurs from the outer wood layers, with the surface moisture content remaining at a constant level within the fiber saturation range (stage 2 in Figure 4.2). The temperature of the material remains constant during this process phase and equals the wet-bulb temperature (t_{wet}) (segment AB in Figure 4.1). The drying rate is also constant and is determined by the intensity of water evaporation from the surface. Therefore, this stage is referred to as the constant drying rate period. During this phase, the partial pressure of vapor at the wood surface equals the saturation pressure and depends only on the temperature of the wood surface.

In the drying of sawn timber, the constant drying rate period is very short. It is more noticeable in assortments that contain a large proportion of sapwood and that are dried from the green state [2]. Also when drying thin assortments (e.g., veneer), the duration of this phase accounts for a considerable percentage of the total duration.

The constant drying rate period lasts as long as the surface is supplied with free water and the velocity of free water movement equals the evaporation rate. Also during this phase there are no stresses in the wood and therefore no risk of cracking (Figure 4.3 a).

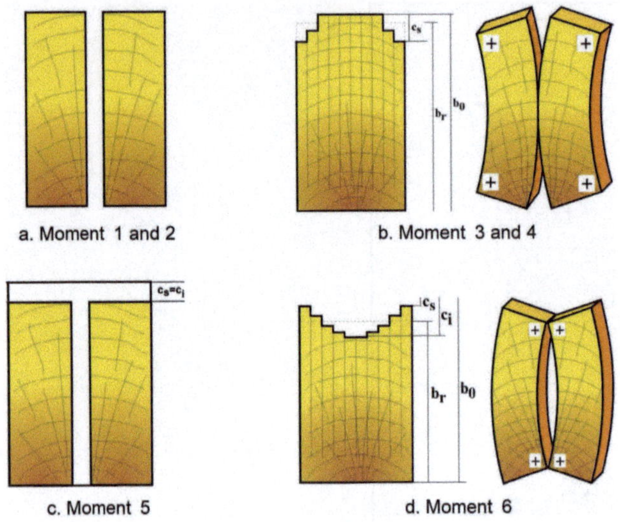

Figure 4.3 Visualization of the internal stresses at the selected moments 1–6 (see Figure 4.2) during wood drying (source: [3])

Over time, as the path of free water from increasingly deeper layers to the surface becomes longer, the partial vapor pressure at the wood surface becomes lower than the saturation pressure. Consequently, the moisture content of the wood surface falls below the fiber saturation range (stage 3 in Figure 4.2) and the falling drying rate period begins (section BC in Figure 4.1).

The average moisture content of the material ($\bar{u}$) at the end of the constant-rate stage and at the beginning of the falling-rate stage is referred to as the critical moisture content (u_{gr}). The critical moisture content is the wood moisture content below which the drying climate can be tightened without problems, without causing cracks [4]. In practice, the critical moisture content is a substitute value for the fiber saturation point (u_{fs}), which is difficult to determine with accuracy. Moreover, the appropriate moment for tightening the drying climate also depends strongly on the wood species. There are wood species that dry easily, e.g. fir, spruce, pine. For these, the critical moisture content can be set much higher than the fiber saturation point ($u_{gr\_resinous\ wood}$ = 45 %) and thus the drying time is significantly reduced. For hard-to-dry woods, on the other hand, e.g. oak, the critical moisture content is much lower (u_{gr_oak} = 27 %). The higher density and the greater shrinkage coefficients, as well as the tendency toward high growth stresses, recommend this cautious drying method for hardwood.

The falling-rate drying phase is characterized by the formation of two zones in the wood:

- an outer zone, with a moisture content below the fiber saturation range;

- an inner zone, where free water is still present and which can therefore be referred to as the capillary zone.

Under the action of capillary forces, free water is drawn to the contact surface of the two regions, where it partially converts to vapor and partially moves as a liquid through the micro-capillary system.

The outer zone is now characterized by the complete absence of free water. The movement of bound water in this zone occurs only under the action of the moisture gradient according to Fick's law (Equation 3.12). The moisture distribution across the wood thickness during this process phase is illustrated by curves 3 and 4 in Figure 4.2. The middle sections of these conventional curves are shown as dashed lines. In reality, the moisture distribution in the capillary zone cannot be strictly uniform, because the permeability of the micro-capillary structure of the wood, determined by the position and degree of filling of the pores, is inhomogeneous within the material volume.

As soon as the surface moisture content falls below the fiber saturation point (stage 3 in Figure 4.2), the surface layers tend to shrink, but the inner layers resist. When the connections between the layers are broken (Figure 4.3 b), it can be observed that the outer teeth shorten (c_s = surface shrinkage), while the height of the inner teeth remains unchanged (b_0). The entire uncut specimen has a real dimension (b_r), where $b_r < b_0$. Consequently, the surface layers are subjected to tensile stresses, which is also demonstrated by the tendency of the two halves to deform outward (Figure 4.3 b, right). In this stage, surface cracks may occur. This happens only when the stresses exceed the maximum tensile strength of the wood species in question.

If one continues to follow the kinematics of water removal, one arrives after some time at the moment when the entire amount of free water has been removed (moment 5 in Figure 4.2) and moisture transport continues exclusively by diffusion.

The internal stresses equalize as soon as the wood can shrink uniformly across the entire thickness (surface shrinkage c_s = shrinkage of the inner layers c_i) (Figure 4.3 c). Unfortunately, this safe, crack-free drying stage does not last until the end of the process.

If the wood were an ideal elastic material, the internal stresses occurring in it at moments 3 and 4 would gradually be reduced as moisture loss decreases and would finally disappear completely upon final moisture equalization. In reality, the stresses in the wood disappear at an intermediate stage of the process (moment 5), but they subsequently increase again toward the end of drying, with a changed sign. The cause of this phenomenon is the development of residual deformations in the wood.

In the initial phase of the process, when the heated wet wood exhibits high plasticity, residual tensile deformations occur under the influence of the tensile stresses acting in the surface layers, and residual compressive deformations remain in the inner layers under the influence of compressive stresses. Furthermore, during the drying of the wood, the elastic deformations are partially converted into permanent deformations as moisture decreases.

As a result, toward the end of the process (stage 6 in Figure 4.2), the shrinkage of the surface layers (c_s) becomes smaller than the shrinkage of the inner layers (c_i) (Figure 4.3 d). In other words, the nominal dimension of the surface layers will be greater than the nominal dimension of the inner layers, and consequently compressive forces occur on the outside and tensile forces on the inside of the wood, which is also reflected in the corresponding stress relief: the two halves bend inward (Figure 4.3 d, right). In this stage, internal cracks can occur in the wood.

Drying continues at an ever slower rate. The intensity of water loss gradually decreases according to a logarithmic curve (see curve segment BC in Figure 4.1), and the moisture gradient across the wood thickness becomes progressively smaller ($\Delta u_6 < \Delta u_5 < \Delta u_4$ – see Figure 4.2).

The temperature of the material gradually increases and tends toward the end of the process to reach the ambient temperature (t_{air}) (see Figure 4.1). The wood moisture content also tends to reach the equilibrium moisture content (*EMC*) to be reached. In practice, the drying process is not carried out until the equilibrium moisture content is reached (this would take too much time), but ends somewhat earlier, when the required final moisture content (u_e) is reached.

In summary, there are two critical stages in fresh air/exhaust air drying at temperatures below 100 °C:

- Stages 3–5, in which the surface layers are under tensile stress and the risk of surface cracks is high; in this stage it is extremely important to dry at a moderate temperature and, above all, at a high relative humidity in order to avoid these cracks;
- Stages 6–7, in which the inner layers are under tensile stress and the risk of internal cracks is high; although the risk of cracking in this stage is actually caused by too aggressive drying in the previous stage (3–5), it is advisable to dry carefully in this phase as well. This drying phase is controlled by the drying gradient (*DG*):

$$DG = \frac{\bar{u}}{\mathrm{EMC}} \tag{4.1}$$

$\bar{u}$ average wood moisture content of the values indicated by the wood moisture measuring points at a specific point in time [%]
EMC equilibrium moisture content at the same point in time [%]

If the condition is set that the drying gradient must be constant, the equilibrium moisture content can be controlled as a function of the decreasing wood moisture content, according to the relationship:

$$\mathrm{EMC} = \frac{\bar{u}}{DG}\,[\%] \tag{4.2}$$

where the *DG* value is chosen depending on the wood species (e.g., *DG* = max. 3.5 for softwoods and *DG* = 1.5 for oak).

A summary of the most important stages during wood drying, with emphasis on the critical stages, and some recommendations for controlling the drying process to avoid cracks, are presented in Figure 4.4.

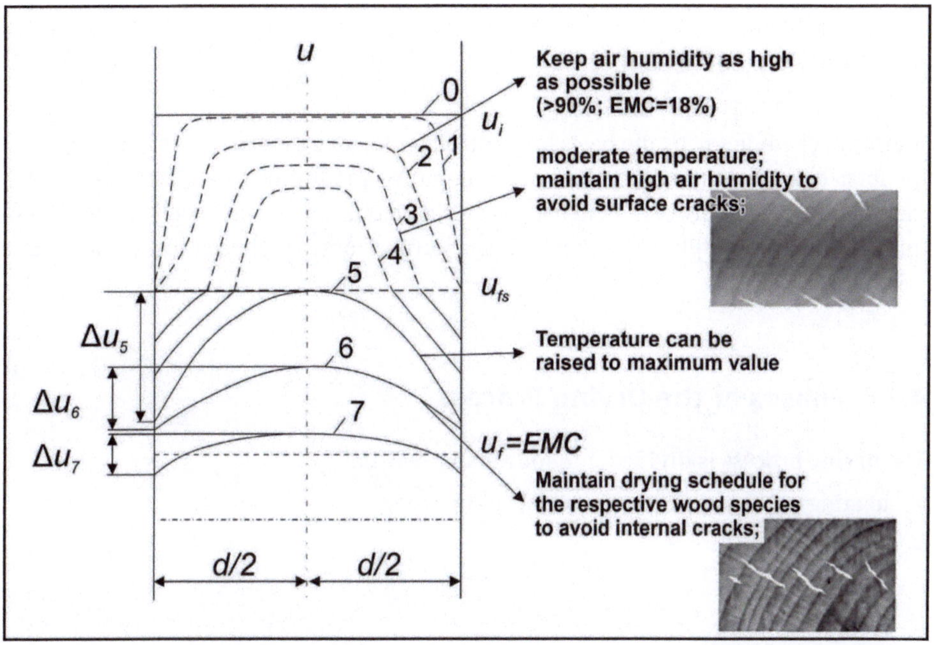

Figure 4.4 Critical stages in lumber drying and preventive measures

Unfortunately, it is impossible to completely avoid the occurrence of internal stresses in wood during conventional fresh air/exhaust air drying, because their cause is the unavoidable shrinkage. This involves both free shrinkage and restrained shrinkage, which leads to residual stresses [5]. Drying stresses can be minimized by using specimens with small dimensions. The thicker the wood pieces, the more problems are caused by the moisture gradient between the surface and the core. By applying adapted temperatures and by humidifying the drying medium accordingly, it is also possible to reduce internal stresses even in these difficult assortments.

The analytical calculation of wood residual stresses during drying is in itself a particularly complex problem, given the highly specific properties of wood, which vary even within the same tree and are therefore difficult to capture in mathematical models. In recent years, remarkable progress has been made in this direction through the numerous research works dedicated to this topic and with the help of modern computer technology, through the development of increasingly sophisticated models for simulating the mechanical behavior of wood during drying, which come ever closer to real behavior. Such models are discussed in more detail in Chapter 5.

By correctly drawing up the drying schedule, it is possible to control the drying process in such a way that the stresses that occur do not exceed the resistance limit of the wood and that the drying proceeds without defects.

4.2 Drying Schedules

A drying schedule forms the basis for controlling the drying process. It is drawn up in the form of a table in which values of temperature and humidity are assigned to each phase of the drying process, which are considered optimal for the wood assortment in question and for the phase in question. The purpose of each phase must be taken into account.

4.2.1 Phases of the Drying Process

The drying process is divided into four main phases:

- heating of the wood and the water in the wood;
- pure drying;
- conditioning;
- cooling.

Each of them has a precisely defined purpose.

The **Heating phase** serves the gradual adaptation of the wood to the drying environment. In this phase, the temperature is raised to a moderate value (t_1) corresponding to the first drying stage. This value varies depending on the wood species and ranges from 30 °C for sensitive wood species (e.g. oak) to 60 °C for easy-to-dry wood species (e.g. spruce).

It is important that the heating rate (temperature rise per unit of time) is constant and does not exceed 5 °C/h, so that the heating occurs uniformly across the entire thickness of the piece and the temperature difference between the surface and inner layers remains relatively small. In particular, frozen wood must be heated very slowly to a temperature of 20 °C, otherwise high stresses will develop in the wood.

In this phase, the relative humidity (φ) plays a particularly important role. Through abundant spraying of steam or water, a high value (φ = 85 – 95 %) must be achieved, corresponding to an equilibrium moisture content of approximately 18 %, in order to prevent case hardening of the surface. An exception to this rule is formed by light-colored wood species (e.g. maple, alder). To avoid discoloration, in this case the wood must be dried below the fiber saturation point as quickly as possible, and therefore heating in this case is carried out without spraying.

At the beginning of the heating period, until a temperature of 25 – 30 °C is reached in the dryer, only the fans and the heating elements operate. Only after reaching this temperature does the corresponding humidification of the air take place (faster for pre-dried species, later for those with high initial moisture content).

During the heating phase, the wood moisture content should not decrease. It can sometimes increase slightly.

The transition to the next phase of the process only occurs when the temperature sensor on the supply air side indicates the planned value (t_1).

The actual drying has as its goal the controlled reduction of the wood moisture content from the initial value (u_a) to the desired final value (u_f). If the initial wood moisture content is higher than the fiber saturation moisture content, then the wood still contains free water and the actual drying is divided into two distinct stages:

- I – drying from the initial moisture content to the fiber saturation point, or more precisely to the so-called limit moisture content;
- II – drying to the desired final moisture content.

The limit moisture contentvalues (u_{gr}) for the most commonly used domestic wood species are listed in Table 4.1.

Table 4.1 Control Data for Conventional Lumber Drying (source: [4])

Group	Wood species	t_1 [°C]	EMC_1 [%]	u_{gr} [%]	t_2 [°C]	DG
		at $u \geq u_{gr}$			at $u < u_{gr}$	
1	Spruce, fir, pine	60	13	45	90	3.5
2	Alder, linden, poplar, willow	50	14	39	80	2.8
3	Chestnut, Larch	50	15	37	80	2.5
4	Pear tree, Cherry tree, Walnut tree, Birch	50	16	35	80	2.2
5	Beech	45	17	34	80	2.0
6	Hornbeam	40	15	30	75	2.0
7	Oak	30	18	27	70	1.5
8	Ash, maple, beech (when a light-colored finish is required)	45	13	32	60	2.5

In the first stage of the actual drying (from u_i to u_{gr}), the drying schedule must ensure gentle drying conditions in order to avoid the development of high stresses in the wood. For this reason, the temperature is not yet increased at this point, but remains at the value reached at the end of the heating period (t_1). The relative humidity is reduced slightly and then kept constant at a value of φ = 60 – 90 %, so that the equilibrium moisture content is set to EMC_1 = 13 – 18 % is established (the lower values in the given interval correspond to the drying of softwood and coniferous wood, and the higher values to the drying of sensitive species such as oak). These drying conditions must be maintained until all wood moisture measurement points reach the limit moisture content.

Notes:

- The values given in Table 4.1 correspond to a gentle drying schedule. To achieve the best possible drying quality, it is recommended to apply these reference values for the first batch. In the case of a successful drying run, a more aggressive schedule may be tried, but under no circumstances should the drying gradient be increased to compensate for a lower temperature.
- For groups 1–6, the temperature t_1 may be increased by 10 °C for wood species with thicknesses below 40 mm.
- Furthermore, for groups 1–6, the drying gradient (*TG*) during the stage of moisture reduction below 15% may be increased by a factor of (1.5 – 2) to accelerate the process.

In this stage, only the fans operate continuously. The heating, the spraying, and the vents are opened and closed periodically as needed to achieve and then maintain the required parameters (t_1 and EMC_1).

From the point at which all wood moisture measurement points indicate the fiber saturation value, the second stage of pure drying can begin. This is characterized by an increase in temperature (to t_2) and a gradual reduction in air humidity. The reduction in air humidity is controlled by the so-called drying gradient.

The drying gradient (*DG*) is an indicator of the severity of the drying regime: the greater the difference between the wood moisture content and the equilibrium moisture content, the higher the *TG*value, and the more severe the drying conditions. It is used in the range below the fiber saturation point/limit moisture content. Its value depends largely on the wood species. The values in Table 4.1 serve as a guide.

As already indicated in Equation 4.1, *DG* is defined as the ratio of the wood moisture content at a given point in time to the equilibrium moisture content at the same point in time. If the wood moisture content at a given point in time is known, and *TG* as a constant value from Table 4.1, the equilibrium moisture content corresponding to the respective point in time can be calculated using Equation 4.2.

Thus, in the second stage of pure drying, the air temperature is gradually increased to the maximum permissible value for the respective wood species (t_2 – see Table 4.1), and at the same time, as the wood moisture content decreases, the air humidity is continuously reduced so that the drying gradient remains constant.

In this stage, the fans operate continuously and the heating registers operate until the operating temperature is reached. To reduce air humidity, the ventilation flaps are opened periodically, but only after the intended temperature has been reached, since the air humidity decreases anyway due to the heating. The pure drying phase is considered complete when the desired final moisture content is reached.

The **conditioning** has as its main objective the equalization of the final moisture content within the charge and the relief of internal stresses in the wood (Figure 4.5). This phase is optional when drying thin wood species (d <40 mm), but should definitely be carried out for $d \geq$ 40 mm.

The conditioning of the wood is achieved by spraying. In this process, the moisture content of the surface layers increases slightly, while the inner layers continue to dry due to the accumulated heat. This reduces the moisture difference across the cross-section of the wood.

In this phase, the temperature is usually reduced by approximately 10 – 15 °C compared to the temperature at the end of pure drying (t_2). The fans and the spraying system are in operation. The heating is switched off and the vents are closed.

The conditioning phase can be terminated either depending on a desired moisture difference across the cross-section of the wood (e.g., max. 5 %) or depending on time.

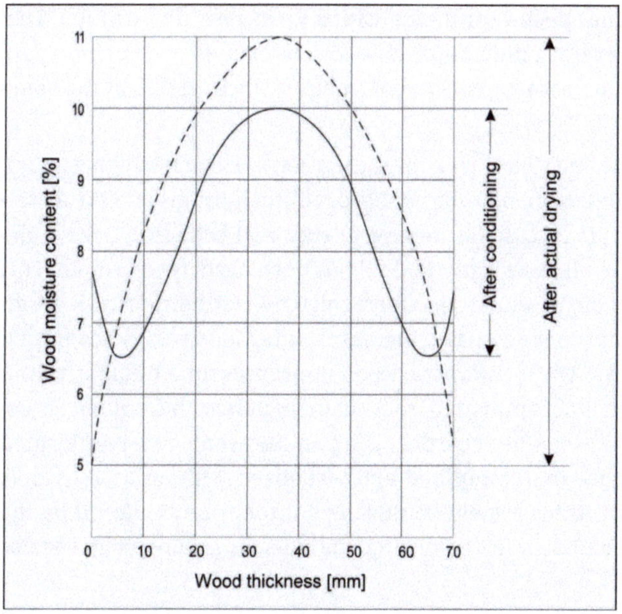

Figure 4.5 Reduction of the moisture gradient across the wood thickness by conditioning (based on [6])

The **cooling** is the last phase of the drying process and aims at the gradual reduction of the wood temperature in order to avoid thermal shock when unloading the drying chamber.

In this phase, heating and humidification are completely stopped. The ventilation flaps are opened to the maximum and, with the help of the fans, the stacks are "flushed" with cold and dry fresh air in order to reduce the wood temperature to approximately 30 °C.

4.2.2 Preparation of the Drying Schedule

Drying schedules can be established depending on the control moisture content, on time, or on internal stresses.

The first option is most commonly used in practice and offers the advantage that the process control is oriented towards the individual behavior of each individual charge. Time is not a control parameter in this case.

Time-dependent drying schedules offer great convenience in the operation of drying plants, but do not always guarantee the corresponding quality, since the process control follows a fixed program in a rigid manner, in which the temperature and air humidity change at fixed time intervals, regardless of whether the wood moisture content has reached the threshold value specific for the transition to the next phase or not. This type of drying schedule is therefore only applicable for companies with many years of experience that dry the same assortments (wood species, thicknesses) in the same plant year after year.

The third category, drying schedules based on internal stresses, are of the highest quality, as they enable the timely detection of unfavorable drying conditions and the avoidance of serious drying defects (e.g., case hardening, cracks, and collapse). For the detection of internal stresses, the ultrasonic method [7] has been used, for example. This is based on the analysis of acoustic emissions during drying. As the moisture gradient across the cross-section of the wood parts increases, internal stresses develop in the wood, causing microcracks. These microcracks do not represent a qualitative impairment, and their formation is accompanied by acoustic signals with frequencies between 50 Hz and 1.5 MHz. These can be recorded using an ultrasonic receiver located inside the plant. The signals are then amplified and visualized with an oscilloscope. The most difficult problem with this system is filtering out the sounds caused by the microcracks, given that other bodies in the drying chamber (e.g., fans, metal walls, etc.) also emit acoustic signals.

Another method for measuring internal stresses is based on a microstructured pressure-measuring instrument made of silicone, which is inserted into a cylindrical Teflon sleeve in the wood [8], or an extensometer [9], which can measure stresses in the longitudinal direction. If the sensor is placed in a borehole at a certain depth in the wood, it can directly measure the compressive stress caused by the shrinkage of the hole diameter. This type of sensor has the potential to be used in industrial drying control.

At present, however, drying schedules are mainly established in tabular form depending on the control moisture content (Table 4.2). In this process, values of the control parameters (temperature, equilibrium moisture content, drying gradient) are assigned to each process phase in such a way that each phase can fulfill its purpose and the wood reaches the desired final moisture content at the end.

Table 4.2 Drying Schedule as a Function of the Control Moisture Content - Template (based on [4])

Phase	Wood moisture content [%]	Temperature [°C]	Equilibrium moisture content [%]	Drying gradient
Heating phase	u_i	t_1	18	–
Pure drying	from u_i to u_{gr}	t_1	EMC_1	–
	from u_{gr} to 20	t_2	$[(u_{gr}+20)/2]/DG$	DG
	from 20 to 15	t_2	$[(20+15)/2]/DG$	DG
	from 15 to u_f	t_2	$[(15+u_f)/2]/DG$	DG
Conditioning	u_f	t_2–10	u_f	–
Cooling	u_f	30	–	–

u_i – measured initial moisture content [%]; u_f – desired final moisture content [%]; u_{gr} – limit moisture content [%]; t_1, t_2, EMC_1, DG – control data (see Table 4.1)

Based on these reference values, a drying schedule for larch wood was determined as an example (Table 4.3). This applies to timber with a thickness of more than 40 mm that is to be dried to a final moisture content of 10%. For thinner assortments, the maximum temperature can be increased to 90 °C and the conditioning phase can be omitted.

In the reference literature, e.g. [10–14], various methods for determining the drying schedule are proposed, and every kiln manufacturer has its own programs integrated into the control software of its equipment, e.g. [15, 16]. All determination methods are based on a grouping of wood species according to difficulty class, with "optimal" values recommended for each group. What differs from one literature source to another are the values of these parameters. In the past, higher temperature values were recommended for the same timber assortment. When these are combined with low air humidity and/or a high drying gradient, this leads to a severe drying process, in which a shorter time but also a high risk of cracking is to be expected. In order to conserve wood resources and reduce drying defects as much as possible, milder drying conditions are applied nowadays.

Table 4.3 Drying Schedule for 48 mm Thick Larch Wood

Phase	Wood moisture content [%]	Temperature [°C]	Equilibrium moisture content [%]	Drying gradient
Heating phase	70	50	18	–
Pure drying	70 to 37	50	15	–
	37 to 20	80	11.4	2.5
	20 to 15	80	7	2.5
	15 to 10	80	5	2.5
Conditioning	10	70	10	–
Cooling	10	40	–	–

Further examples of drying schedules for various European wood species are listed in Appendix D.

4.3 Calculation of Drying Duration

In accordance with the four phases of the process, the total drying time (D) consists of the following four components:

- Duration of the heating phase (D_h);
- Duration of the pure drying phase (D_{drying});
- Duration of the conditioning phase (D_{cond}) (optional for timber assortments under 40 mm thickness, strongly recommended above 40 mm thickness);
- Duration of the cooling phase (D_{cool}).

$$D = D_h + D_{drying} + D_{cond} + D_{cool} \text{ [h]} \quad (4.3)$$

The **duration of the heating phase** depends on the area of the heating system and the parameters of the heat transfer medium. In general, the following rule of thumb can be applied:

D_h = 1 h/cm thickness for softwoods and soft hardwoods

and

D_h = 1.5 h/cm thickness for hardwoods.

The **duration of the actual drying phase** depends on numerous factors, such as:

- wood species (density);
- initial and final moisture content of the wood;

- thickness of the boards;
- edging of the boards;
- temperature and humidity of the drying environment;
- drying gradient;
- air velocity;
- operating mode of the heating system (h/day);
- desired quality;
- condition of the system, etc.

The moisture decrease during drying follows a logarithmic function; therefore, the theoretical calculation of the drying duration is based on Kollmann's formula [13]:

$$D_{\text{drying}} = \ln \frac{u_{\text{i}}}{u_{\text{f}}} \text{ [h]} \quad (4.4)$$

In order to account for all the above-mentioned influences, the obtained value is then multiplied by coefficients that take into account the material and medium properties.

Corresponding to the two stages of actual drying, the duration D_{drying} consists of two components:

- $D_{\text{dry\_1}}$ – duration of the first drying stage (from u_{i} to u_{gr}) and
- $D_{\text{dry\_2}}$ – duration of the second drying stage (from u_{gr} to u_{f}).

The two components are calculated using the following equations:

$$D_{\text{dry\_1}} = \ln \frac{u_{\text{i}}}{u_{\text{gr}}} \cdot f_1 \cdot f_{2'} \cdot f_3 \cdot f_{4'} \cdot f_5 \cdot f_6 \text{ [h]} \quad (4.5)$$

and

$$D_{\text{dry\_2}} = \ln \frac{u_{\text{gr}}}{u_{\text{f}}} \cdot f_1 \cdot f_{2''} \cdot f_3 \cdot f_{4''} \cdot f_5 \cdot f_6 \cdot f_7 \text{ [h]} \quad (4.6)$$

u_{i} initial moisture content [%]
u_{gr} critical moisture content [%] (see Table 4.1)
u_{f} final moisture content [%]
f_1–f_7 coefficients depending on wood species (Table 4.4), thickness (Table 4.5), edging (Table 4.6), temperature (Table 4.7), velocity (Table 4.8), operating time (Table 4.9) and drying gradient (Table 4.10)

Remarks:

1. The coefficient f_4 has different values in Equation 4.5 and Equation 4.6, because the temperature in the two stages of pure drying is different.
2. When calculating the duration of the first drying phase ($D_{\text{dry\_1}}$), the coefficient f_7 is not taken into account, because at wood moisture contents above the limit moisture content the drying gradient (*DG*) is a fictitious quantity.

Table 4.4 Values of the Wood Species Coefficient f_1 (source: [4])

Wood species	f_1
Softwood with $\rho_0 \leq 500$ kg/m^3 (fir, pine, spruce)	19
Softwood with $\rho_0 > 500$ kg/m^3 (larch)	26
Soft hardwood with $\rho_0 \leq 450$ kg/m^3 (poplar, alder, lime, willow)	20
Hardwood with $\rho_0 = 450 - 650$ kg/m^3 (beech, oak, ash, maple, chestnut, cherry, walnut, elm)	30
Hardwood with $\rho_0 > 650$ kg/m^3 (black locust, hornbeam)	40

Table 4.5 Values of the Thickness Coefficient f_2 (based on [4])

Thickness [mm]	f_2
15	0.60
20	0.80
25	1.00
30	1.20
35	1.40
40	2.02
50	2.41
60	5.76
70	7.84
80	10.24
90	12.96
100	16.00
120	23.04
140	31.36

Table 4.6 Values of the Coefficient f_3 Depending on the Type of Trimming (source: [4])

Type of trimming	f_3
Cut pieces	0.70 - 0.90
Edged timber	1.00
Unedged timber	1.10 - 1.30

Table 4.7 Values of the Temperature Coefficient f_4 (based on [4])

Temperature [°C]	f_4
30	8.000
35	5.956
40	4.560
45	3.564
50	2.828
55	2.270
60	1.837
65	1.495
70	1.222
75	1.000
80	0.818
85	0.669
90	0.544
95	0.441

Table 4.8 Values of the Air Velocity Coefficient f_5 (based on [6])

Air velocity [m/s]	f_5
0,5	1.25
1,0	1.15
1,5	1.08
2,0	1.00
2,5	0.98
3,0	0.94
3,5	0.90
4,0	0.88
4,5	0.84
5.0	0.80

Table 4.9 Values of the Coefficient f_6 Depending on the Operating Time of the Drying Chamber (source: [4])

Operating time [h/day]	f_6
24	1.00
16	1.17
12	1.26
8	1.30

Table 4.10 Values of the Drying Gradient Coefficient f_7 (based on [4])

Drying gradient	f_7
1.5	2.0
2.0	1.5
2.5	1.2
3.0	1.0
3.5	0.86
4.0	0.75

The duration of the actual drying is then obtained as the sum of the two components:

$$D_{\text{drying}} = D_{\text{dry\_1}} + D_{\text{dry\_2}} \text{ [h]} \tag{4.7}$$

The **duration of the conditioning phase** is:

$$D_{\text{cond}} \approx 0{,}1 \cdot D_{\text{drying}} \text{ [h]} \tag{4.8}$$

D_{drying} Duration of actual drying [h]

The determining factor for setting this duration is the desired quality (uniformity of final moisture content).

The **duration of the cooling phase** is:

$$D_{\text{cool}} = D_{\text{h}} \text{ [h]} \tag{4.9}$$

All determined values are generally rounded up to the next whole number and then summed (Equation 4.3) to obtain the total duration of the drying process.

In addition, the **handling time** (D_{loading}) for loading and unloading the drying plant must also be added:

$$D_{\text{loading}} = 0.33 \text{ h/m}^3 \tag{4.10}$$

It must be emphasized that the mathematical determination of the drying time is only an estimate. Accurate drying times can only be determined in practice.

What is the occupancy time of a drying chamber that dries 100 m^3 of unedged beech wood with a thickness of 40 mm from 60 % initial moisture content to 10 % final moisture content?

Solution:

$D_h = 1.5\ h/cm \cdot 4\ cm = 6\ h$

$D_{dry\_1} = \ln(60/34) \cdot 30 \cdot 2.02 \cdot 1.3 \cdot 3.564 \cdot 1 \cdot 1 = 174\ h$

$D_{dry\_2} = \ln(34/10) \cdot 30 \cdot 2.02 \cdot 1.3 \cdot 0.818 \cdot 1 \cdot 1 \cdot 1.5 = 118\ h$

$D_{drying} = D_{dry\_1} + D_{dry\_2} = 292\ h$

$D_{cool} = 0.1 \cdot D_{drying} = 29\ h$

$D_{cool} = D_h = 6\ h$

$D_{loading} = 0.33\ h/m^3 \cdot 100\ m^3 = 33\ h$

$D_{total} = D_h + D_{drying} + D_{cond} + D_{cool} + D_{loading} = 366\ h = 15.25$ days

4.4 The Drying Diagram

The drying diagram (Figure 4.6) is the graphical representation of the drying schedule. It shows the chronological progression of the most important drying parameters (wood moisture content, temperature, and equilibrium moisture content) throughout the entire drying process.

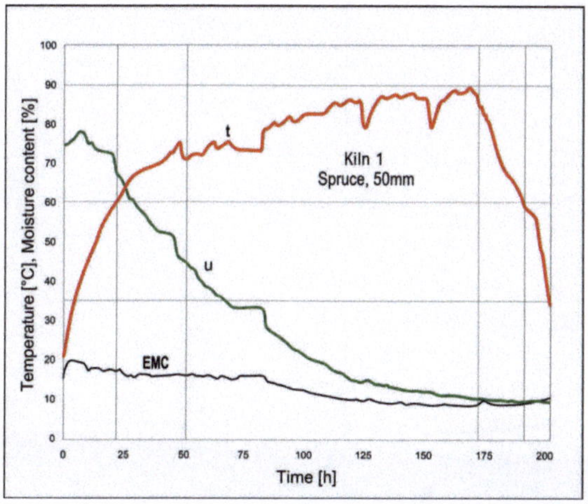

Figure 4.6 Drying diagram for spruce, 50 mm, from an initial moisture content u_i = 75% to a final moisture content u_f = 10% (based on [18])

The drying diagram shows whether the process was carried out correctly. Any deviation of the drying curves from the theoretical course indicates possible measurement or control errors, which may be the cause of certain drying defects. Therefore, the drying diagram is an important element for process analysis and for its optimization. The example in Figure 4.7 for the same wood assortment (spruce, 50 mm) shows numerous deviations in all process phases:

- During the heating phase, the target value of the equilibrium moisture content *EMC* = 18% was not reached, but only 14%; the most probable cause is insufficient water pressure in the spray system;
- During the pure drying phase above the fiber saturation point (45%), the target value of the temperature (60 °C) was not reached; the cause was identified as insufficient boiler capacity. The *EMC*-value remained smaller than the target value (11.9% instead of 13%), which again is attributable to the incorrect water pressure in the spray system;
- During the pure drying phase below the fiber saturation point, the same deficiencies can be observed: the temperature is lower than the target value (64 °C instead of 90 °C). In addition, an unintentional spray event at the moment 105 h caused a sudden drop in temperature and an increase in air humidity and wood moisture content;
- During the conditioning phase, the temperature remains below the target value (62 °C instead of 80 °C), but the spray system switches on correctly.

In general, it can be noted that the charge was controlled too mildly, which led to an unjustified extension of the drying time.

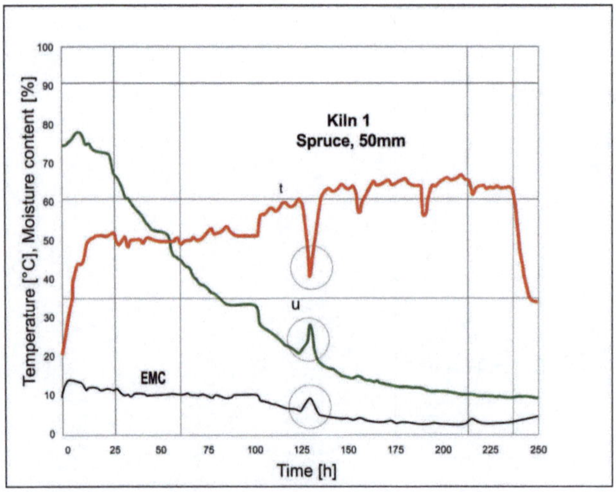

Figure 4.7 Drying diagram showing deviations from the normal course of drying

References

[1] Marinescu, I. (1979): *Uscarea lemnului*. Editura Tehnică, Bucharest, Romania

[2] Perré, P. (1993): *Fundamentals of Wood Drying*. COST A.R.BO.LOR, ENGREF, Nancy, France

[3] Câmpean, M. (2003): *Heat Treatments of Wood*. Verlag der Transilvania University, Brasov, Romania

[4] Trübswetter, T. (2009): *Holztrocknung. Verfahren zur Trocknung von Schnittholz – Planung von Trocknungsanlagen*. 2nd ed., Carl Hanser Verlag, Leipzig, Germany

[5] Avramidis, S.; Lăzărescu, C.; Rahimi, S. (2023): *Basics of Wood Drying*. Chapter 13 in *Handbook of Wood Science and Technology*. Springer International Publishing, Cham, p. 679 – 706

[6] Câmpean, M. (1997): *Tratamente termice ale lemnului. Uscarea cherestelei*. Verlag der Transilvania University, Brasov, Romania

[7] Becker, H. F. (1982): *Schallemissionen während der Holztrocknung*. In: Holz als Roh- und Werkstoff, 40, p. 345 – 350

[8] Allegretti, O.; Ferrari, S. (2008): *A Sensor for Direct Measurement of Internal Stress in Wood During Drying: Experimental Tests Toward Industrial Application*. In: Drying Technology, 26(9), p. 1150 – 1154. *https://doi.org/10.1080/07373930802266256*

[9] Sanchez, J. Y. C. S.; Lima, J. T.; da Silva, J. R. M.; Dias Soares, B. C. (2022): *Application of a Strain Gauge to Assess Drying Stresses in Normal and Tension Wood of Corymbia citriodora*. In: Maderas. Ciencia y Tecnología, 24. *http://dx.doi.org/10.4067/s0718-221x2022000100421*

[10] Boone, S. C.; Bois, C.; Wengert, E. (1988): *Dry Kiln Schedule for Commercial Woods: Temperate and Tropical*. USDA Forest Products Laboratory, GTR-57, Madison, USA

[11] Denig, J.; Wengert, E. M.; Simpson, W. T. (2000): *Drying Hardwood Lumber*. USDA Forest Service, Forest Products Laboratory, Madison, WI, USA

[12] Dumitrescu, N. St. (1969): *Uscarea lemnului*. Editura Tehnică, Bucharest, Romania

[13] Kollmann, F. F. P. (1951): *Technologie des Holzes und der Holzwerkstoffe. Band 1*. 2nd ed., Springer Verlag, Berlin/Göttingen/Heidelberg

[14] Langendorf, G.; Eichler, H. (1982): *Holzvergütung*. 3rd ed., VEB Fachbuchverlag, Leipzig, Germany

[15] Hildebrand, R. (Hrsg.) (1979): *Schnittholztrocknung*. 4th ed., Oberboihingen, Germany

[16] Cividini, R. (2001): *Conventional Kiln-Drying of Lumber*. NARDI S.p.A., Chiampo, Italy

[17] Lohmann, U. (1998): *Holz Handbuch*. 5th ed., DRW Verlag, Leinfelden-Echterdingen, Germany

[18] Bedelean, B.; Campean, M. (2013): *Tratamente termice ale lemnului. Îndrumar pentru lucrări practice*. Verlag der Transilvania University, Brasov, Romania

[19] Janik, W. (1965): *Handbuch der Holztrocknung*. Fachbuchverlag, Leipzig, Germany

5 Modeling of Lumber Drying

The modeling of the drying process is one way to determine the behavior of a piece of wood in a specific environment with a certain degree of accuracy [1]. Scientists have designed various mathematical models to develop a useful tool for the design, analysis, and optimization of drying systems. The most important performance indicators for optimization are usually: shorter drying times, lower energy consumption, and better quality of the dried material.

For a long time, modeling focused mainly on the phenomenon of moisture transfer [2]. Since Tuttle presented the first model theory of the drying process in 1925, numerous models have been developed [1–5], employing various modeling methods, such as:

- Models based on the moisture diffusion equation, and models based on the law postulated by Adolf Fick in 1855, which generally refers to the phenomenon of diffusion [1]. These models describe the movement of water in wood from the interior to the surface. In addition to Fick's law, Fourier's law was also used, which represents the fundamental equation of heat conduction in solids [1, 4].
- Empirical models, which were determined experimentally and describe a relationship between input variables (initial moisture content, wood species, air parameters, etc.) and the drying rate or moisture content at a specific point in time.
- Models based on fundamental equations of moisture and heat transfer. These models account for the simultaneous transfer of mass and heat during the drying period, with or without integration of the airflow into the system. These models are created by applying FEM (Finite Element Method) or CFD methods (Computational Fluid Dynamics).
- Models based on artificial intelligence techniques, e.g., artificial neural networks, which are capable of determining a relationship between the input and output variables of the process much more efficiently than regression equations [2, 5].

The results of the modeling efforts were materialized through the creation of theoretical models, which were implemented in various drying process analysis and optimization programs such as Torksim [1, 6] and others.

5.1 Models Based on the Moisture Diffusion Equation

Fick's law of moisture diffusion in wood is presented in its simplified form in Equation 5.1 [1]:

$$\frac{\partial u}{\partial t} = \frac{\partial}{\partial x}\left(D\frac{\partial u}{\partial x}\right) \tag{5.1}$$

u moisture content
t time
x spatial coordinate
D diffusion coefficient

Numerous studies have been conducted to determine the diffusion coefficient (D) of some wood species in the three principal directions for various moisture ranges and different temperature values.

Equation 5.1 can only be solved if boundary conditions are defined for the volume under consideration and an interaction between the piece of wood and the environment is established. The simplest alternative is to assume that the surface of the piece of wood almost instantaneously reaches an equilibrium moisture content that depends only on the ambient parameters. This means that the mass transfer resistance at the surface is neglected. However, it is desirable not to neglect this resistance and to account for the introduction of an approximation in the modeling [1].

The analytical solution of Equation 5.1 is presented in Equation 5.2, which takes into account the external mass transfer resistance. The moisture flux per unit area considered is:

$$\frac{\dot{m}}{A} = S(u_{\text{surf}} - u_{\text{EMC}}) \tag{5.2}$$

u_{surf} surface moisture content
u_{EMC} equilibrium moisture content corresponding to the environment in which the part is located
S surface emission coefficient

Newman presented in 1931 both the analytical solutions for Equation 5.1 and Equation 5.2 as well as the numerical values for a unidirectional flow between two planes. This study was the starting point for numerous efforts to determine the S- and D-coefficients [1].

Equation 5.1 cannot describe the actual drying process without incorporating temperature. Theoretically, the equation for determining the temperature field in solids is derived from Equation 5.1by replacing moisture with temperature (T) and the diffusion coefficient with the thermal diffusivity coefficient (a), thereby obtaining the Fourier equation. However, incorporating the Fourier equation alone does not achieve a complete interaction between the moisture and temperature fields. To overcome this disadvantage, mass and energy balance equations must be determined for each element of the investigated volume. In this way, the moisture and temperature fields are coupled. The boundary conditions are defined by Equation 5.3 and Equation 5.4 [1].

$$\frac{\dot{m}}{A} = \beta(c_{surf} - c_{air}) \qquad (5.3)$$

Here, c_{air} and c_{surf} are the vapor concentrations in the air and in the air layer that is in equilibrium with the moisture content at the surface, and β is the mass transfer coefficient.

$$\frac{\phi}{A} = \alpha(T_{air} - T_{surf}) \qquad (5.4)$$

ϕ/A heat flux per unit area considered
α heat transfer coefficient
T_{air} air temperature
T_{surf} – temperature at the surface of the wood material

For a temperature level below 80 °C at atmospheric pressure, a model based on mass and energy balances is sufficient [1]. For slow drying processes and slow temperature changes, the presence of temperature gradients in the wood can be neglected. It is assumed that the temperature in the material is constant but variable over time. For models developed for temperatures near or above the boiling point of water, the manner of moisture movement in the wood must also be taken into account by considering, in addition to the moistureand temperature gradients, a pressure gradient as well.

The similarity of Fick's and Fourier's equations, but also between Equation 5.3 and Equation 5.4, suggests that within the common boundary layer there is a relationship between the mass transfer coefficient (β) and the heat transfer coefficient (α). If at least one of the coefficients is known (usually the coefficient α), the other transfer coefficient can be determined using Equation 5.5.

$$\frac{\alpha}{\beta} = c \cdot \rho \qquad (5.5)$$

c specific heat of moist air in the boundary layer
ρ bulk density

This coupling is referred to as the mass-heat analogy.
The heat transfer coefficient (α) can be determined using Equation 5.6 (Dittus-Boelter equation).

$$Nu = 0{,}023\, Re^{0{,}8}\, Pr^{1/3} \qquad (5.6)$$

Nu, Re and *Pr* Nusselt, Reynolds and Prandtl number

This equation has some limitations, for example the fact that it assumes that the airflow occurs in a smooth channel. In reality, the value of the heat transfer coefficient (α) is much higher, due to factors that contribute to enhancing air turbulence (gaps between the wood pieces, separation of the airflow by the edges of the wood pieces, sudden contractions and expansions of the flow cross-section, etc.). Further information on the characteristics of heat transfer in a drying plant is provided in [1].

5.2 Empirical Models

One of the first empirical models was developed by Bramhall in 1976 [2]:

$$\frac{d\overline{M}}{d_t} = \frac{P_{DS\_dry} - P_{DS\_wet}}{R} \tag{5.7}$$

P_{DS_dry} saturation pressure of water vapor at dry-bulb temperature [Pa]
P_{DS_wet} saturation pressure of water vapor at wet-bulb temperature [Pa]
R resistance to moisture displacement [Pa kg kg^{-1}]

The moisture movement resistance is calculated using Equation 5.8 or Equation 5.9:

$$R = c_1 \ \exp(-c_2\overline{M}) \tag{5.8}$$

$$R = c_3\,\overline{M}^{-c_4} \tag{5.9}$$

c_1, c_2, c_3 and c_4 experimentally determined constants

Another empirical model was developed by Tschernitz and Simpson [2]:

$$\overline{\Phi} = exp\left(\frac{-b \cdot t}{l^n}\right) \tag{5.10}$$

t drying time [days]
l wood thickness [cm]
b, n empirical parameters

Olek et al. (1994) proposed a new empirical model for the drying of European beech (*Fagus sylvatica*) [2]:

$$-\frac{d\overline{M}}{dt} = a_1 + \frac{a_2}{(100\,M)^n + a_3} \tag{5.11}$$

a_1, a_2, a_3 and n are the parameters determined graphically on the basis of experimental data. Their values vary depending on the drying conditions.

In the literature, in addition to empirical models, graphical-analytical methods are also presented [2]. One such model, developed at the Moscow State Forest University, is given in Equation 5.12 represented and facilitates the determination of the drying time:

$$\tau = \frac{0{,}233}{D_m}\left[1 + \frac{\pi^2}{4\,Bi^2}\right]\ln\left[B\frac{(u_{surf} - \text{EMC}) + 0{,}81(u_0 - u_{surf})}{\overline{u} - \text{EMC}}\right] \tag{5.12}$$

D_m average moisture diffusion coefficient [m^2/s]
Bi Biot mass transfer number
B coefficient,
u_{surf} surface moisture [%]
$\overline{u}$ average moisture across the thickness of the wood piece at time D
EMC corresponding equilibrium moisture content [%]
u_0 average moisture content (measured inside the wood piece) at the beginning of drying [%]

In the case of a drying schedule for beech wood comprising three phases, Equation 5.13 can be used:

$$\tau = C\left[\frac{0{,}81(u_{01}-\text{EMC}_1)}{u-\text{EMC}_1}\ln\frac{u_{02}-\text{EMC}_2}{u_2-\text{EMC}_2}+\ln\frac{u_{03}-\text{EMC}_3}{u_3-\text{EMC}_3}\right] \tag{5.13}$$

C is a coefficient that depends on the thickness of the wood, the width of the stack, and the air velocity in the drying chamber. The value of this coefficient is determined using specific nomograms [2]
u_{01} wood moisture content at the beginning of the phase (%)
EMC average equilibrium moisture content corresponding to the relative humidity

Kayihan (1993) developed an empirical model that accounts for the influence of external conditions (via a constant drying rate coefficient) and the internal moisture transport process (via a diffusion parameter *b*) to estimate the rate of change of the average wood moisture content [2]:

$$-\frac{d\bar{u}}{d_t} = \frac{b\bar{u}}{\left[1+(b\bar{u}/a)^3\right]^{1/3}} \tag{5.14}$$

The drying behavior is influenced by the parameter (*b*), which also takes into account the temperature of the material. The value of the parameter *b* is calculated using Equation 5.15, as a function of temperature *T* [K].

$$b = b_0\ \exp\left(-\frac{2{,}700}{T}\right) \tag{5.15}$$

Keey and Pang (1994) analyzed the drying kinetics of pine wood (*Pinus Radiata*) using a characteristic drying curve and moisture movement analysis [2]. The developed model assumes the existence of a constant drying rate period, namely from the initial moisture content (u_0) to the first critical drying point (u_1). This constant drying period applies only to sapwood. The drying rate (j_{w1}) can be calculated using Equation 5.16

$$j_{w1} = \varphi K_0\left(H_s - H_g^0\right) \tag{5.16}$$

$H_s - H_g^0$ moisture potential [%]
K_0 mass transfer coefficient
φ relative humidity [%]

The first drying period is applied from the wood moisture content u_1 to the wood moisture content u_2 (second critical point) and the corresponding drying rate changes from j_{w1} to j_{w2}.

The normalized moisture value (Φ_1) and the relative drying rate (f_1) are determined by the absolute moisture content (u) and the actual drying rate (j_w) expressed.

$$\Phi_1 = \frac{u-u_2}{u_1-u_2} \tag{5.17}$$

$u_2 < u < u_1$

$$f_1 = \frac{j_w - j_{w2}}{j_{w1} - j_{w2}} \tag{5.18}$$

$j_{w2} < j_w < j_{w1}$

The second phase of the reduction in drying rate occurs when the wood reaches the equilibrium moisture content and drying ends. The normalized moisture value (Φ_2) and the relative drying rate (f_2) are calculated using Equation 5.19 and Equation 5.20.

$$\Phi_1 = \frac{u - u_{\mathrm{EMC}}}{u_1 - u_{\mathrm{EMC}}} \tag{5.19}$$

$$u_2 > u > u_1$$

$$f_2 = \frac{j_{\mathrm{w}}}{j_{\mathrm{w2}}} \tag{5.20}$$

$$j_{\mathrm{w2}} > j_{\mathrm{w}} > 0$$

The parameters u_1, u_2 and j_{w2} are based on the properties of the respective wood species.

Currently, it cannot be claimed that a perfect model exists, as all models still exhibit variability in prediction. This is mainly due to the coefficients of the mathematical equations, the simplifying assumptions underlying the development of the model, and the different methods used to calculate mass and heat transfer [2].

5.3 Models Based on Fundamental Equations of Moisture and Heat Transfer

Wood can be regarded as a porous, hygroscopic medium consisting of elements in solid, liquid, and gaseous states. Therefore, transfer mechanisms must be developed separately for each phase, taking into account the complex interaction between them. Since the variability of wood is very complex, the equations must be based on conservation laws at the macroscopic level. To date, several equations have been developed to simulate the drying process. The first significant difference lies in the number of state variables used to describe the environment [7]:

- Mathematical models with a single state variable such as wood moisture content or another equivalent quantity;
- Mathematical models with two variables such as wood moisture content and temperature;
- Mathematical models with three variables such as wood moisture content, temperature, and ambient pressure.

Currently, model 1 is most commonly used (Equation 5.21 to Equation 5.23).

Moisture conservation equation:

$$\frac{\partial}{\partial t}\left(\varepsilon_{\mathrm{w}}\rho_{\mathrm{w}} + \varepsilon_{\mathrm{g}}\rho_{\mathrm{v}} + \overline{\rho}_{\mathrm{b}}\right) + \nabla \cdot \left(\rho_{\mathrm{w}}\overline{v}_{\mathrm{w}} + \rho_{\mathrm{v}}\overline{v}_{\mathrm{g}} + \overline{\rho_{\mathrm{b}}v_{\mathrm{b}}}\right) = \nabla \cdot \left(\rho_{\mathrm{g}}\overline{\overline{D}}_{\mathrm{eff}}\nabla\omega_{\mathrm{v}}\right) \tag{5.21}$$

Energy conservation equation:

$$\frac{\partial}{\partial t}\left(\varepsilon_w\rho_w h_w + \varepsilon_g(\rho_v h_v + \rho_a h_a) + \overline{\rho}_b\overline{h}_b + \rho_o h_s - \varepsilon_g P_g\right)$$
$$+\nabla\cdot\left(\rho_w h_w\overline{v}_w + (\rho_v h_v + \rho_a h_a)\overline{v}_g + h_b\overline{\rho_b v_b}\right) = \nabla\cdot\left(\rho_g\overline{\overline{D}}_{eff}(h_v\nabla\omega_v + h_a\nabla\omega_a) + \overline{\overline{\lambda}}_{eff}\nabla T\right) + \Phi \tag{5.22}$$

Air conservation equation:

$$\frac{\partial}{\partial t}(\varepsilon_g\rho_a) + \nabla\cdot(\rho_a\overline{v}_g) = \nabla\cdot\left(\rho_g\overline{\overline{D}}_{eff}\nabla\omega_a\right) \tag{5.23}$$

In Equation 5.21, the velocities of the gas and liquid phases are given by the generalized Darcy's law:

$$\overline{v}_l = -\frac{\overline{\overline{K}}_l\overline{k}_l}{\mu_l}\nabla\varphi_l \tag{5.24}$$

where: $\nabla\varphi_l = \nabla P_l - \rho_l g\nabla\chi$.

The following boundary conditions are formulated for the drying of external surfaces:

$$J_w|_{x=0^+}\cdot\hat{n} = h_m c M_v \ln\left(\frac{1-x_\infty}{1-x_v|_{x=0^+}}\right) \tag{5.25}$$

$$J_e|_{x=0^+}\cdot\hat{n} = h(T|_{x=0^+} - T_\infty)$$

$$P_g|_{x=0^+} = P_{atm}$$

J_w and J_e represent the total moisture flux and the total enthalpy at the outer surface.

The two-equation model for heat and mass transfer (Model 2) can be regarded as a simplified version of Model 1, since it considers only two variables, namely temperature and moisture content. Assuming that the capillary pressure depends only on the moisture content (and not also on temperature), both the gas and liquid phases can be described by diffusion mechanisms involving the so-called "pseudo-diffusion coefficients" (K_v and K_l). Since this model cannot account for the internal pressure driving the movement of water in the liquid and gaseous phases, it is not suitable for drying processes with intensive moisture movement. However, due to the differences between the enthalpies of the gas and liquid phases, this model accounts for the latent heat of vaporization and thus incorporates the most important characteristic of drying, namely the coupled transfer of heat and mass.

Water conservation equation:

$$\frac{\partial}{\partial t}(\rho_0 X) = \nabla\cdot\left(\rho_v\overline{\overline{\boldsymbol{K}}}_v\nabla X\right) + \nabla\cdot\left(\rho_l\overline{\overline{\boldsymbol{K}}}_l\nabla X\right) \tag{5.26}$$

Energy conservation equation:

$$\frac{\partial}{\partial t}[\rho_0(h_0 + X\cdot h_l)] + \nabla\cdot(\rho_l h_l\overline{q}_l + \rho_v h_v\overline{q}_v) = \nabla\cdot\left(\overline{\overline{\lambda}}_{eff}\nabla T\right) + \Phi \tag{5.27}$$

with $\overline{q}_v = \nabla\cdot\left(\rho_v\overline{\overline{\boldsymbol{K}}}_v\nabla X\right)$; $\overline{q}_l = \nabla\cdot\left(\rho_l\overline{\overline{\boldsymbol{K}}}_l\nabla X\right)$

Boundary conditions

$$J_{\mathrm{w}}|_{x=0^+} \cdot \hat{n} = h_{\mathrm{m}} c M_{\mathrm{v}} \ln\left(\frac{1 - x_{\infty}}{1 - x_{\mathrm{v}}|_{x=0^+}}\right) \tag{5.28}$$

$$\mathrm{J}_{\mathrm{e}}|_{\mathrm{x}=0^+} \cdot \hat{\mathrm{n}} = \mathrm{h}(\mathrm{T}|_{\mathrm{x}=0^+} - \mathrm{T}_{\infty})$$

$$P_{\mathrm{g}}|_{x=0^+} = P_{\mathrm{atm}}$$

Model 3 with a single equation for heat and mass transport [6] can be derived from Model 2 by assuming that the wood temperature is constant and equal to the air temperature in the facility. Furthermore, a distinction between liquid and gas flow is no longer necessary, and the use of a global diffusion coefficient becomes possible.

Water conservation equation:

$$\frac{\partial X}{\partial t} = \nabla \cdot \left(\overline{\overline{\boldsymbol{K}}} \nabla X\right) \tag{5.29}$$

Boundary conditions

$$\boldsymbol{J}_{\mathrm{w}}|_{x=0^+} \cdot \hat{\boldsymbol{n}} = \alpha(c_{\mathrm{v}}|_{x=0^+} - c_{\infty}) \tag{5.30}$$

Since the goal of drying is to remove moisture from wood, the use of a single-equation model can be attractive, as it focuses on the movement of moisture within the medium and at its surface. According to Perre and Turner [6], the single-equation model is widely used in research.

The second model, based on two equations, represents a considerably more realistic simulation tool for the drying process [6]. This model is suitable for simulating drying at moderate temperatures or for online monitoring of the drying process. However, this model also has some limitations, such as the fact that it cannot simulate the drying process at high temperatures, vacuum drying, microwave heating of wood, etc.

Model 1, based on three equations, assumes that heat and mass transport also occur in the longitudinal direction. This model has been successfully used to simulate several drying configurations, taking into account both the thickness and the length of the material [6].

Table 5.1 presents a comparison of studies available in the literature with regard to the type of model used.

Depending on the number of dependent variables, the model size, and the model type, the time required for simulation can vary between a few seconds and a maximum of one hour [7].

Table 5.1 Types of Models Used to Simulate the Drying Process (adapted from [6])

Dependent variables, spatial dimension of the model	Model 3 – one variable wood moisture content	Model 2 – two variables Wood moisture content and temperature	Model 1 – three variables Wood moisture content, temperature and pressure
1D (thickness)	It is simple, but does not account for all physical phenomena (e.g., coupling of mass and heat transfer). Therefore, the results are not very meaningful.	Suitable for drying at low temperatures (<100 °C)	It enables detailed modeling of physical phenomena, but does not provide a realistic result for internal pressure. For high-temperature drying, the results are not very meaningful.
2D (thickness and width)	It correctly describes the cross-sectional shape, but does not account for all physical phenomena (e.g., coupling of mass and heat transfer). Therefore, the results are not very meaningful.	Correctly describes the shape of the cross-section, suitable for drying at low temperatures (<100 °C)	It enables detailed modeling of physical phenomena, but does not provide a realistic result for internal pressure. For high-temperature drying, the results are not very meaningful.
2D (thickness and length)	The additional computational effort compared to the 1D model is not justified. It does not account for all physical phenomena (e.g., coupling of mass and heat transfer). The results are not very meaningful.	Significantly higher computational effort compared to the 1D model	Enables detailed modeling of physical phenomena. Provides a good description of the analyzed geometry. Suitable for all drying processes. Accounts for the influence of internal pressure.
3D	Complex numerical calculations are not justified, as they do not account for all physical phenomena. The results are not very meaningful.	It provides a good description of the analyzed geometry and can be useful for calculating the global deformations of all boards.	The best description at the macroscopic level both in terms of geometry and the physical phenomena occurring.

Regarding the modeling of the material stack, the literature recommends actually modeling the stack geometry by considering an equivalent airflow model (Figure 5.1).

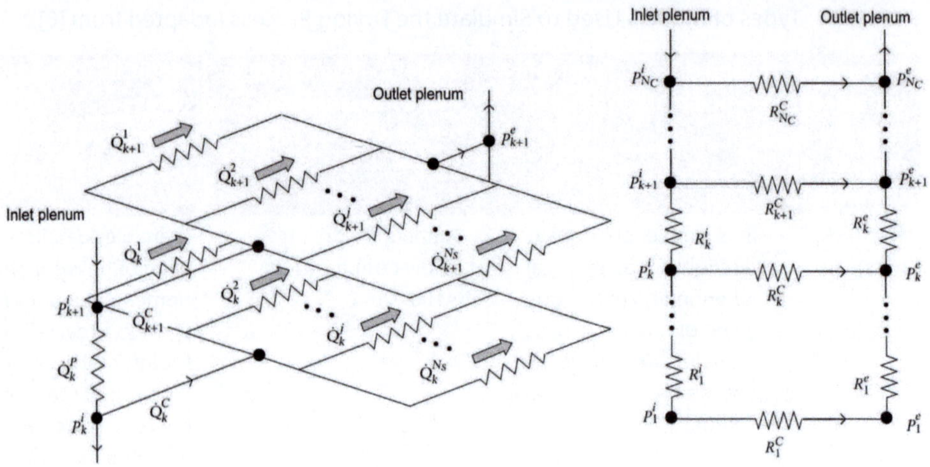

Figure 5.1 Equivalent airflow circuit between two pieces of wood (left) and equivalent airflow circuit over the height of the material stack (right) (source: [8], with permission)

5.4 Drying Models Based on Artificial Intelligence

Artificial neural networks (ANNs) are a tool that can be used to model the drying process. Wu and Avramidis [9] used this modeling technique to estimate the final moisture content (M_f) of the material to be dried by taking into account the initial moisture content (M_i), density (GM) and drying time (Figure 5.2).

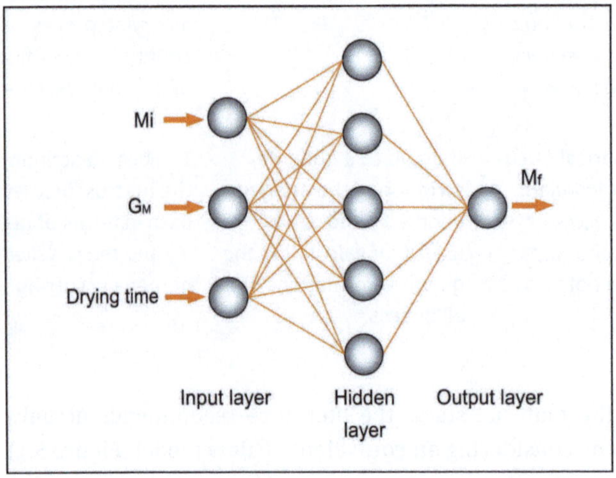

Figure 5.2 ANN model architecture for estimating the final moisture content (source: [9], with permission)

Zhang et al. [10] developed an ANN model, which can convert the structure of a drying schedule into a mathematical model that can be used to control the drying process in order to describe the relationship between temperature, relative humidity and wood moisture content. Three wood species were considered in the study: *Fraxinus mandshurica, Betula platyphylla* and *Xylosma racemosum.*

Ceylan [11] used this modeling technique to develop a model for estimating the moisture content of poplar and pine wood in the dryer. The input variables considered are air temperature, relative humidity and stack weight.

Watanabe et al. [12] developed an ANN model for estimating the final moisture content of Japanese cedar (*Cryptomeria japonica*) specimens during fresh air/exhaust air drying. The input variables of the developed ANN model are initial moisture content, basic density, annual ring orientation, annual ring width, heartwood proportion and color difference in the CIE-L*a*b* system.

Also for Japanese cedar specimens, Watanabe et al. [13] investigated the possibility of determining the stress level at the surface of wood during drying using near-infrared spectroscopy (NIR) in combination with artificial neural networks. The input of the ANN model consisted of the NIR spectrum, which was subsequently analyzed by a genetic algorithm to determine the input variables of the model.

Chai et al. [14] present an ANN model that estimates the variation of wood moisture content during HF vacuum drying, based on drying time, position of measurement points, material temperature, and pressure. Pine samples (*Pinus sylvestris* var. mongholica Litv.) were used for the study.

The study conducted by Fu et al. [15] presents an ANN model that can estimate the elastic strain (ES), the most important parameter related to drying stresses, of birch wood samples (*Betula platyphylla* Suk). The input parameters considered were temperature (*T*), wood moisture content (*MC*), relative humidity (*RH*), and distance from the pith (*DP*) (Figure 5.3).

Laaroussi et al. [16] used machine learning techniques to estimate the average wood moisture content in a fresh air/exhaust air drying chamber, such as Random Forest (RF), Extra Trees (ET), Adaboost (AB), Gradient Boosting (GB), Support Vector Regressor (SVR), Long Short-Term Memory (LSTM), and a combination of a convolutional network and a bidirectional LSTM (CBLSTM) algorithm. The input data considered include, among others, temperature and relative humidity, the phase of the drying process, and seasonal variables in which the drying is carried out. In the end, 20 of the 39 considered input variables remained. Of all the algorithms analyzed, CBLSTM was the one that led to the best estimation of the final moisture content.

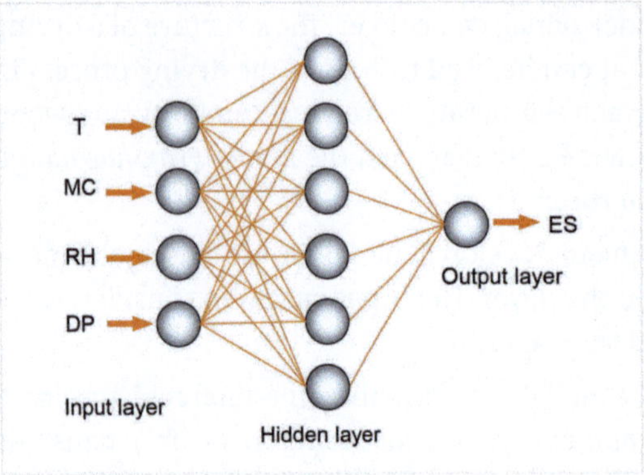

Figure 5.3 Neural network architecture for estimating elastic strains (source: [15], with permission)

Rahimi and Avramidis [17] developed three artificial neural networks, namely the multilayer perceptron (MLP) neural network, the radial basis function (RBF) neural network, and the group method of data handling (GMDH), to create a prediction model (Figure 5.4) that combines the material properties (density ρ_b, initial weight w_i, initial moisture content M_i, and target moisture content M_t) with the final moisture content (M_f) of the material. To improve the performance of the models, the authors also performed a Monte Carlo simulation. The GMDH-based model showed the best performance in estimating the final moisture content, followed by the MLP model and then the RBF model. With regard to the Monte Carlo simulation, the most efficient solution was achieved in the case of the RBF model. Of all the variables analyzed, the initial moisture content and the basic density had the greatest influence on the final moisture content. The wood species analyzed was hemlock.

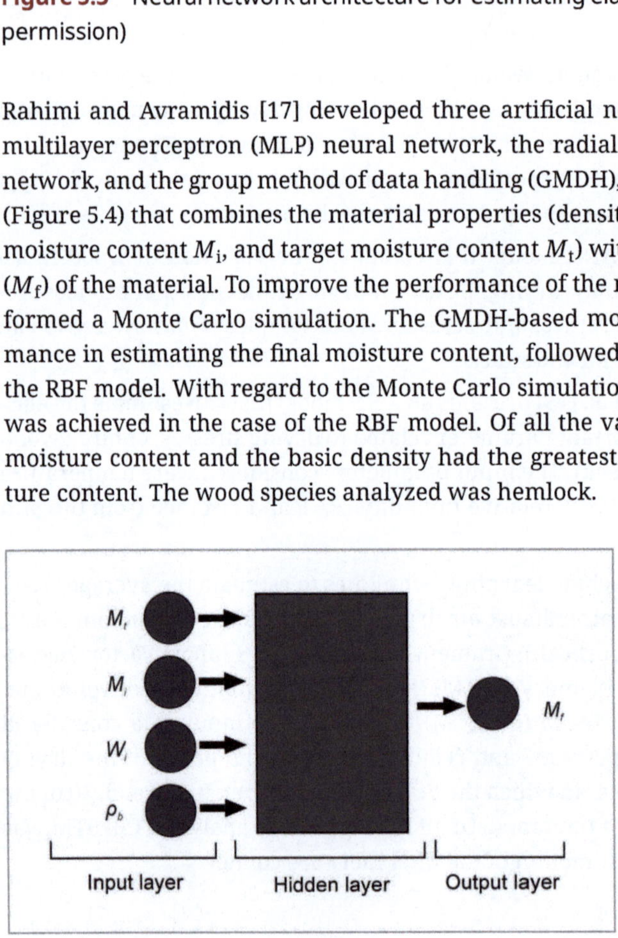

Figure 5.4 Structure of the input layers of the ANN model for estimating the final moisture content (source: [17], with permission)

Rahimi et al. [18] also developed a gradient boosting machine learning model for the analysis and optimization of the drying process of hemlock wood. The input variables of the developed model are the initial moisture content, initial weight, density, conditioning parameters, and the storage method after drying. It was found that, of all these variables, the initial weight has the highest degree of correlation with the final moisture content of the material.

Chai and Li [19] developed a neural network for estimating drying quality, based on the wood drying rate and the degree of longitudinal crack formation.

5.5 Modeling of Drying Stresses

Most mathematical models assume that the total strain rate ε is the sum of several mechanisms occurring during drying [1]:

$$\dot{\varepsilon} = \dot{\varepsilon}_{\mathrm{e}} + \dot{\varepsilon}_{\mathrm{u}} + \dot{\varepsilon}_{\mathrm{ve}} + \dot{\varepsilon}_{\mathrm{ms}} \tag{5.31}$$

Indices:
e elastic strain
u swelling and shrinkage
ve viscoelasticity
ms mechanosorptive load

$$\dot{\varepsilon}_{\mathrm{ms}} = m(\sigma - E_{\mathrm{ms}}\varepsilon_{\mathrm{ms}})|\dot{u}_{\mathrm{b}}| \tag{5.32}$$

m a constant
σ stress
E_{ms} constant related to the maximum stress that can be reached at a given stress level
$\dot{u}_{\mathrm{b}}$ bound water content

For wood drying, it has been found that 1D computer models provide useful information for the development and analysis phase of drying systems. 2D and 3D models, which often exhibit a grain slope, are extremely complicated. The most complicated part in simulating drying stresses is knowing the actual value and variability of the wood properties required for the models used. Therefore, it is nearly impossible to accurately estimate crack development in the drying material [1]. Through modeling, however, only the degree of probability of drying cracks can be determined – a useful tool in the analysis and optimization phase of drying schedules.

In the literature, several mathematical formulations and examples of the use of the numerical code TransPore for estimating drying stresses are presented [20]. For example, in Figure 5.5 we can observe the moisture and stress fields at different time intervals. After a drying time of six hours, the outer part of the profile enters the hygroscopic range, which leads to the development of shrinkage stresses in the surface area of the material. As a result, the inner regions are subjected to compression. At the end of the drying process (after 60 hours), on the other hand, the shrinkage stresses in the inner region can be noticed: the stress distribution is reversed (Figure 5.5).

Salinas et al. [22] developed a one-dimensional model to simulate the stresses occurring during drying of pine wood as a function of the moisture gradient. The resulting model is a set of partial differential equations that were subsequently solved using the finite volume method. After validation of the model with experimental data, it was found that the model can be used to study the phenomenon of drying stresses, which is difficult to evaluate experimentally.

Moutee et al. [23] developed a rheological model that can be used to estimate the development of stresses and deformations during wood drying. This model is based on the shrinkage coefficients, the instantaneous relationship between stress and strain, rheological deformation (time-dependent creep), and mechano-sorptive creep. The mathematical relationship is based on an elasto-viscoplastic model that takes into account the moisture gradient in the wood, the influence of external loads, and the stress- and time-dependent viscoplastic threshold. After validation, it was found that the model can estimate the reversible stress phenomenon, the residual stress, and the maximum stress across the material thickness.

Pang [24] developed a one-dimensional model for estimating the drying stresses of pine wood based on shrinkage coefficients, instantaneous stress–strain relationships, mechano-sorptive creep, time-dependent creep, and the influence of temperature. The hardening behavior of wood in the plastic range and the differences between increasing and decreasing stresses were also taken into account. The developed model can estimate stresses when the mechanical and rheological properties are known.

In addition to classical viscoelastic creep, most mathematical models are based on mechano-sorptive creep at various moisture contents. Further information on the developed models is given in the literature [25, 26].

One-dimensional (1D) models for calculating drying stresses can provide useful information in the development and optimization phase of drying systems, while two-dimensional (2D) or three-dimensional (3D) models that take fiber orientation into account are quite complicated to apply, as they require knowledge of all wood properties and their variability. Therefore, it is nearly impossible to accurately predict the stresses occurring during the drying process. The models developed so far serve merely as tools to indicate a probability for the risk of crack formation [1].

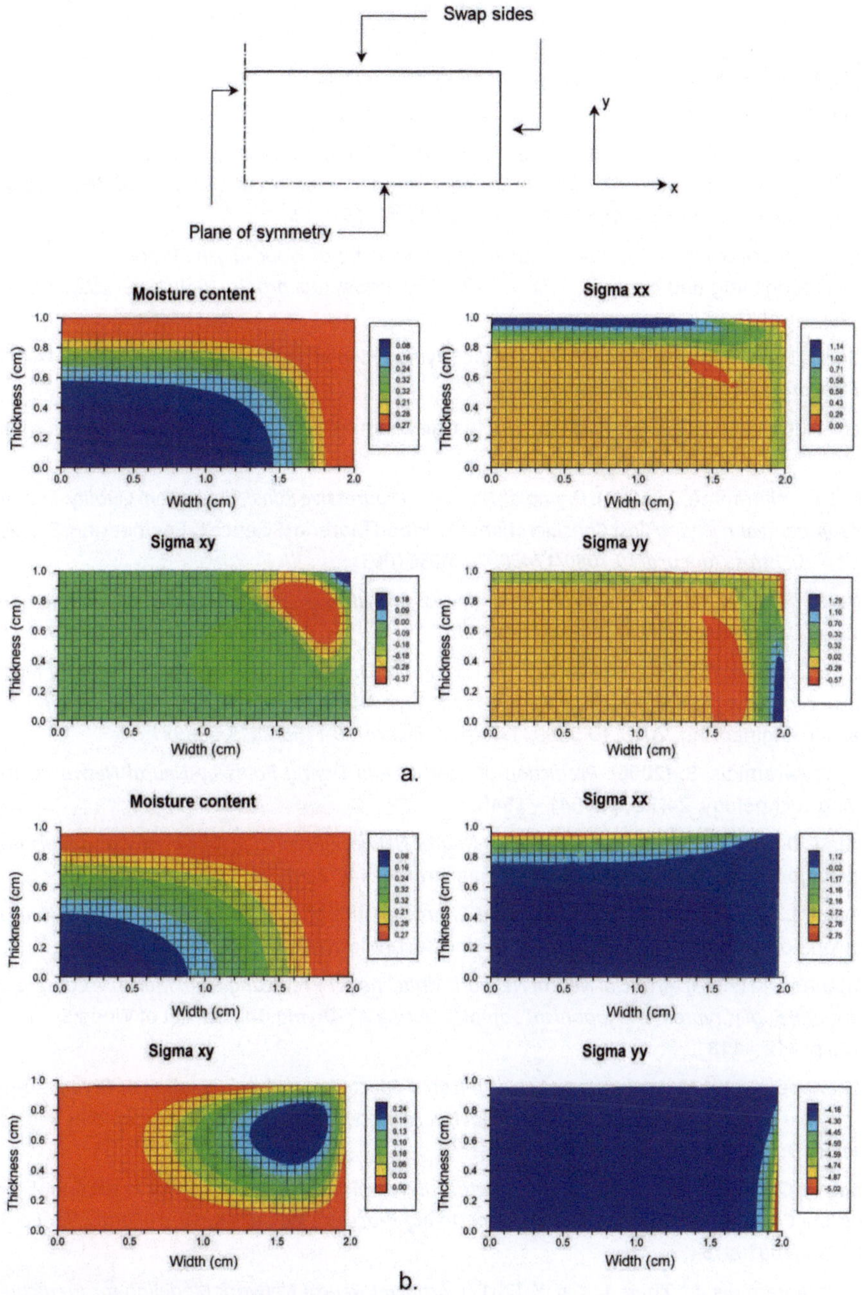

Figure 5.5 Moisture and stress fields after 6 and 60 hours (t = 80 °C and t_{wet} = 60 °C) (source: [21], with permission)

References

[1] Salin, J. G. (2010): *Problems and Solutions in Wood Drying Modelling: History and Future*. In: Wood Material Science and Engineering, 5, p. 123 – 124

[2] Avramidis, S.; Lăzărescu, C.; Rahimi, S. (2023): *Basics of Wood Drying*. Chapter in Niemz, P.; Teischinger, A.; Sandberg, D. (Hrsg.): *Springer Handbook of Wood Science and Technology*. Springer, Cham. *https://doi.org/10.1007/978-3-030-81315-4_13*

[3] Čiegis, R.; Starikovičius, V. (2002): *Mathematical Modeling of Wood Drying Process*. In: Mathematical Modelling and Analysis, 7(2), p. 177 – 190. *https://doi.org/10.1080/13926292.2002.9637190*

[4] Tuttle, F. (1925): *A Mathematical Theory of the Drying of Wood*. In: Fuel, 200(5), p. 609 – 614. *https://doi.org/10.1016/S0016-0032(25)90713-2*

[5] OpenAI (2023): ChatGPT (Mar 14 Version) [Large Language Model] . *https://chat.openai.com/chat*

[6] Salin, J. G.; Wamming, T. (2008): *Drying of Timber in Progressive Kilns: Simulation, Quality, Energy Consumption and Drying Cost Considerations*. In: Wood Material Science & Engineering, 3(1–2), p. 12 – 20. *https://doi.org/10.1080/17480270802561003*

[7] Perré, P.; Turner, I. (2007): *Coupled Heat and Mass Transfer*. In: Perré, P. (Hrsg.): *Fundamentals of Wood Drying*. A.R.BO.LOR., Nancy, France

[8] Vaz, M.; Zdanski, P. S. B.; Cerqueira, R. F.; Possamai, D. G. (2013): *Conjugated Heat and Mass Transfer in Convective Drying in Compact Wood Kilns: A System Approach*. In: Advances in Mechanical Engineering, 2013, ID 538931. *https://doi.org/10.1155/2013/538931*

[9] Wu, H.; Avramidis, S. (2006): *Prediction of Lumber Kiln Drying Rates by Neural Networks*. In: Drying Technology, 24(12), p. 1541 – 1545

[10] Zhang, D.; Liu, Y.; Cao, J.; Sun, L. (2008): *Neural Network Prediction Model of Wood Moisture Content for Drying Process*. In: Scientia Silvae Sinicae, 44(12), p. 94 – 98

[11] Ceylan, I. (2008): *Determination of Drying Characteristics of Timber by Using Artificial Neural Networks and Mathematical Models*. In: Drying Technology, 26, p. 1469 – 1476

[12] Watanabe, K. (2013): *Artificial Neural Network Modeling for Predicting Final Moisture Content of Individual Sugi (Cryptomeria japonica) Samples During Air-Drying*. In: Journal of Wood Science, 59(2), p. 112 – 118

[13] Watanabe, K. (2014): *Application of Near-Infrared Spectroscopy for Evaluation of Drying Stress on Lumber Surface: A Comparison of Artificial Neural Networks and Partial Least Squares Regression*. In: Drying Technology, 32(5), p. 590 – 596

[14] Chai, H.; Chen, X.; Cai, Y.; Zhao, J. (2019): *Artificial Neural Network Modeling for Predicting Wood Moisture Content in High Frequency Vacuum Drying Process*. In: Forests, 10, 16. *https://doi.org/10.3390/f10010016*

[15] Fu, Z.; Avramidis, S.; Zhao, J.; Cai, Y. (2017): *Artificial Neural Network Modeling for Predicting Elastic Strain of White Birch Disks During Drying*. In: European Journal of Wood and Wood Products, 75, p. 949 – 955. *https://doi.org/10.1007/s00107-017-1183-x*

[16] Laaroussi, M.; Benabbou, L.; Ouhimmou, M.; Abasian, F.; Haddad, S. (2022): *Predicting the Wood Mean Moisture Content in a Conventional Kiln-Based Drying Process: A Data-Driven Approach*. In: International Federation of Automatic Control (IFAC), 55(10), p. 1447 – 1452

[17] Rahimi, S.; Avramidis, S. (2022): *Predicting Moisture Content in Kiln-Dried Timbers Using Machine Learning*. In: European Journal of Wood and Wood Products, 80, p. 681 – 692. *https://doi.org/10.1007/s00107-022-01794-7*

[18] Rahimi, S.; Nasir, V.; Avramidis, S.; Sassani, F. (2023): *The Role of Drying Schedule and Conditioning in Moisture Uniformity in Wood: A Machine Learning Approach*. In: Polymers, 15, 792. *https://doi.org/10.3390/polym15040792*

[19] Chai, H.; Li, L. (2023): *Prediction of Wood Drying Process Based on Artificial Neural Network*. In: BioResources, 18(4), p. 8212 – 8222

[20] Perré, P.; Passard, J. (2007): *Stress Development*. In: Perré, P. (Hrsg.), *Fundamentals of Wood Drying*. A.R.BO.LOR., Nancy, France

[21] Perré, P.; Passard, J. (2004): *A Physical and Mechanical Model Able to Predict the Stress Field in Wood Over a Wide Range of Drying Conditions*. In: Drying Technology, 22(1–2), p. 27 – 44. *https://doi.org/10.1081/DRT-120028202*

[22] Salinas, C.; Chavez, C.; Ananias, R. A.; Elustondo, D. (2015): *Unidimensional Simulation of Drying Stress in Radiata Pine Wood*. In: Drying Technology, 33(8), p. 996 – 1005. *https://doi.org/10.1080/07373937.2015.1012767*

[23] Moutee, M.; Fortin, Y.; Fafard, M. (2007): *A Global Rheological Model of Wood Cantilever as Applied to Wood Drying*. In: Wood Science and Technology, 41, p. 209 – 234. *https://doi.org/10.1007/s00226-006-0106-5*

[24] Pang, S. (2000): *Modelling of Stress Development During Drying and Relief During Steaming in Pinus Radiata Lumber*. In: Drying Technology, 18(8), p. 1677 – 1696. *https://doi.org/10.1080/07373930008917806*

[25] Elustondo, D.; Matan, N.; Langrish, T.; Pang, S. (2023): *Advances in Wood Drying Research and Development*. In: Drying Technology, 41(6), p. 890 – 914. *https://doi.org/10.1080/07373937.2023.2205530*

[26] Yin, Q.; Liu, H.-H. (2021): *Drying Stress and Strain of Wood: A Review*. In: Applied Sciences, 11, 5023. *https://doi.org/10.3390/app11115023*

6 Lumber Drying Methods

6.1 Historical Development

Wood has accompanied mankind through all its stages of development. The necessity of drying it before use has been known since antiquity. Furniture parts from the tombs of Egyptian pharaohs support this claim. After being stored for 2000 years in hermetically sealed chambers beneath the desert sand, the excavated wooden objects (statues, chests, cabinets) from the pharaonic era showed not the slightest tendency to crack, flake, or have their joints come apart.

Hesiod (750 BC), Vitruvius (50 BC) and Pliny (50 AD) provide written evidence of contemporary knowledge about how wood should be stacked and dried. Open-air drying, flue gas drying and sand drying were the only techniques applied until the early modern period.

It seems that the first patent related to wood drying was filed in England in 1636 and relates to the drying of wood for ship and boat building [1].

The first drying plants were also built near ports. These were simple, without fans and with direct heating using exhaust gases produced by the combustion of wood waste (Figure 6.1).

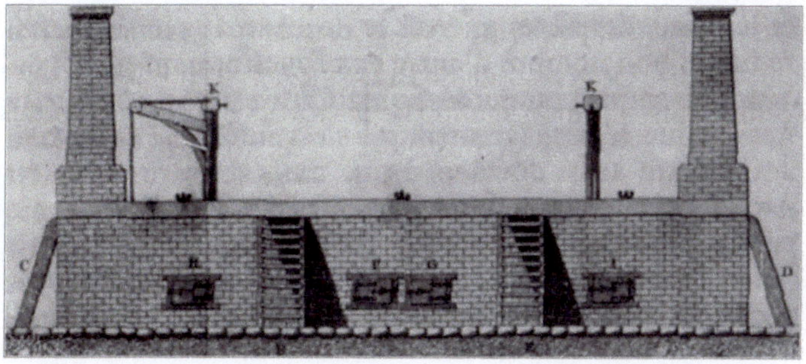

Figure 6.1 Wood drying kiln from the 18th century (based on [1])

The first dryers with heating elements (usually placed in the lower part of the unit) were built at the beginning of the 18th century and were based on the principle of natural convection (Figure 6.2).

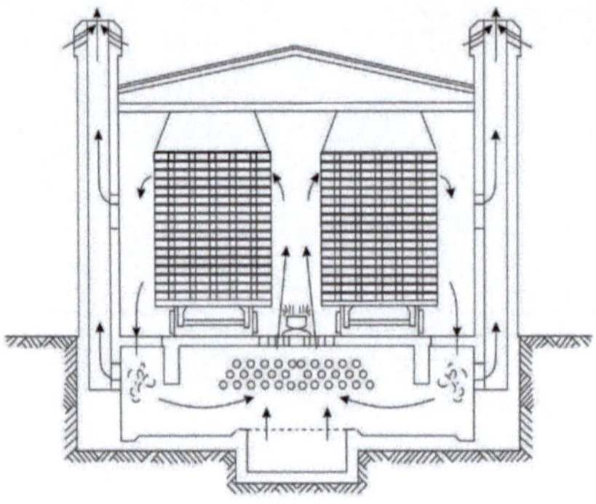

Figure 6.2 Drying kiln with natural convection (based on [1])

Due to their economy, safety and ease of operation, these systems were popular in Europe for a long time.

But as soon as the advantages of artificial ventilation became known around 1850, wood drying entered a new stage. The performance of installations equipped with fans increased considerably, both in terms of drying time and drying quality. It was recognized how important uniform air circulation is as a prerequisite for high-quality drying.

During the First World War, developed in the USA and later imported to Europe, the technique of artificial drying with warm air (which we today refer to as "conventional" drying or fresh air/exhaust air drying) spread rapidly. The first installations were masonry dryers (Figure 6.3), the transition to metal constructions came later. Like today's dryers, these systems were equipped with heat exchangers, humidification pipes, fans and air ducts. The motors driving the fans were located outside the system to avoid being damaged by the corrosive environment inside the drying system.

For the generation and guidance of air through the stacks, various variants were tested (Figure 6.4). It was concluded that the best variant is the one with axial, individually driven fans placed above the intermediate ceiling. This ensures uniform air circulation through the stacks and enables a clear separation of the overpressure/underpressure zones in the dryer. Therefore, this variant remains the most widely used one to this day.

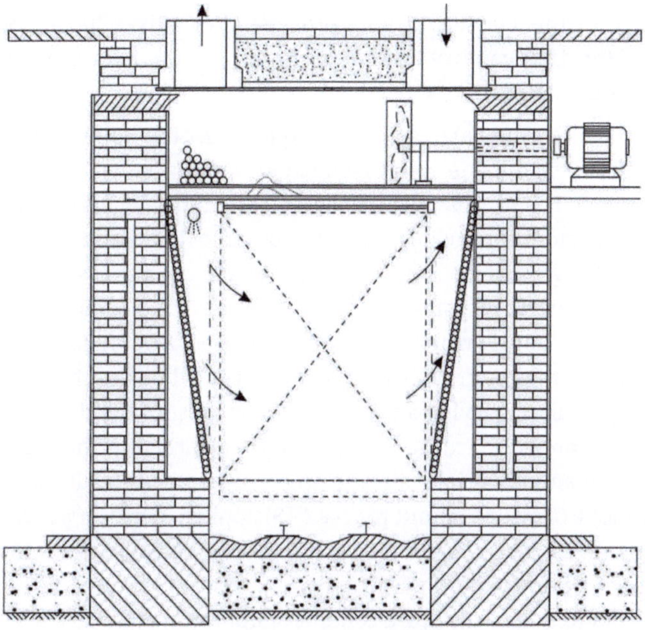

Figure 6.3 Masonry drying system (based on [2])

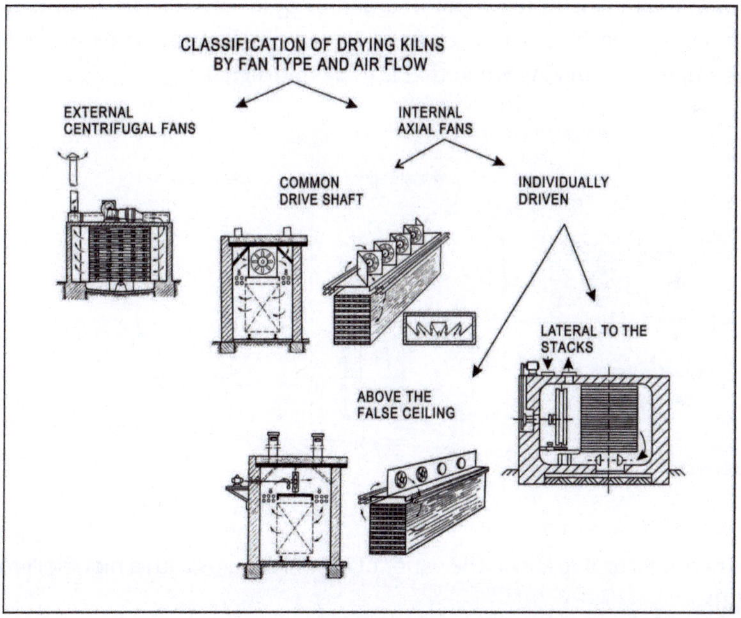

Figure 6.4 Types of dryers with different types of fans and air guidance

Parallel to the development of this classical drying process, alternatives have also emerged [2, 3]. The end of the 19th century and the beginning of the 20th century were productive periods of innovation:

1865 Robbins [4] patents drying in the vapor of organic liquids with a boiling point above 100 °C (e.g. xylene 135 °C or toluene 110 °C), which are not miscible with water (so that separation is easier). The process is carried out in an autoclave performed (Figure 6.5). An evaporator provides for the heating of the organic liquid and its conversion into steam. The hot steam is then passed over the wood in the autoclave. The drying can take place at atmospheric pressure or under vacuum. The mixed vapors of chemicals and water are continuously withdrawn and condensed. After xylene and water are separated, the water settles by gravity and is disposed of, while the chemical is recycled through a storage tank for drying and reused. After most of the drying is completed, a final vacuum is applied to extract the substance adsorbed in the wood and to complete the drying. On an industrial scale, the process was first introduced in 1942 in the USA under the name "Taylor-Colquitt process" [5] applied, where it proved its effectiveness in drying railway sleepers and telegraph poles. The wood is t = 150 °C in an autoclave filled with xylene vapor at atmospheric pressure. Vacuum drying brings the impregnated wood to a final moisture content of approximately 15%. Oak railway sleepers require 12 – 16 h to reach a wood moisture content of 40%, and 200 mm thick pine poles require 10 h to go from 90% to 35% moisture content. Small surface checks are common. Drying in the vapor of organic liquids can also be applied to beech wood. 26 mm thick beech boards dried from 28% to 17% in approximately 14 h [2]. The process is suitable for drying wood that is to be impregnated.

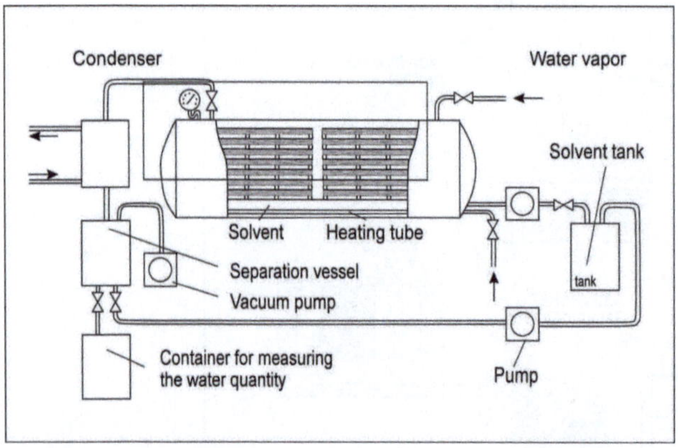

Figure 6.5 Schematic of a drying kiln in the vapor of organic liquids with a high boiling point (based on [2])

1867 Allen and Campbell [6] patent high-temperature drying in superheated steam. Tiemann [7] uses the same principle in a plant where the movement of steam was ensured by four pairs of steam spray lines and could also be reversed. Basically, water vapor is superheated when its temperature exceeds that of saturated steam at the relevant pressure. When the wet-bulb temperature (t_{wet} – see also Section 2.2.2) at atmospheric pressure is below 100 °C, both air and water vapor are present in the chamber. Once the wet-bulb temperature reaches 100 °C and the dry-bulb temperature is above 100 °C, pure superheated steam remains in the chamber. Both variants are referred to as "high-temperature drying." The main advantage of this process is the reduction of drying time compared to drying at temperatures below 100 °C. 25 mm thick softwood can be dried to 10% in 24 hours [8]. However, it is a severe drying process with frequent checking. After an establishment period up to the 1950s, new physical insights regarding this process led to a revival and industrial application, especially in regions that dry large quantities of softwood [9–12].

1879 drying in hot oils is patented in England under the name "Boulton process" [13] and two years later in the USA. The principle consists of immersing the wood in a hydrophobic liquid (Figure 6.6), which has a higher boiling point than water. This is heated to approximately 120 °C, high enough to evaporate the water. At the same time, the oil penetrates the emptied wood cavities. The drying is very severe and casehardening cannot be avoided. The process is also used today as a drying and wood protection method against mold and insects for wood used in outdoor conditions.

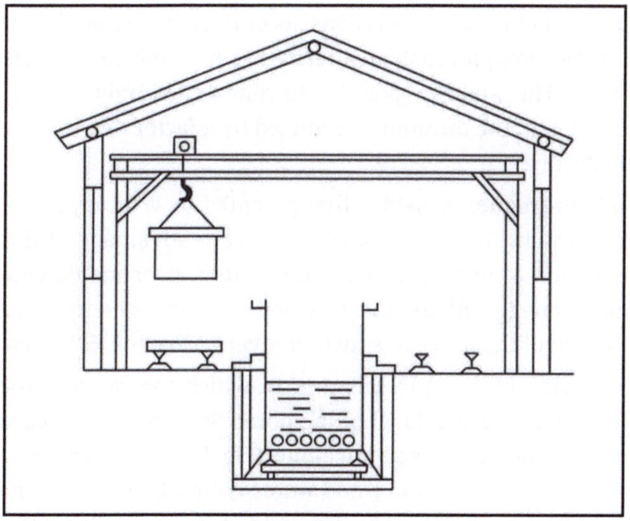

Figure 6.6 Schematic of a drying kiln in hot oils (based on [14])

1893 Howard receives the first patent for vacuum drying of sawn timber. The process was intensively investigated, but without achieving the expected spectacular results in terms of drying time reduction. Vacuum has only a very minor influence on moisture diffusion through wood. The removal of water from the surface is associated with a cooling effect, which slows down diffusion. Therefore, heat must be supplied to the wood [13]. Initially, attempts were made to heat the wood between vacuum periods at atmospheric pressure (this process is known as "discontinuous vacuum drying"). Better results were achieved with continuous heating under uninterrupted vacuum. Only after 1965 were combined processes (vacuum press drying, vacuum drying in superheated steam, vacuum high-frequency drying) applied industrially. These are described in more detail in Section 6.6 discussed.

1893 Lyon patents ozone drying [1]. The presence of ozone in warm and humid air leads to condensation of the vapors contained in the drying medium, thereby maintaining moisture on the wood surface and preventing case-hardening. A further advantage of ozone drying is the oxidation of secondary compounds contained in the wood (e.g., resins), achieving the effect of artificial aging. Although successfully promoted in France during the years 1925 – 1930, experiments conducted a few years later in India with native species did not confirm the same good results.

1914 an electrical drying process called the Nodon process is tested in France, advertised as a "rapid process for drying and protecting wood." The process is based on the transfer of heat from electrodes in the form of wire meshes, which are placed between rows of wood during stacking and connected to a power source (direct or alternating current). The treatment lasts 1 – 2 days and is considered an effective method for accelerating air drying. A further development of this process is found in induction drying, proposed and tested by Romanovski in Russia. This consists of introducing mild steel rod grids between the stacks, which are placed in an alternating electromagnetic field at industrial frequencies (50 Hz). This makes it possible to remove not only free but also bound water from the wood, and the duration is reduced by a factor of two compared to fresh air/exhaust air drying.

1928 the drying of wood in a high-frequency field is first patented by Whitney in the USA and a year later by John Gray in Great Britain [7]. The main advantage of this process is the substantial reduction of drying times. Thick timber assortments with high density and wood species prone to collapse can be dried with high quality using this process. A detailed description of the process is further given in Section 6.7 given.

1933 Stamm [15] patents the drying of wood in solvents, in which the wood in the green state is exposed to a hot solvent with a low boiling point. Stamm initially used paraffin as the solvent, but this was then replaced by acetone, which has an even lower boiling point (56.2 °C). The mixture of solvent and wood water is circulated. When the mixture contains too much water, the solvent is recovered by distillation. Using this process, ponderosa pine was dried from an initial moisture content of 150% to a final moisture content of 12% in 24 h. The process can also be used for the extraction of extractives from wood (e.g., tannins or polyphenols).

1934 Schwalbe and Bartels design the first plant for drying round timber using solar energy [13].

1935 Rawling [16] publishes a classification of various wood species according to the penetration depth of infrared radiation. *Pinus cembra* L. and *Acer pseudoplatanus* L. were found to be particularly permeable. Norway spruce is cited as the most permeable wood species, with a penetration depth of 6 mm, followed by maple and poplar (4 mm) and jaw (3 – 4 mm) [13]. The aim of this research was to apply this heating principle as a drying method to wood. Later experiments by Keylwerth [17] showed that drying with IR radiation is only suitable for thin wood assortments (veneers).

1937 Grau develops the idea of "oscillating drying"", to accelerate open-air drying [2]. The boards (with a mass of approximately 10 t) were stacked vertically and balanced at an angle of 20° to the vertical, with 15 oscillations/minute or an air velocity of 4 – 7 m/s in the middle of the height of the stack (Figure 6.7). The drying time for 25 mm thick spruce boards amounts to 4.5 days from 90% down to 30% wood moisture content, and a further 9.5 days down to 20% wood moisture content. Compared to open-air drying of the same sawn timber assortment, the drying time is reduced by 33 – 66%, with very low energy consumption (0.25 kWh/m^3 for a large swing and 0.10 kWh/m^3 for smaller swings).

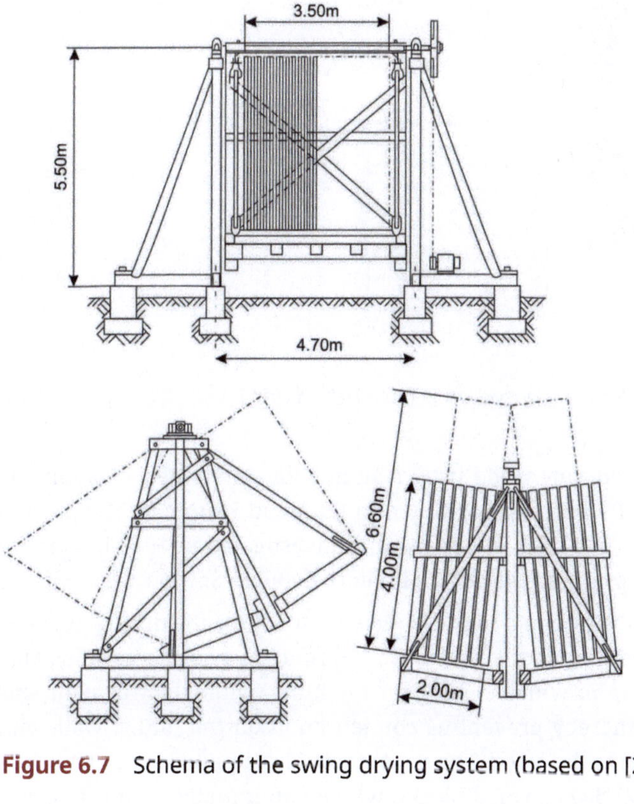

Figure 6.7 Schema of the swing drying system (based on [2])

1943 Kastmark [18] takes up the idea of using centrifugal force when drying. He is building the first plant of this kind in Sweden (Figure 6.8 a) and tests the correlation between different wood moisture levels and the rotational speed. The process proves to be applicable only for drying in the moisture range above the fibre saturation point. A moisture reduction from 55% to 35% of wood preheated to 40°C was achieved in 4 h. However, the final moisture content was uneven, with higher values at the ends of the pieces, which is presumably attributable to the formation of a "jacket" of air-saturated air between the stack and the circular housing. To remedy this deficiency, Eisenmann [19] attempted to improve Kastmark's design by changing the shape of the housing to one with a square cross-section and baffles for air guidance (Figure 6.8 b). Although it was not applied on an industrial scale, the process was not forgotten. 65 years later, Krämer [20] builds a carousel for drying firewood logs based on this principle.

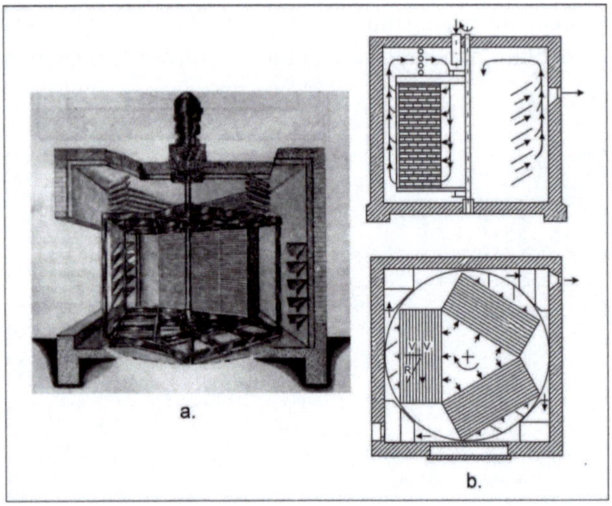

Figure 6.8 Centrifugal dryers (a – based on [1]; b – based on [2])

In 1945, the first patent for drying wood in a vacuum with simultaneous heating by high-frequency current (HFV process) is granted to Luth and Krupnick. The process was also proposed in 1948 by Miller in Canada and investigated in 1949 by Murata and Iso in Japan [21]. This process is discussed in more detail in Section 6.8 depicted.

In 1963, Brown, Boveri & Cie design a high-frequency tunnel for the drying of beech wood (Figure 6.9). The generator operated at 13.56 MHz and the power was adjustable between 2 and 25 kW. A fan provided cooling for the tubes, while the warm air was directed into the tunnel, thereby preventing condensation on the tunnel walls and also transferring some heat to the wood. The system was also equipped with air shafts for removing moist air [22]. Beech wood blocks, which had initially been brought to a moisture content below the fibre saturation range (approximately 20%), could be dried within four hours in the HF field to a final moisture content of approximately 9% with almost no defects.

Figure 6.9 HF dryer by Brown, Boveri & Cie (source: [1])

1965 the Italian Vincenzo Pagnozzi patents vacuum press drying [23]. The method (Figure 6.10) is based on the concept of accelerating vacuum drying through direct heat transfer by contact between heated plates and wood. The process is particularly applicable to diffuse-porous hardwood species (e.g. beech).

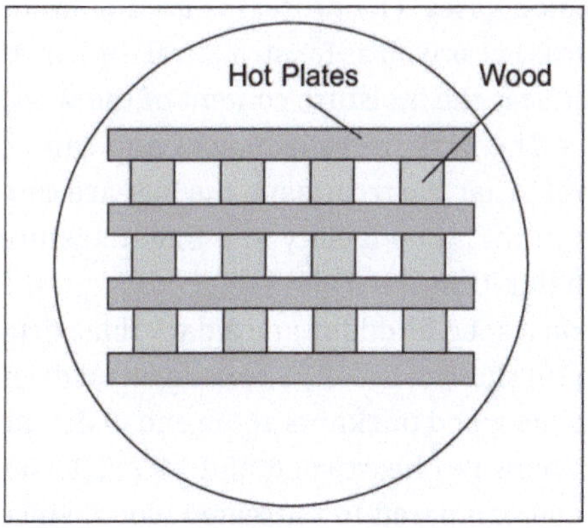

Figure 6.10 Principle of vacuum press drying

In the late 1960s, freeze and sublimation drying was introduced in the USA investigated [24], in the hope of reducing shrinkage and collapse when drying very sensitive wood species. This is actually a variant of vacuum drying, in which the freshly cut wood is first frozen at a temperature of –30 °C in order to convert the free water in the lumen of the cells into ice. This freezing phase is followed by a vacuum phase at approximately 0.27 bar, whereby the ice is directly converted into steam (sublimation phase). The removal of water in gaseous form from the wood without passing through the liquid phase protects the wood and makes the drying of sensitive wood species, or of archaeological wood, possible, thanks to the low temperatures.

In the late 1960s and early 1970s, the new technology of microwave heating and microwave drying also tested in the area of wood materials, particularly as a continuous process for thinner materials (veneers, thin sawn timber, pencils) [25]. The first production-ready, continuous microwave hot-air dryer was built by Cryodry Corp. With a microwave power of 50 kW and a frequency of 915 MHz, it was developed to equalize the moisture content of softwood veneer [26].

1971 Bror Häger patented [27] the so-called "Royal process", a combination of drying in hot oils and vacuum drying. The process is applied to sawn timber in the green state. The wood is stacked in an autoclave filled with oil with a high boiling point (150 – 400 °C). The oil is heated to 60 – 90 °C while simultaneously a vacuum is created (0.03 – 0.16 bar). The negative pressure leads to water evaporation. The drying time is 4 – 6 h for 25 mm thick wood and 6 – 8 h for 50 mm thick wood. A similar process was developed in Germany 30 years later, under the name "Oil Heat Treatment" (OHT™). This process uses vegetable oils heated to a higher temperature (180 – 220 °C), which leads not only to drying but also to changes in the chemical composition (thermal modification) of the wood.

1976 in the USA, drying with air dehumidification by condensation patented. This method is described in detail in Section 6.4 described.

1991, the Ecological Drying System (EDS) is developed in Japan [28] and advances the idea of thermally treating wood with combustion gases. The process is used both for unbarked tree trunks (small diameters) (Figure 6.11) as well as for structural timber. At high ambient temperature (t = 70 – 200 °C) reduces the moisture content of the wood to approximately 40 – 50 % within 3 – 5 days. The bark prevents cracks and can be removed much more easily after this treatment. After the treatment, the logs are converted into sawn timber, which can be dried in the same facility to a final moisture content of 8 – 10 %, with the drying time ranging from 3 – 7 days depending on the wood species. In this process, the wood takes on a specific odour and a dark (blackish) colour. Studies with EDS-treated beech wood [29] showed that EDS-treated wood dries more uniformly. The moisture gradient across the wood thickness at the end of drying is 14 % lower in EDS-treated beech wood without red heartwood and 21.7 % lower in EDS-treated beech wood with red heartwood compared to untreated wood. Since beech wood with red heartwood is known to be particularly inhomogeneous and dimensionally unstable, the benefit of the EDS treatment becomes even more apparent.

Figure 6.11 EDS dryer for undebarked logs and sawn timber (source: [29], with permission)

1992 the so-called "Incubation/Decompression" (I/D) procedure patented [30], characterized by alternating overpressure and normal pressure phases. The drying process takes place in an autoclave. During the "incubation phase", the pressure rises to 10 bar (or more, up to 40 bar). After an "incubation phase", a "decompression phase" follows, i.e. a relaxation of the pressure to atmospheric value (1 bar). One cycle lasts approximately 20 minutes. Without any supply of thermal energy, the air humidity in the free water range can be reduced. For drying below the fibre saturation point, a supply of heat is absolutely necessary. Despite the relatively high investment costs, the payback period is estimated at less than two years [31].

In the 1990s, an intensive investigation of vacuum high-frequency drying also begins in the USA and Canada. Avramidis et al. [32–35] investigate in a new HFV chamber with 23 m^3 capacity to dry various Canadian softwood species at a frequency of 3 MHz and a maximum power of 260 kW. With RF technology, it was possible to dry thick wood pieces that cannot be dried in conventional drying kilns, as hot air dryers would require excessively long drying times and cause unacceptable defects. By using appropriate drying schedules, not only cracks but also discolorations can be avoided.

Also in the 1990s, microwave drying examined more closely. Research by Antti et al. [36–38] aims to overcome the problems of field distribution and power intensity in the development of a continuous microwave dryer. Multimode applicators are proposed

[39] to further develop this technology for industrial application. A further improvement of this concept is the additional application of vacuum to achieve even higher drying rates and to improve the quality of the dried material [40].

2001 Brunner patents a plant based on the principle of vacuum drying in superheated steam [41], which is considered the "drying process of the future", thanks to the much shorter drying times and the better quality for all types of wood [42].

In 2008, Brodie [43] published the results of his investigations regarding the combination of microwave heating with solar drying. Experiments with short, 30 mm thick *Populus alba*-Wood parts that were first heated in a microwave oven showed a 17% shorter drying time compared to untreated samples dried in the same experimental solar installation. According to model results, the reduction should amount to up to 60% for thicker wood. Further experimental tests with 30 mm thick samples of *Eucalyptus regnans* reduced the drying time by 33% compared to solar drying without microwave heating. Based on the experimental results of this research, this combined method is particularly recommended for hardwoods with low permeability and medium to high density, with thicknesses exceeding 30 mm.

The advances recorded after the year 2000 do not relate to new processes, but rather to optimizations of the existing ones, and the development of increasingly smarter control and regulation systems.

In summary, regarding the historical development of drying methods, it can be highlighted that open-air drying – the oldest wood drying method – is still used today, as a cheaper alternative to artificial drying, and also in situations where the wood must be dried with as little stress as possible (e.g. in the case of musical instruments).

In artificial drying, more than 40 different methods have been tested for wood drying so far. Some aimed at reducing the duration, others at reducing energy consumption, others at utilizing ecological energy sources, and still others at improving drying quality in certain aspects.

Every process has its advantages and disadvantages. Some are only applicable under very specific conditions and therefore never achieved widespread industrial use. Conventional drying (fresh air/exhaust air process) has held the top position for decades with approximately 85% of industrially deployed installations. Condensation drying, vacuum drying and high-frequency drying are also of industrial significance. Their principles will therefore be explained in more detail below.

6.2 Air Drying (Natural Drying)

The natural drying consists of properly stacking the wood outdoors or under open sheds and protecting it from direct exposure to sunlight and precipitation until it reaches the equilibrium moisture content in the given environment. Wind speed plays an essential role in air drying, as it ensures ventilation of the stacks and removal of moisture from the wood surface.

The final moisture content, which the wood can reach through natural drying, depends on the season and local climatic factors that determine the wood equilibrium moisture content. Based on the average temperature and humidity values for Germany, the graph in Figure 6.12 was constructed.

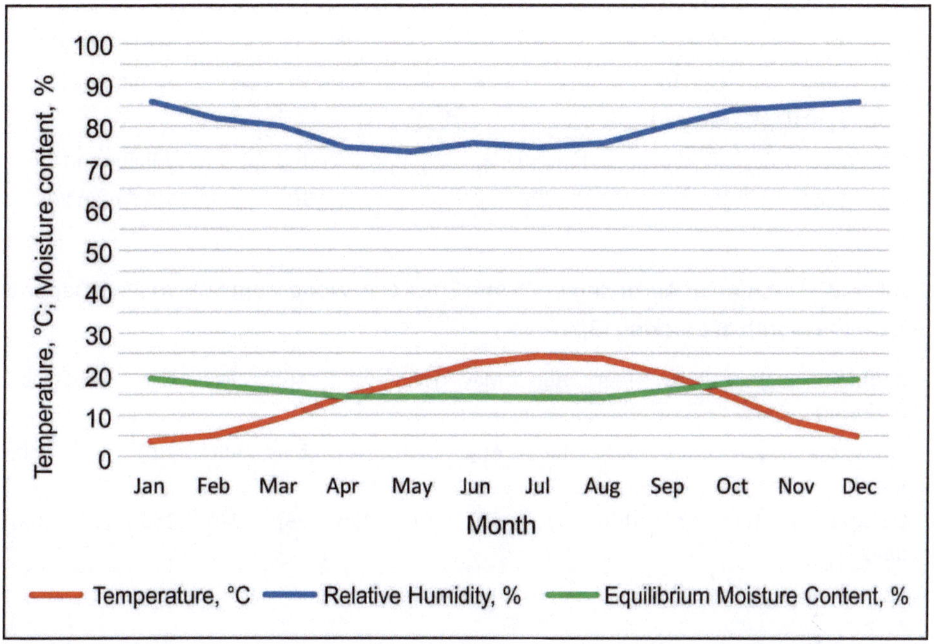

Figure 6.12 Average climatic conditions in Germany (based on data from the German Weather Service, *https://www.dwd.de*)

It can be noted that the equilibrium moisture content in Germany lies in the range of 14.3 – 18.5 % and therefore practically coincides with the range applicable to Central Europe (15 – 18 %). The minimum value is reached in the months of July and August and the maximum value in the month of January. The mean value is 16.2 %.

The time required to reach the equilibrium moisture content during air drying depends significantly on the wood species, thickness, season, and climatic conditions of the respective geographic area (annual average values of temperature and relative humidity, altitude, direction and intensity of prevailing winds, intensity of fog, average annual precipitation, solar radiation, etc.). Reference values for the duration of air drying in Central Europe are given in Table 6.1. For softwoods, the drying time in the open air is on the order of weeks to months, whereas for hardwoods it can exceed three years, e.g. for oak, bearing in mind that from November to March air drying practically stagnates (Table 6.2).

Table 6.1 Duration of Air Drying in the Temperate Climate Zone [44]

Period	Drying time, in days, for a wood thickness of:			
	≥ 18 mm	19 - 37 mm	38 - 75 mm	> 75 mm
Wood species: Softwood				
Oct - Apr	15 - 25	26 - 35	36 - 50	Minimum 51
May - Sep	10 - 18	19 - 28	29 - 35	Minimum 36
Wood species: Hardwood				
Oct - Apr	30 - 50	50 - 70	70 - 100	Minimum 100
May - Sep	20 - 36	35 - 56	55 - 70	Minimum 70

Table 6.2 Duration of Air Drying of 25 mm Thick Oak Wood from 50% to 25%, Depending on the Month of Stacking [2]

Beginning of drying	Jan	Feb	Mar	Apr	May	Jun	Jul	Aug	Sep	Oct	Nov	Dec
End of drying	Jun	Jun	Jun	Jul	Aug	Sep	Nov	May	May	Jun	Jun	Jun
Duration, days	160	140	120	110	90	70	120	250	240	240	220	180

Prolonged storage usually leads to quality deterioration (graying, surface and end checks, case hardening, deformations, fungal and insect attack). To prevent these, the following rules should be observed as far as possible:

- the lumber yard should be located on elevated ground, exposed to the wind and protected from obstacles (trees, tall buildings);
- the ground should be flat and dry; water-permeable clay or loam soils are well suited [45]; if the ground is swampy, it must be drained; unevenness must be covered with soil, gravel, or slag; vegetation must be destroyed; a slight slope (1:1,000) is advantageous for rainwater runoff;
- fresh lumber must never, even temporarily, be left unstacked;
- the stacks are placed not directly on the ground, but at a distance of 50–60 cm on stacking supports (Figure 6.13), to avoid moisture absorption from the ground;
- the stacks must be covered to avoid direct contact with precipitation and sunlight; the stack roof must be secured against wind uplift and should have a slope of approximately 3% so that precipitation can drain off; the roof should extend beyond

the stack length by approximately 0.5 m so that water does not drain off in the immediate vicinity of the stack (Figure 6.14);

- for optimal space utilization, multiple stacks can be overlapped to a total height of up to 8 m, provided the total height does not exceed three times the width of the stack; however, the load-bearing capacity of the floor must also be taken into account to prevent any risk of sinking;
- for long-term outdoor storage of timber and especially for assortments with a thickness of more than 40 mm, water-repellent end-grain coatings with paraffin or wax dispersions are recommended to prevent premature drying of the sawn edges and thus reduce the formation of end checks.

Each wood species has its own characteristics: pine, beech, alder, and maple are susceptible to blue stain and should therefore be stacked as quickly as possible after sawing. Larch, ash, and oak must dry slowly and should therefore be stacked tightly and well protected from rain. Lime and alder require airy storage without direct sunlight [46].

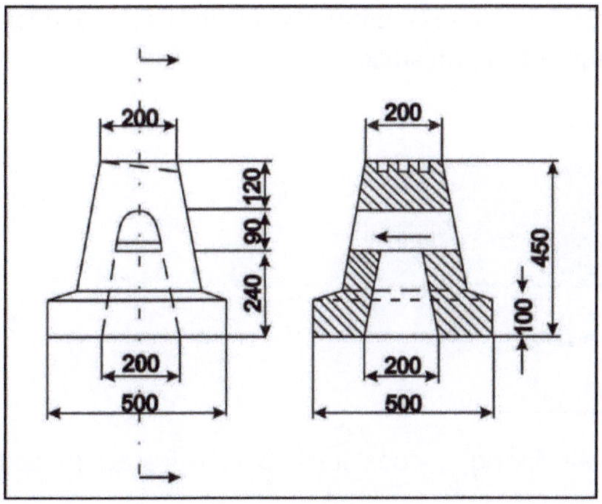

Figure 6.13 Concrete stacking blocks (source: [2], with permission)

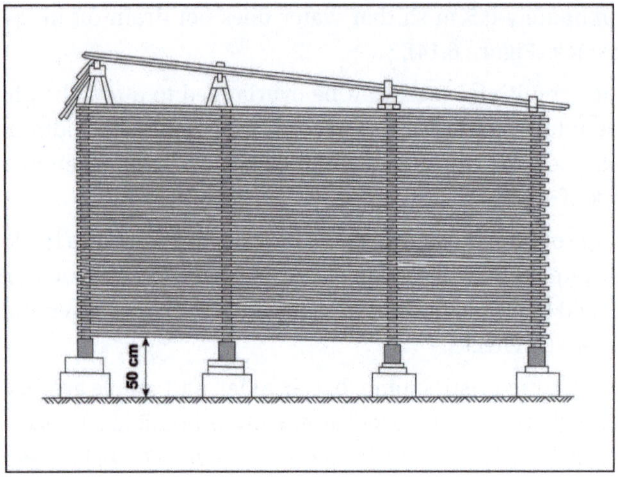

Figure 6.14 Open-air stack with roof (source: [2], with permission)

With regard to open-air drying, sawn timber is most commonly stacked as a box stack (Figure 6.15 a) or as a block stack (Figure 6.15 b) on stickers.

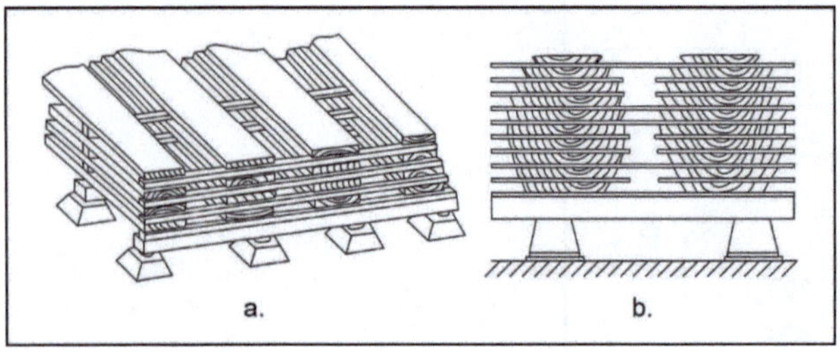

Figure 6.15 Stacking forms for open-air drying: a – box stack; b – block stack (based on [45])

Special stacking forms (without stickers), such as the flat stack for parquet raw strips (Figure 6.16) or for resonance wood strips for musical instruments (Figure 6.17), are also encountered.

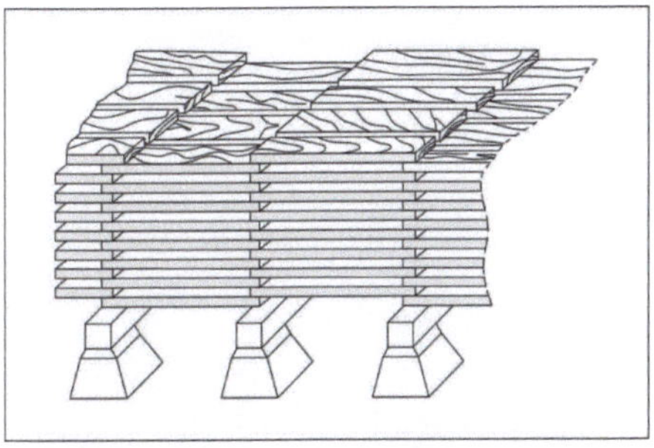

Figure 6.16 Flat stack without stickers for parquet strips (based on [45])

Figure 6.17 Flat stack for semi-finished products (sides) of musical instruments (source: [47], with permission)

When removing free water from wood, air velocity is an essential factor. An increase in air circulation speed within the stack during open-air drying is possible by:

- inclined stacking (in the form of an "inverted V") (Figure 6.18 a) of freshly sawn boards, especially recommended for wood species with high susceptibility to mold. This stacking form should only be maintained for 5–10 days; otherwise there is a risk of longitudinal deformation of the pieces, particularly with long and thin material;
- horizontal triangular stacking (Figure 6.18 b) creates better air circulation between the boards than compact stacking, but requires more space;
- box stacking with air shaft (Figure 6.18 c).

The use of mobile fans, arranged to the side of the stacks (Figure 6.19), can also reduce the duration of air drying.

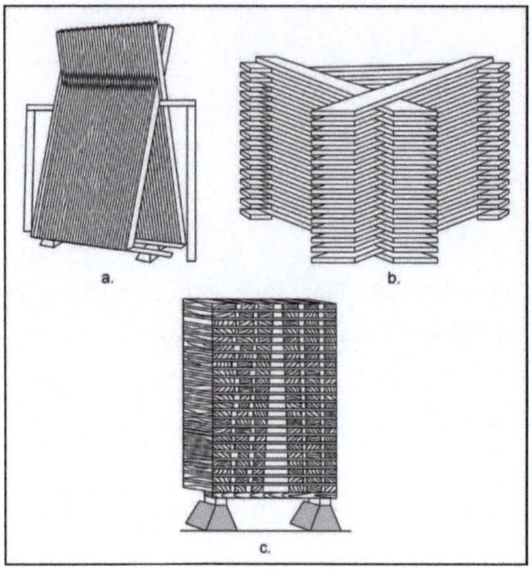

Figure 6.18 Special stacking arrangements for accelerating air drying: a – inclined stacking, b – triangular stacking, c – air-basket stacking (source: [2], with permission)

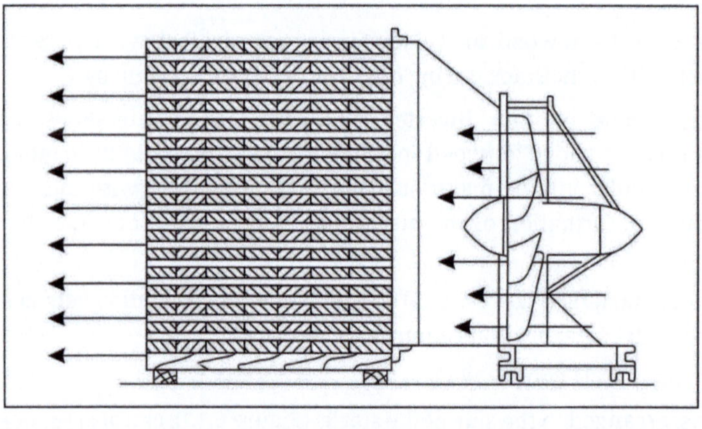

Figure 6.19 Acceleration of air drying by laterally arranged fans (source: [2], with permission)

Another way to accelerate drying is the use of solar energy, which provides an additional contribution of thermal energy necessary to break the bonds between moisture and the wood substance. The simplest form of application is a wood shed whose south wall is painted black to capture solar energy and raise the temperature inside the shed by a few degrees. A more advanced solution is the use of transparent plastic panels on the outside of the roof and the west, east, and south walls, along with black-lacquered corrugated aluminum sheet panels on the inside. The solar rays passing through the film heat the aluminum sheets while simultaneously preventing the heat from escaping to the outside.

The main advantages and disadvantages of air drying are summarized in Table 6.3.

Table 6.3 Advantages and Disadvantages of Air Drying

Advantages	Disadvantages
▪ Simplicity ▪ Minimal investment ▪ Eco-friendly ▪ No energy consumption	▪ Long drying duration and the associated immobilization of important material stocks and large space requirements ▪ Uneven reduction of wood moisture content in the stack (no possibility of control) ▪ Impossibility of reducing wood moisture content below 12 – 15 % in temperate climate zones ▪ Risk of wood infestation by fungi and insects

For this reason, natural drying is no longer used as the sole drying method (down to service moisture content), except in warm climate zones. In Europe, it is mainly practiced as pre-drying, particularly for hardwood (e.g., beech parquet strips, oak boards) and for tonewood for musical instruments. However, once the wood moisture content reaches the fiber saturation point, the wood should be artificially dried to the desired final value in order to save time and to achieve a uniformly distributed final moisture content.

6.3 Conventional Drying

Conventional drying was and remains the best-known and most widely used drying method. It is a convective evaporative drying process in a steam–air mixture, which changes its temperature and relative humidity during the course of the process. Drying takes place at temperatures below 100 °C and at atmospheric pressure.

The **principle** of conventional drying (Figure 6.20) consists in the controlled circulation of air heated by the heating element by means of axial fans through the wood stacks. The air velocity within the stack should be 2 – 3 m/s. The warm air ensures a continuous transfer of heat to the wood material while simultaneously absorbing the moisture removed from it. To prevent premature drying of the wood surface, it is necessary, especially at the beginning of the process, to spray. Cold, decalcified water or steam can be injected into the chamber through spray nozzles for this purpose. When the air becomes too humid, a portion of the recirculated air is discharged to the outside through air ducts equipped with adjustable dampers , while an equal amount of fresh air is simultaneously drawn into the chamber. This means that dehumidification of the air takes place by exchange with the external environment, whereby the thermal energy of the expelled exhaust air is lost. For a better energy balance, a heat exchanger can be used in the ventilation circuit (Figure 6.21).

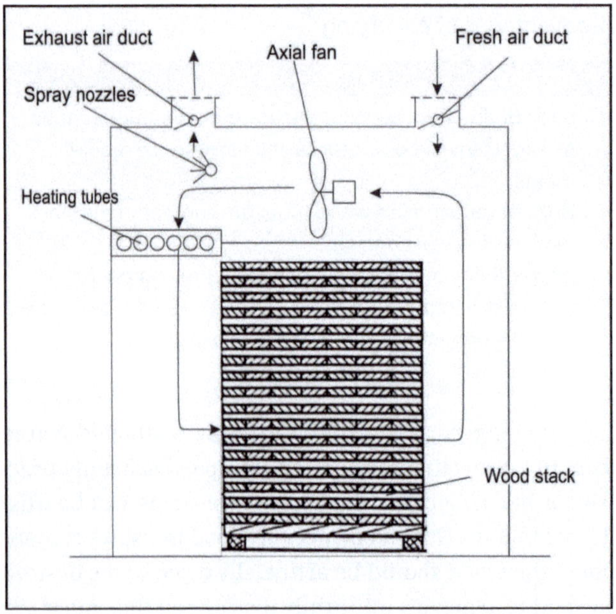

Figure 6.20 Principle of conventional drying (based on [48])

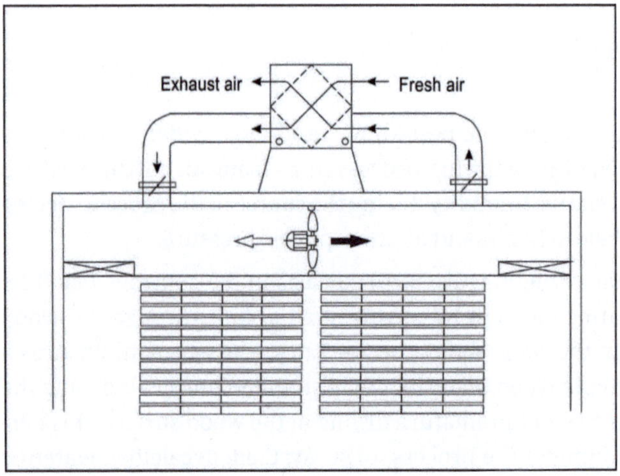

Figure 6.21 Partial heat recovery through plate heat exchangers in the fresh air/exhaust air circuit (based on [49])

The **performance** of conventional drying can be evaluated as follows:

- The duration is satisfactory for drying softwoods but relatively long for drying hardwoods, especially when artificial drying starts from a high initial moisture content of the material. Table 6.4 provides reference values for the duration of conventional drying of sawn timber, from an average initial moisture content of 60% to a final moisture content of approximately 12%.

Table 6.4 Drying Times for Conventional Drying of European Wood Species (source: [2])

Thickness [mm]	Drying time in days				
	Spruce, fir	Pine, poplar	Birch, alder	Beech, ash, larch	Oak
19 - 22	1 - 2	2	3	5	10
24 - 30	2 - 3	2 - 3	4	6	12
30 - 40	3	3	5	6 - 8	15
40 - 50	3 - 4	4	6	10	20
50 - 60	4 - 5	5	8	12	25
60 - 70	6	6	10	15	30
70 - 80	6 - 8	8	12	20	-

- The specific energy consumption (without heat recovery) is 280 – 330 kWh/m^3 for softwoods and 530 – 560 kWh/m^3 for hardwoods [50]. Within the total energy consumption, the thermal energy required for heating accounts for approximately 70% and the electrical energy required for air circulation accounts for approximately 30%.
- When correct drying schedules and good control are applied, the drying quality is very good for softwood and satisfactorily good for hardwoods. Cracks and uneven final moisture content are the most common drying defects in hardwood, but also a high moisture gradient between the surface layers and the interior of the wood in thick structural timber assortments of any wood species.
- The costs of conventional drying double the costs of fresh wood (e.g., if 1 m^3 of fresh wood costs 300 euros, then the same 1 m^3 of dried wood will cost approximately 600 euros).

The **main advantages and disadvantages** of conventional drying are summarized in Table 6.5.

Table 6.5 Advantages and Disadvantages of Conventional Drying

Advantages	Disadvantages
■ Universal method that can be used to dry wood of any type and thickness, from any initial moisture content, no matter how high, to a final moisture content ≥ 6 % ■ Short drying times (1 – 7 days) for softwood of medium thickness (< 60 mm) ■ Good quality when drying softwood	■ High energy consumption associated with high costs at a relatively poor efficiency, due to the constant exchange of air with the outside air ■ Drying defects when drying conditions are too severe ■ Requires a boiler system

The **application range** of conventional drying is very wide. It is particularly advantageous for drying softwood with thicknesses below 60 mm. For hard/dense wood species and especially for thick assortments, the use of other methods should be considered.

6.4 Dehumidification Drying

Drying with air dehumidification by condensation is, like the previous method, a convective evaporation drying in a steam-air mixture, which takes place at temperatures below 100 °C and at atmospheric pressure.

The difference between the two methods lies in the way the air is dehumidified. In dehumidification drying, the warm exhaust air is no longer blown outside, but is directed into a unit connected to the dryer (condensation device) and cooled there (heat pump principle). The water vapor contained in the air condenses and is discharged. The thus dehumidified air is reheated and returned to the dryer. The closed-loop air circulation saves energy. However, this involves exclusively electrical energy.

The **principle** of dehumidification drying is illustrated in Figure 6.22. After flowing through the lumber stacks, the air laden with water vapor is drawn into the connected condensation unit.

There, the air first passes through a cooling coil, which cools the air to a temperature below the dew point. As a result, the water vapor contained in the air transitions to a liquid state and is collected in a tank outside the dehumidification unit. Since the amount of condensate corresponds to the amount of moisture removed from the wood, this provides a way to monitor the decrease in wood moisture content without sensors.

The heat released by the air during cooling is absorbed by the refrigerant in the cooling coil. This evaporates and is then compressed by a compressor and pumped into a second heat exchanger. This acts as a heating battery for the dehumidified air, which thereby recovers part of the heat lost during dehumidification. The heat recovery efficiency is 25 – 30 %. To heat the dehumidified air to the operating temperature, an additional electric heating battery is required.

The thus reheated and dehumidified air is then blown by an axial fan into the drying chamber, where it resumes its circulation through the stacks.

Since the heat pump only starts operating at an air temperature of approximately 30 °C, it is necessary to ensure that the drying chamber is equipped with its own heating register.

The condensation unit can be placed directly inside the drying chamber (Figure 6.23 left), but due to the risk of corrosion, it is recommended to install it outside the drying chamber (Figure 6.23 right). All its components must be made of aluminum or stainless steel.

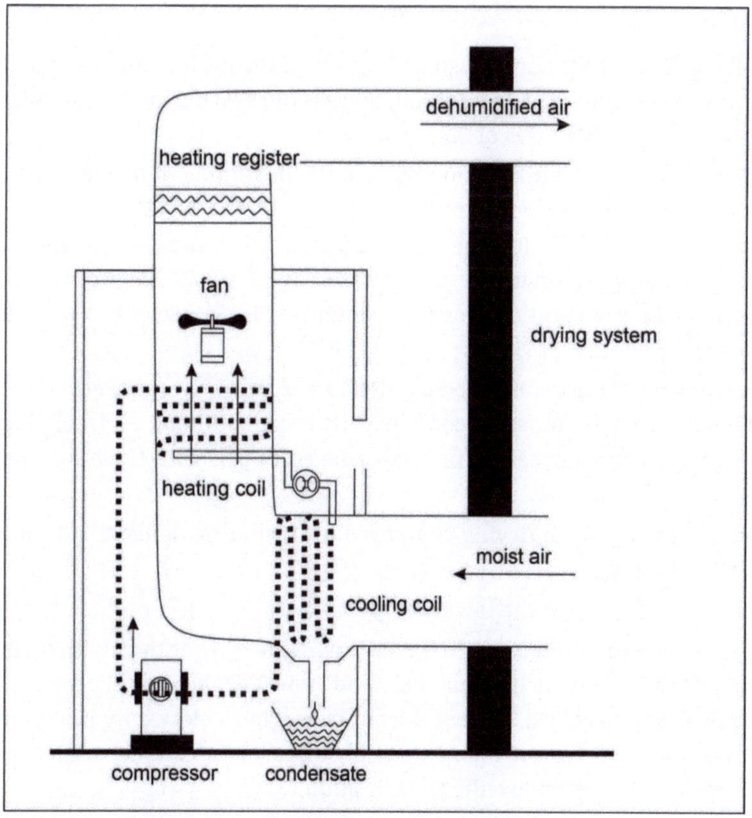

Figure 6.22 Principle of dehumidification drying (source: [51], with permission)

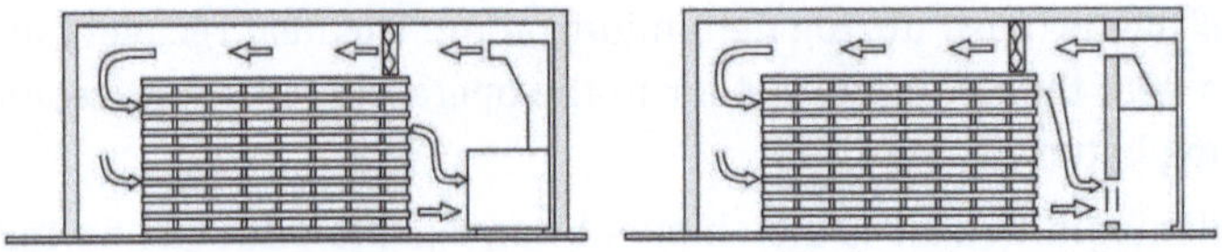

Figure 6.23 Placement of the condensation unit: a – inside the dryer; b – outside the dryer (source: [2], with permission)

The fundamental characteristic of this method is the relatively low drying temperature. For technical reasons related to the operating conditions of the heat pump, this is a maximum of 60 °C. Due to this fact, drying proceeds slowly and is not recommended for wood species prone to blue stain (e.g., pine) and also not for light-colored species that are to retain their color (e.g., maple, alder, ash, beech). Because if these are exposed to a very humid climate with low temperature over a longer period of time, undesirable discoloration occurs.

The specific installed power of the compressor is 0.17 – 0.22 kW/m^3 for softwoods and 0.33 – 0.44 kW/m^3 for hardwoods [49]. In addition, there is the specific installed power of the supplementary heating batteries of 0.3 – 0.5 kW/m^3.

The maximum yield of the dehumidification unit is at the beginning of drying, when the free water is removed from the wood, and reaches values of 100 – 130 l/h. As soon as the wood moisture content falls below the fiber saturation point, the yield decreases and the specific energy consumption increases by a factor of 6, from 0.3 kW/kg of evaporated water (above the fiber saturation moisture content) to 1.9 kW/kg of evaporated water, at final moisture contents < 15%.

In order to achieve final moisture contents below 15% under more favorable economic conditions, condensation drying must be combined with conventional recirculating air drying. In this case, the condensation unit must be equipped with fresh air and exhaust air ducts.

The **performance** of condensation drying compared to fresh air/exhaust air drying can be summarized as follows:

- The drying time is longer due to the lower temperature.
- The specific energy consumption is approximately two times lower than with fresh air/exhaust air drying (Table 6.6). It should be noted, however, that this energy consumption is exclusively electrical, so that in countries where electricity costs are much higher than those of thermal energy, the drying costs for condensation drying are higher despite the lower specific consumption.
- The drying quality is very good due to the gentle drying process, even for hardwood.
- The costs of condensation drying down to final moisture contents ≥ 15% are equal to the costs of fresh air/exhaust air drying. The advantage of energy savings is usually offset by the extended drying time and the higher costs of electrical energy.

The **main advantages and disadvantages** of condensation drying are summarized in Table 6.7.

Table 6.6 Specific Energy Consumption in Condensation Drying [46]

Wood species	Wood thickness [mm]	Wood moisture content [%]		Specific energy consumption [kWh/m³]
		u_i	u_f	
Spruce; fir	30	70	12	110 - 130
Pine	50	60	8	110 - 130
Beech	35	30	12	120 - 150
Oak	25	25	11	120 - 150
	40	30	12	140 - 160
	65	80	10	340 - 380

Table 6.7 Advantages and Disadvantages of Condensation Drying [46]

Advantages	Disadvantages
▪ Gentle drying of heavy and difficult-to-dry hardwoods thanks to low temperature ▪ Up to 50 % energy savings when drying heavy woods compared to conventional drying ▪ No emission pollution of the environment	▪ Longer drying times and therefore higher capital commitment than with conventional drying ▪ Final moisture contents below 15 % require long drying times ▪ Risk of discoloration in wood species with high initial moisture content due to slow drying, because of the climate favorable to fungi ▪ Condensate drainage must comply with environmental protection regulations

Area of application: Although wood of any species and thickness can be dried with this process, economic analyses have shown that the application of this process is optimal under the following conditions:

- For hardwoods with a thickness greater than 40 mm and for softwoods with a thickness greater than 120 mm;
- For wood species that must be dried to a final moisture content of more than 15 %, since below this value the duration increases significantly;
- When the company does not have its own thermal energy source.

The process is not suitable for wood species that are susceptible to blue stain or prone to discoloration.

6.5 High-Temperature Drying

Drying at temperatures above 100 °C is generally referred to as high-temperature drying (HT drying). Flue gas, hot air–superheated steam mixtures, pure superheated steam, or hot oil can serve as drying agents.

Principle. Increasing the temperature above 100 °C accelerates drying, but also leads to a sharp decrease in the equilibrium moisture content (Table 6.8), the higher the psychrometric difference ($t - t_{wet}$) is.

Table 6.8 Dependence of the Equilibrium Moisture Content of Wood on Temperature During High-Temperature Drying in a Steam-Air Mixture [2]

Dry-bulb temperature [°C]	Wet-bulb temperature [°C]	Wood equilibrium moisture content [%]
110	100	8.0
110	80	3.0
110	60	1.5
120	100	4.3
120	80	1.6
120	60	0.8
130	100	3.0
130	80	1.2
130	60	0.5

If, for example, one heats rapidly to 120 °C in the usual manner (at atmospheric pressure), even when t_{wet} = 100 °C, this corresponds to severe drying, since the wood tends toward an equilibrium moisture content of 4.3 %. The result is severe case hardening. The drying process must therefore be controlled differently. The hygroscopic equilibrium at temperatures above 100 °C and various pressure values (Figure 6.24) shows that high-temperature drying can be carried out in a milder manner, either in a steam-air mixture at elevated pressure or in superheated steam with or without admixture of air.

The first variant requires pressure-resistant equipment, which is costly.

Therefore, in practice only **drying in pure superheated steam** experienced industrial application.

In this process, wood is dried in a pure steam atmosphere at temperatures of 100 – 130 °C and atmospheric pressure, while the wet-bulb temperature is kept constant at 98 – 100 °C. Under these conditions, the steam is superheated and can transfer heat to the wood being dried without condensing, until the saturation temperature (100 °C) is reached. Its reheating once again ensures the capacity to absorb moisture. The superheated steam temperature must be carefully increased (in tenths of a degree) above 100 °C (Table 6.9) in order to keep the decrease in wood equilibrium moisture content under control.

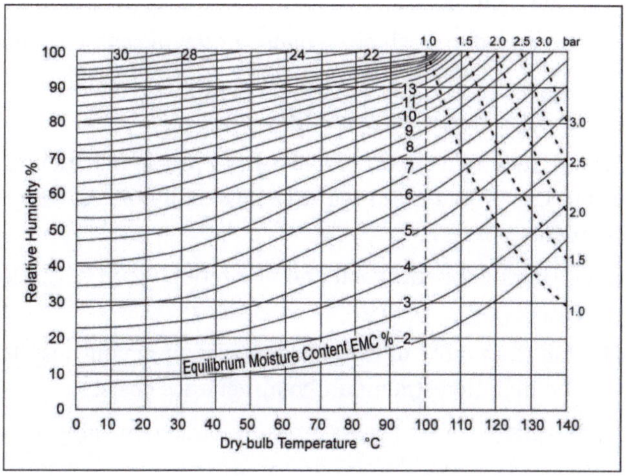

Figure 6.24 Hygroscopic equilibrium at temperatures above 100 °C (adapted from [52])

Table 6.9 Dependence of the Equilibrium Moisture Content of Wood on the Degree of Superheating, in Pure Superheated Steam (source: [2])

Dry-bulb temperature [°C]	Wet-bulb temperature [°C]	Wood equilibrium moisture content [%]
100.2	100	20
105.8	100	10
129.3	100	3

Sufficient heating energy for the evaporation of water must be available throughout the entire drying process.

High-temperature drying allows a greater drying gradient than fresh air/exhaust air drying, i.e., although the conditions are more severe, the risk of cracking is lower, thanks to the higher temperatures and high humidity, which impart greater plasticity to the wood. The drying rate is 3 to 5 times higher than with fresh air/exhaust air drying.

An increased risk of cracking exists only during the heating-up phase, when wood temperatures are still well below 100 °C. With pre-dried boards and thicker timber, spraying must be carried out at the beginning of the heating-up phase in order to displace the air still present in the chamber.

Admixtures of air result in a significant reduction of the wet-bulb temperature and intensify the drying conditions. This leads to severe heating of the wood surface and ultimately to the undesirable case-hardening.

The **performance** of HT drying compared to fresh air/exhaust air drying can be summarized as follows:

- The drying time is shorter due to the higher temperature.
- The heat requirement is lower than with fresh air/exhaust air drying, but achieving a flow temperature above 100 °C is difficult and requires powerful boilers, which only a few wood processing plants possess.
- The drying quality is satisfactory only for softwood and plantation wood (e.g., Radiata pine, Pitch pine).
- The costs of HT drying are higher than those of fresh air/exhaust air drying despite the reduction in time, due to the high investment and maintenance costs.

The **main advantages and disadvantages** of HT drying are summarized in Table 6.10.

Table 6.10 Advantages and Disadvantages of Superheated Steam High-Temperature Drying

Advantages	Disadvantages
▪ Very short drying times, approximately 40 – 60 % of the drying times for fresh air/exhaust air drying ▪ Simple control of the drying process by regulating the temperature ▪ No surface cracks, little tendency to warp	▪ Suitable only for permeable wood species (softwood, soft hardwood) ▪ High stress on the material (frequent internal cracks and cell collapse) due to severe drying conditions ▪ Surface discoloration (can be eliminated by planing) ▪ Requires high heating capacity of the dryer (equipment must be fitted with a larger heating surface than in conventional drying) ▪ Requires high air velocities, up to 6 m/s (equipment must be fitted with more powerful fans than in conventional drying)

The **optimal application range** is in countries that dry large quantities of softwood. In Europe, the process is currently used for drying softwood middle layers for laminated wood.

6.6 Vacuum Drying

Principle. The main characteristic of vacuum drying is the reduction of pressure inside the drying system (usually an autoclave) down to 0.5 – 0.08 bar. The reduction in pressure leads to a reduction in the boiling point (see Figure 2.15) of the water contained in the wood. Due to the vacuum condition, the vapor pressure of the water in the wood is higher than in the surrounding environment. As a result, the water in the wood evaporates and flows to the surface in the form of water vapor. The flow velocity is greater the lower the pressure.

Therefore, vacuum drying is characterized by shorter drying times than fresh air/exhaust air drying, but also by better drying quality, because drying takes place at lower temperatures than in atmospheric pressure systems, thus protecting the wood from cracking.

However, the low pressure also has a disadvantage, namely a negative effect on heat transfer, since vacuum is a very poor conductor of heat. If one considers that the removal of water from the surface of the material leads to significant cooling of the wood, after some time the drying process will stagnate if sufficient heat is not continuously supplied to the wood.

There are three ways to continuously heat the wood while the vacuum pump is also operating:

- by contact – the so-called vacuum press method (or ”plate method”);
- by convection in a superheated steam atmosphere – the so-called vacuum superheated steam method;
- by means of a high-frequency field – the so-called vacuum high-frequency method.

6.6.1 Vacuum Press Drying

Principle. In vacuum press drying (Figure 6.25), the boards are stacked on a trolley without stickers. An aluminum plate is placed below and above each layer of wood. The plates are heated with hot water by means of flexible hoses. The cooling plates ensure condensation of the moisture escaping from the wood, so that the air extracted by the vacuum pump contains as little moisture as possible.

After the wood stacks have been introduced into the autoclave, the wood is first heated to approximately 50 °C at atmospheric pressure, after which the vacuum pump is activated and generates a negative pressure of 0.08 bar. This low pressure is maintained until the end of the drying process.

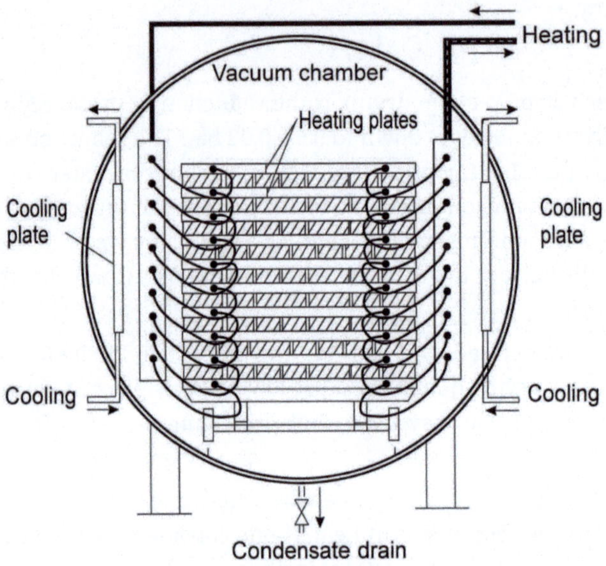

Figure 6.25 Principle of vacuum press drying (based on [53])

The process is controlled in accordance with the decrease in wood moisture content by controlling the temperature of the plates (which increases as the wood moisture content decreases).

An essential prerequisite for the success of drying with this method is that the wood pieces have the same thickness in order to ensure uniform contact between the plates and the material over the entire surface.

Research on drying times with this method has indicated significantly faster drying rates for hardwoods than with fresh air/exhaust air drying (e.g., 20 – 50 % shorter drying time for 40 mm thick oak wood [54]).

The **main advantages and disadvantages** of vacuum press drying are summarized in Table 6.11.

Table 6.11 Advantages and Disadvantages of Vacuum Press Drying

Advantages	Disadvantages
▪ Short drying times ▪ Very good wood quality; gentle drying ▪ Low energy consumption	▪ Relatively long loading and unloading time due to heating plates ▪ Reduction in wood thickness (wood compression) under the weight of the heating plates ▪ Slight darkening of the wood surface

The **optimal area of application** of this process is in the drying of diffuse-porous hardwoods (e.g., beech). Thanks to direct contact heat transfer, the drying time is very short: 5 to 20 times shorter than with fresh air/exhaust air drying, and the drying quality is very good thanks to the gentle drying regime (low pressure, maximum temperature 70 °C).

6.6.2 Vacuum Drying in Superheated Steam

Principle. Drying in a mixture of air and superheated steam at low pressure is the most modern variant of vacuum drying, which has made it possible to build systems with greater capacity than the plate process, with diameters of up to 4 m and lengths of up to 30 m. These are equipped similarly to fresh air/exhaust air dryers (Figure 6.26). The stainless steel autoclave is equipped with fans and heating pipes. It also has a cooling system for condensing water vapor. Unlike conventional dryers, the fans have a higher volumetric flow rate, as they must be capable of generating velocities of 20 – 40 m/s. The heating surface is also larger (approximately double) than in a fresh air/exhaust air dryer. The material is stacked using normal stacking stickers.

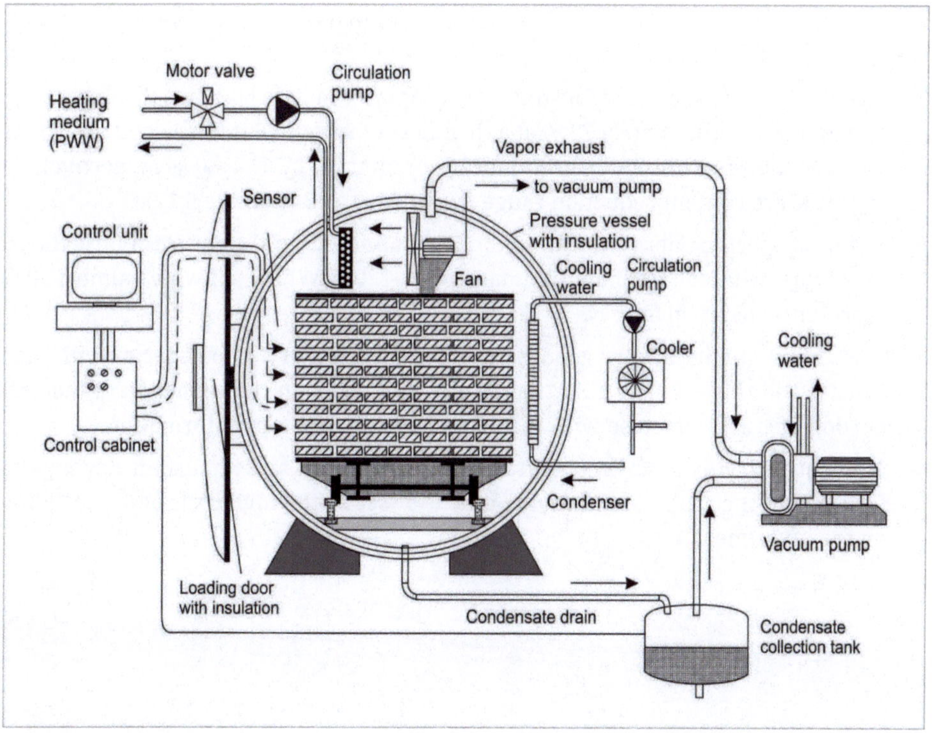

Figure 6.26 Principle of vacuum superheated steam drying (based on [53])

After loading the stacks, a vacuum (approximately 0.1 bar) is applied immediately for 80 minutes. The temperature is then raised to the boiling point of water. The evaporation of water from the wood and, if necessary, additional spraying creates the superheated steam environment. The drying rate in this case depends only on the difference between the temperature in the system and the boiling point of water (dependent on the negative pressure).

The drying process is controlled according to the wood moisture content. The measured parameters are: the temperature of the wood, the pressure inside the system, the equilibrium moisture content, and the wood moisture content. The control variables are the temperature and the pressure of the environment.

The greatest challenge with this process is achieving uniform drying throughout the entire system. Two conditions are required for this:

- the velocity of the drying medium through the stacks should be as high as possible;
- no temperature fluctuations may occur within the system, since at low pressures small temperature fluctuations lead to a significant drop in relative humidity and thus also in equilibrium moisture content, which can cause serious drying defects (internal checks).

The **performance** that can be achieved with this process, which can be successfully used for drying both softwoods and hardwoods, compared to fresh air/exhaust air drying, can be summarized as follows:

- The drying time is reduced for softwood by approximately 60 % at a thickness of 52 mm and by approximately 72 % at a thickness of 90 mm. For hardwood (e.g., oak), the reduction in duration is even more pronounced, by 75 – 80 % respectively in the moisture content reduction range from 60 % to 20 % (Table 6.12).
- The total specific energy consumption is comparable to that of fresh air/exhaust air drying, with values of approximately 150 – 200 kWh/m^3, of which almost 50 % represents electrical energy.
- The drying quality is very good, even for difficult-to-dry wood species with high density and thick timber assortments, not only through the reduction of the proportion of cracks, but also with regard to discoloration and deformation.
- The cost of drying 1 m^3 is comparable to that of fresh air/exhaust air drying. Unfortunately, the reduction in drying time is offset by the high costs of depreciation on the investment.

Table 6.12 Drying Times for Vacuum Superheated Steam Drying for Various Wood Species

Wood species	Thickness [mm]	Initial moisture content [%]	Final moisture content [%]	Duration [h]	Source
Spruce (*Picea abies*)	23	77	12	73.5	[55]
	48	49	13	127	
Spruce (*Picea abies*)	52	80	20	58	[42]
	90	80	20	108	
Oak (*Quercus robur*)	27	80	8	288	[42]
	52	80	8	624	
Oak (*Quercus robur*)	27	50	10	230	[56]
	54	100	10	560	
Red oak (*Quercus rubra*)	50	63.3	7.8	684	[57]
White oak (*Quercus alba*)	50	53.1	14.5	519	[58]

The **main advantages and disadvantages** of vacuum superheated steam drying are summarized in Table 6.13.

Table 6.13 Advantages and Disadvantages of Vacuum Superheated Steam Drying

Advantages	Disadvantages
▪ Shorter drying times than with fresh air/exhaust air drying ▪ Very good wood quality, even for hardwoods with ring-porous structure; no cracks, no discoloration ▪ Lower energy consumption than with the press drying method ▪ Less effort in loading and unloading the system than with the press drying method	▪ Expensive equipment ▪ Sensitive sensors ▪ Demanding maintenance of the equipment ▪ High cooling water consumption ▪ Resin exudation when drying resinous softwoods (e.g. pine)

Optimal area of application. The above-mentioned advantages classify the vacuum superheated steam process as an advantageous alternative to fresh air/exhaust air drying, particularly when the specification to be dried includes valuable wood species (e.g. oak, exotic wood species) or thick structural timber assortments that cannot be dried to quality by other methods.

From an economic point of view, vacuum processes have proven themselves in the drying of hardwoods of any thickness and softwoods with thicknesses exceeding 120 mm, whereby the drying time can be reduced by 50 – 75 % compared to conventional drying.

Due to the high costs of the equipment, the choice of this method is only recommended when there is the possibility of operating the equipment at maximum capacity. In this regard, there are two main directions of economical application of vacuum drying:

- When the company already has a drying department with fresh air/exhaust air systems in which various specifications are dried (both softwood and hardwood species), but which can no longer cover the entire required annual capacity, so that the acquisition of a vacuum drying system for a relatively small, constant quantity of valuable hardwoods would solve the problem of insufficient capacity while simultaneously enabling an improvement in the quality of these difficult-to-dry species;
- When the operation has a large quantity of slow-drying assortments for which a particular drying quality is required (e.g. oak, large thicknesses).

6.7 Drying in a High-Frequency Field

Principle. High-frequency drying (HF) utilizes the effect of heating the moisture in a wet body (in this case wood) when it is exposed to an electromagnetic high-frequency field (f > 100 MHz), due to dielectric losses (see also Section 2.1.8).

Since the thermal energy is generated inside the wood, where the moisture content is highest, a temperature gradient is created in the wood that is oriented from the interior of the wood toward the surface, thereby accelerating the movement of water in the wood. This has a positive effect on drying times, which are considerably shorter in HF drying than in the case of fresh air/exhaust air drying. The process is only suitable for permeable wood species (e.g. softwoods, beech without red heart, birch, poplar, walnut), because if the wood is not sufficiently permeable (e.g. in tylosis-containing species such as white oak, black locust, ash, elm), the pressure and temperature inside the wood rise so sharply that internal cracks and collapse can occur.

In industrial HF drying, two different frequency ranges can be distinguished (Table 6.14): radio frequencies below 100 MHz, in which open circuits are used, and microwaves with frequencies above 500 MHz, in which waveguides are used for energy transmission.

Table 6.14 High-Frequency Bands for Industry (source: [59])

	Frequency [MHz]	Wavelength [m]
Radio frequencies	13.56	22.11
	27.12	11.05
	40.68	7.37
Microwaves	915	0.328
	2,450	0.122
	5,800	0.052

Due to the low penetration depth, microwaves are more suitable for drying thin wood assortments (e.g., veneer).

Radio frequency dryers are generally used as tunnel systems (Figure 6.27). The metal belt, which serves to transport the wood parts, is simultaneously the lower (grounded) electrode. The upper electrode made of perforated aluminum sheet is height-adjustable. This allows the drying process to be indirectly controlled, since the energy transferred to the wood depends on the thickness of the air layer between the upper electrode and the wood. The final moisture content is determined by the feed rate control. In order to achieve the most uniform possible reduction in wood moisture content, pre-drying of the wood to approximately 40% is necessary.

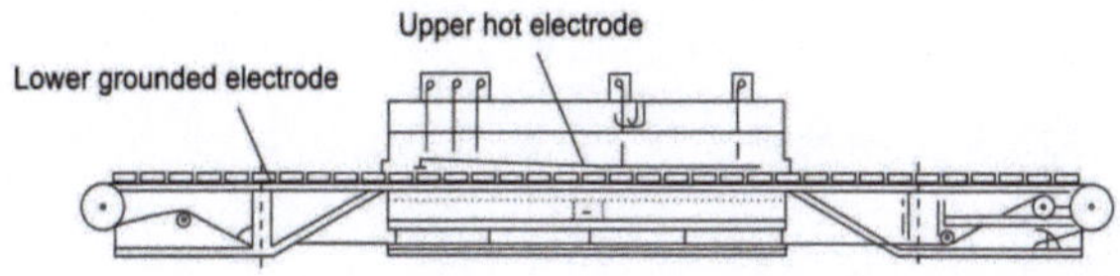

Figure 6.27 Schematic diagram of a radio frequency dryer for wood (based on [22])

The **main advantages and disadvantages** of radio frequency drying are summarized in Table 6.15.

Table 6.15 Advantages and Disadvantages of Radio Frequency Drying

Advantages	Disadvantages
▪ Shorter drying times (approximately 5–20% of the time required for fresh air/exhaust air drying) ▪ Very good wood quality for permeable hardwoods; no cracks, no discoloration	▪ Expensive equipment ▪ High energy consumption ▪ Demanding maintenance of the equipment

The **optimal application range** is for permeable wood species (softwoods, beech, birch, poplar, walnut, and others).

6.8 Radio Frequency Vacuum Drying

Even better results with regard to drying time and quality can be achieved by combining vacuum drying with heating in a radio frequency field (Figure 6.28). In this case, the wood is heated dielectrically by using a generator to create a high-frequency electric field between the two electrodes. The negative pressure in the cylinder (p = 0.03 bar) is achieved using a water ring vacuum pump equipped with an air injector. The water vapor extracted from the wood is partly evacuated from the drying chamber by the vacuum pump, then cooled and condensed in a condenser, and partly discharged at the bottom of the system in the form of condensate, which is directed into a water tank by means of a pump.

The drying process is controlled in accordance with the decrease in wood moisture content by adjusting the frequency of the generator.

With drying times on the order of hours, even for hard hardwood species and softwoods more than 100 mm thick [21], radio frequency vacuum (RFV) drying is the fastest drying method currently known for sawn timber.

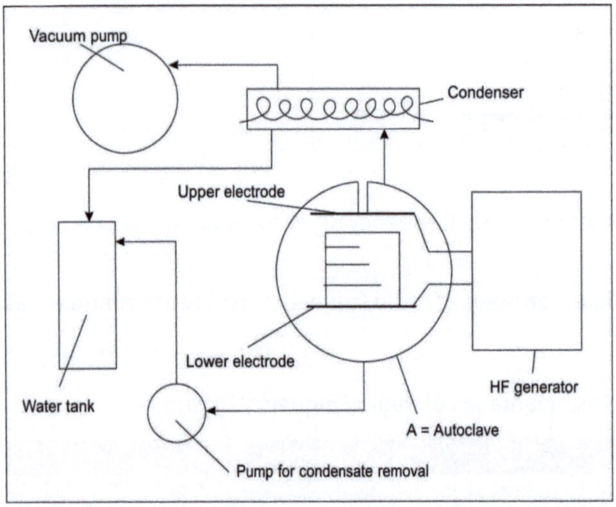

Figure 6.28 Principle of radio frequency vacuum drying

The **performance** of RFV drying compared to fresh air/exhaust air drying can be summarized as follows:

- The drying time is reduced by at least half.
- The specific energy consumption is comparable and amounts to approximately 250 kWh/m^3; however, the energy costs are higher than with fresh air/exhaust air drying, since at least 50 % (depending on the process) of the total energy consumption is electrical.
- In addition to electricity and, in some cases, thermal energy, a considerable consumption of coolant for condensation is required, approximately 0.2 – 5.0 m^3 cold water/m^3_{wood} h.
- The drying quality is very good even for difficult-to-dry wood species and assortments, and is characterized by the absence of cracks, discoloration, and deformation, as well as by the uniformity of the final moisture content throughout the entire thickness of the material.
- The drying costs are strongly influenced by the investment costs (expensive equipment with many components) and the energy costs.

The **optimal application range** is recommended for valuable and collapse-prone wood species, e.g. eucalyptus, which cannot be dried to quality by any other method.

References

[1] Villiere, A. (1966): *Séchage des bois*. Dunod Editeur, Paris, France

[2] Marinescu, I. I. (1980): *Uscarea și tratarea termică a lemnului. Vol. 2*. Editura Tehnică, Bucharest, Romania

[3] Câmpean, M. (2010): *Timber Drying Methods: Passing from History into the Future*. Keynote Address at the 11th International IUFRO Wood Drying Conference „Recent Advances in the Field of Wood Drying", Skellefteå, Sweden

[4] Robbins, L. S. (1865): Improved Process for Preserving Wood. U.S. Patent No. 47 132, 4 Apr. 1865

[5] Hudson, M. S. (1942): Treating Wood and Wood Products. U.S. Patent No. 2 273 039, 17 Feb. 1942

[6] Allen, C. F.; Campbell, L. W. (1867): Apparatus for Drying and Seasoning Lumber by Superheated Steam. U.S. Patent No. 64 398, 1867

[7] Tiemann, H. D. (1944): *Wood Technology*. 2nd ed., New York und Chicago, USA

[8] Mathewson, J. S. (1960): *High-Temperature Drying: Its Application to the Drying of Lumber*. In: Australian Timber Journal, 26(9). Online: *https://www.fpl.fs.usda.gov/documnts/pdf1960/mathe60a.pdf*

[9] Keylwerth, R. (1949): *Grundlagen der Hochtemperaturtrocknung des Holzes*. In: Holz-Zentralblatt, 75, p. 953 – 954

[10] Eisenmann, E. (1949): *Künstliche Schnelltrocknung mit Temperaturen über 100 °C*. In: Holztechnik, 29, p. 151

[11] Kollmann, F.; Malmquist, L. (1952): *Untersuchungen über die Trocknung von Kiefernschnittholz mit erhöhten Temperaturen*. Schwedisches Holzforschungsinstitut, Stockholm, Sweden

[12] Sturany, H. (1952): *Trocknung im reinen Heißdampf*. In: Holz als Roh- und Werkstoff, 10, p. 358 - 362

[13] Kollmann, F.; Côté, W. A. Jr. (1968): *Principles of Wood Science and Technology. Vol. I. Solid Wood*. Springer Verlag, Berlin/Heidelberg/New York

[14] Folomin, A. I. (1955): *High-Temperature Seasoning of Wood in Petrolatum*. In: Derev. Prom., 4(4), p. 3 - 7

[15] Stamm, A. J. (1933): *Method for Simultaneously Seasoning and Treating Water-Swollen Fibrous Materials*. U.S. Patent No. 2 060 902, 1933

[16] Rawling, S. O. (1935): *Infrared Photography*. 2nd ed., London, UK

[17] Keylwerth, R. (1951): *Infrarotstrahler in der Holzindustrie*. In: Holz als Roh- und Werkstoff, 9, p. 224 - 231

[18] Kastmark, C. F. (1943): *Eine neue Verfahrensweise beim Trocknen von Holzwaren*. In: Trävaruindustrien, 9, p. 134

[19] Eisenmann, E. (1950): *Mechanische Holztrocknung mit Trockenzentrifugen*. In: Holz-Zentralblatt, 76(106), p. 1153

[20] Krämer, G. (2004): *Zentrifuge beschleunigt Trocknung von Scheitholz*. In: Holz-Zentralblatt, 76, p. 1076

[21] Resch, H. (2006): *High-Frequency Electric Current for Drying of Wood – Historical Perspectives*. In: Maderas Ciencia y Tecnología, 8(2), p. 67 - 82

[22] Czepek, E.; Sporckmann, H. (1968): *High-Frequency Drying of Beech Wood in a South-German Wood Factory*. In: Elektrowärme International, 26(12), p. 440 - 447

[23] Pagnozzi, V. (1970): Process and Plant for the Vacuum Drying of Wood in the Form of Planks or Laths. U.S. Patent No. 3 521 373, 1970

[24] Erickson, H. D.; Schmidt, R. N.; Laing, J. R. (1968): *Freeze-Drying and Wood Shrinkage*. In: Forest Products Journal, 18(6), p. 63 - 68

[25] Resch, H. (1968): *Über die Holztrocknung mit Mikrowellen*. In: Holz als Roh- und Werkstoff, 26(9), p. 317 - 324

[26] Resch, H.; Lofdahl, C. A.; Smith, F. J.; Erb, C. (1970): *Moisture Leveling in Veneer by Microwaves and Hot Air*. In: Forest Products Journal, 20(10), p. 50 - 58

[27] Häger, B. O. (1971): Wood Treatment Method. U.S. Patent No. 3 560 251, 1971

[28] Ishii, S. (1991): System for Drying Green Wood. JP 6213568 / EP 0480445 / US 5283700

[29] Câmpean, M.; Ishii, S.; Georgescu, S. (2017): *Drying Time and Quality of EDS-Treated Compared to Untreated Beech Wood (Fagus japonica)*. In: Pro Ligno, 13(3), p. 23 - 30

[30] Schill, V.; Korger, V.; Stahl, W. (1992): *Mechanical Drying of Wood – An Alternative to Seasoning Drying*. In: Proceedings of the 3rd Int. IUFRO Wood Drying Conference, Wien, Austria

[31] Stahl, M. (2000): *Das Inkubations-/Dekompressionsverfahren zur Holzentfeuchtung*. Dissertation, TU Karlsruhe, Germany

[32] Avramidis, S.; Zwick, R. L. (1992): *Exploratory Radiofrequency/Vacuum Drying of Three B.C. Coastal Softwoods*. In: Forest Products Journal, 42(7–8), p. 17 - 24

[33] Avramidis, S.; Zwick, R. L.; Neilson, J. B. (1996): *Commercial-Scale RF/V Drying of Softwood Lumber. Part 1: Basic Kiln Design Considerations*. In: Forest Products Journal, 46(5), p. 44 – 51

[34] Avramidis, S.; Zwick, R. L. (1996): *Commercial-Scale RF/V Drying of Softwood Lumber. Part 2: Drying Characteristics and Lumber Quality*. In: Forest Products Journal, 46(6), p. 27 – 36

[35] Avramidis, S.; Zwick, R. L. (1997): *Commercial-Scale RF/V Drying of Softwood Lumber. Part 3: Energy Consumption and Economics*. In: Forest Products Journal, 47(1), p. 48 – 56

[36] Antti, L. (1992): *Microwave Drying of Hardwood: Simultaneous Measurements of Pressure, Temperature, and Weight Reduction*. In: Forest Products Journal, 42(6), p. 49 – 54

[37] Antti, L. (1999): *Heating and Drying Wood Using Microwave Power*. Dissertation, Luleå University of Technology, Sweden

[38] Hansson, L.; Antti, L. (2003): *Design and Performance of an Industrial Microwave Drier for On-Line Drying of Wood Components*. In: 8th IUFRO Wood Drying Conference, Brasov, Romania, p. 156 – 158

[39] Leiker, M.; Adamska, M. A.; Mollekopf, N. (2004): *Microwave Drying of Single Wood Boards in Multimode Cavities*. In: Proceedings of the International Workshop and Symposium on Drying, Mumbai, India

[40] Leiker, M.; Aurich, K.; Adamska, M. A. (2005): *Accelerated Drying of Single Hardwood Boards by Combined Vacuum-Microwave Application*. In: 9th International IUFRO Wood Drying Conference, Nanjing, China, p. 185 – 190

[41] Brunner, R. (2001): Vacuum-Operated Drying Device, Especially for Sawn Timber. International Patent WO 01/59378 A3

[42] Brunner, R. (1999): *Vacuum Drying in Economic Comparison*. In: Holz-Zentralblatt, Nr. 55 – 58, p. 823 – 875

[43] Brodie, G. J. (2008): *Innovative Wood Drying. Applying Microwave and Solar Technologies to Wood Drying*. VDM Verlag Dr. Müller, Saarbrücken, Germany

[44] Câmpean, M.; Marinescu, I. (2012): *Tratamente termice ale lemnului. Uscarea cherestelei* . Transilvania University, Brasov, Romania

[45] Langendorf, G.; Eichler, H. (1982): *Holzvergütung*. 3rd ed., VEB Fachbuchverlag, Leipzig, Germany

[46] Frühwald, A.; Ressel, J. (1982): *Holztrocknung. Teil 1. Grundlagen der Holztrocknung*. Universität Hamburg, Germany

[47] Gliga, V. (2021): *Cercetări privind influenţa unor factori specifici de material asupra calităţii sunetului viorilor din lemn* . Dissertation, Transilvania University, Brasov, Romania

[48] Bollmann, L. (1977): *Leitfaden der Schnittholztrocknung*. Singen, Germany

[49] Trübswetter, T. (2006): *Holztrocknung. Verfahren zur Trocknung von Schnittholz – Planung von Trocknungsanlagen*. Fachbuchverlag Leipzig, Carl Hanser Verlag, Munich, Germany

[50] Jay, S.; Oliver, T. N. (1995): *Timber Drying: Trends in Energy Consumption and Equipment Utilization Within the UK Timber Industry*. In: Proceedings of the International Drying Symposium 1994, p. 663 – 670

[51] Hiziroglu, S. (2017): *Fundamental Aspects of Kiln Drying Lumber*. Oklahoma State University. Online: *https://extension.okstate.edu/fact-sheets/fundamental-aspects-of-kiln-drying-lumber.html*

[52] Eisenmann, E. (1949): *Künstliche Schnelltrocknung mit Temperaturen über 100 °C*. In: Holztechnik, Bd. 29

[53] Lohmann, U. (1998): *Holz Handbuch*. 5th ed., DRW Verlag, Leinfelden-Echterdingen, Germany

[54] Fohr, J. P.; Chakir, A.; Arnaud, G.; du Peuty, M. A. (1995): *Vacuum Drying of Oak Wood*. In: Drying Technology, 13(8–9), p. 1675 – 1675

[55] Trübswetter, T.; Weber, M. (1992): *Untersuchungen an einer DWT Moldrup-Trockenanlage*. In: Holz als Roh- und Werkstoff, 50, p. 286 – 290

[56] Guilman, C.; Baixeras, O.; Jomaa, W. (1996): *Discontinuous and Convective Vacuum Drying of Oak: Experimental and Modeling Studies*. In: Proceedings of the 5th International IUFRO Wood Drying Conference, Québec, p. 147 – 157

[57] Tremblay, C.; Lavoie, V.; Savard, M. (2001): *Drying Performance of 8/4-Inch Red Oak in a Superheated Steam/Vacuum Kiln*. Canadian Forest Service, Forintek Canada Corp.

[58] Tremblay, C.; Lavoie, V.; Savard, M. (2001): *Drying Performance of 8/4-Inch White Oak in a Superheated Steam/Vacuum Kiln*. Canadian Forest Service, Forintek Canada Corp.

[59] Resch, H. (2009): *Drying Wood with High-Frequency Electric Current*. Wood Science and Technology, ISBN 978-0-9817876-1-9

7

Technology of Conventional Kiln Drying

The process of sawn timber drying has several technological phases, which can be divided into three categories (Table 7.1).

Table 7.1 Technological Stages of Sawn Timber Drying

Before drying	During drying	After drying
▪ Sorting ▪ Stacking ▪ Attaching the wood moisture measuring points ▪ Preparation of the drying plant ▪ Loading ▪ Screening off empty spaces ▪ Connecting the wood moisture measuring points ▪ Closing the doors ▪ Entering the initial data of the charge ▪ Creation/verification of the intended drying schedule ▪ Start of drying	▪ Continuous measurement of wood moisture ▪ Continuous measurement of air parameters ▪ Continuous control of air parameters ▪ Transition from one phase to the next ▪ Recording of the drying log ▪ Verification of final moisture content ▪ End of drying	▪ Opening the doors ▪ Unloading the stacks ▪ Conditioning ▪ Quality control ▪ Identification of possible causes if drying defects occur ▪ Optimization

The drying quality depends to a large extent on compliance with the rules specified for each of these phases.

7.1 Sorting and Stacking

Before stacking, the material is sorted so that only boards of the same wood species, the same thickness, as uniform a length as possible, and as homogeneous an initial moisture content as possible are placed in one stack.

Sawn timber is most commonly stacked by hand. In doing so, the boards can also be visually inspected, and visible defects (such as end checks and surface cracks) should be marked with chalk in order to distinguish them from actual drying defects at the end of the process.

However, there are also mechanical stacking systems that can be used to reduce stacking time and to lower labor costs.

Stacking is a technological operation of great importance for drying success. Many drying defects (e.g., scattered final moisture content, longitudinal deformations, cross-bending, certain discolorations) are attributable to non-compliance with the basic rules of stacking.

Correct stacking is that which creates sufficient spaces for air movement between the boards. This ensures both uniform heating of all wood parts and rapid removal of moisture from the wood surface.

Sensitive wood species prone to discoloration, e.g., maple, beech, alder, oak, pine, should be stacked immediately after sawing.

In technical drying, sawn timber is stacked exclusively horizontally, on stickers (Figure 7.1). This stacking method has the advantage of utilizing the dead weight of the wood pieces as a load to prevent deformation during drying, and also creates uniform spaces for air movement between the boards in the stack.

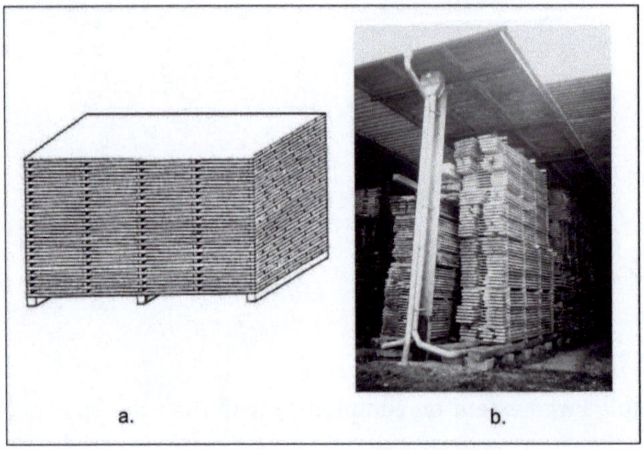

Figure 7.1 Sawn timber stacks prepared for technical drying: a – box stack (for edged lumber) (source: [1]); b – block stack for unedged or one-side-edged lumber (source: [2])

Stacking rules. The following basic rules must be observed when forming a stack:

- a row of stickers is placed between every two rows of wood;
- when viewing the stack from the side (see Figure 7.1 a), the stickers must be positioned strictly vertically one above the other in order to create equally wide air channels between the boards (this rule is extremely important for the uniformity of the final moisture content);
- the stacking stickers are placed flush with both end faces to prevent longitudinal deformation of the board ends (Figure 7.2);
- inside the stack, the stickers are placed at specific intervals depending on the wood thickness (Table 7.2); the thinner the wood pieces, the greater the risk of their longitudinal deformation (see Figure 7.2), and therefore the stickers must be placed closer together; small spacing between stickers is also recommended for wood with diagonal grain or spiral grain;
- in the stack width, the wood pieces should not be placed edge-to-edge, but with small gaps to ensure better ventilation of the stack;
- boards containing the wood core should be placed with the heart side facing up in order to minimize cupping:
- when stacking unedged boards, care must be taken to ensure that the slanted board edges face a uniform direction (Figure 7.3), so as not to disturb or obstruct the airflow; for one-side-edged lumber, the edged side should always face the air inlet side.

Figure 7.2 Longitudinal deformations during wood drying due to incorrect stacking (source: [2])

Table 7.2 Distance between the Stickers in the Stack

Wood thickness [mm]	Distance between two successive stickers [m]
< 30	0.4 - 0.8
30 - 60	0.6 - 1.0
> 60	0.8 - 1.2

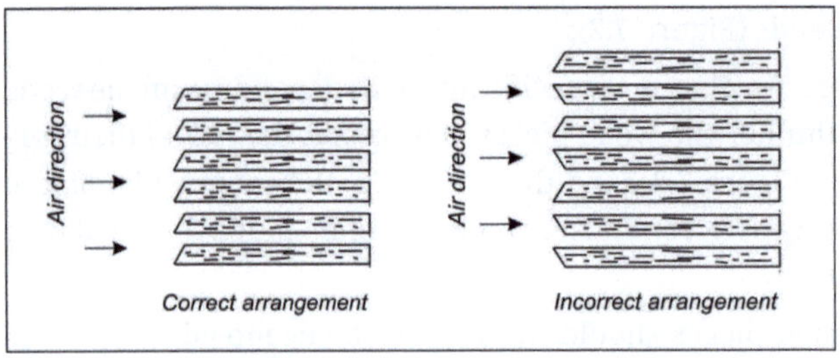

Figure 7.3 Arrangement of the unedged board edges (based on: [3])

Stacking stickers. The stacking stickers are usually made from pre-dried wood. They are typically 1.2 m long (corresponding to the stack width), 30 mm wide and 20 mm thick. To ensure the stability of the stack, the uniform thickness of the stacking stickers with a tolerance of ±1 mm is particularly important. Spruce wood is preferred because it does not stain.

The moisture content of the stacking stickers should not exceed 15 % so that they do not leave marks on the board surface (Figure 7.4), due to the difficulty of removing moisture from the contact area between the sticker and the board.

Figure 7.4 Sticker strips

To avoid this defect in wood species sensitive to discoloration, especially in light-colored woods, e.g. maple, ash, special wooden stickers can also be used to reduce the contact area (Figure 7.5).

The stacking stickers can be reused, provided they have not suffered any damage (crushing, cracks, mold) from previous use.

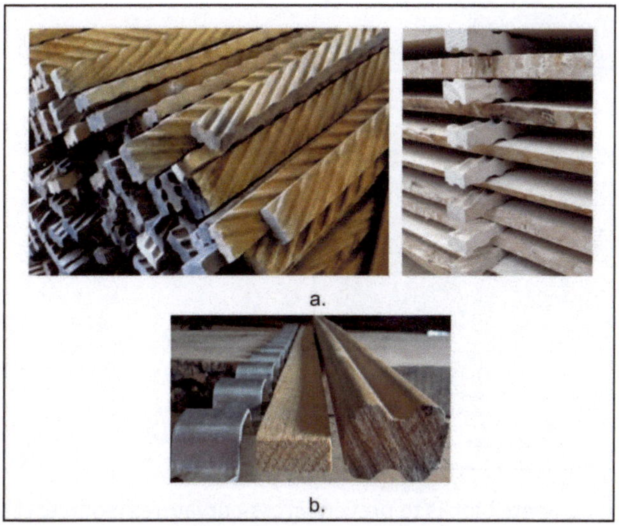

Figure 7.5 Special stacking stickers for reducing the contact area between the sticker and the board surface: a – 'Ventolatten' (source: *https://gruber-saege.at/website/project/ventostapellatten/*); b – corrugated aluminum stickers or milled wooden stickers (source: *https://www.bm-online.de/wissen/technik/qualitaet-ab-lager/*)

Stack loading. Due to shrinkage anisotropy (see also Section 2.1.5), the wood pieces are prone to cross-curvature during drying, especially the boards in the upper part of the stack (Figure 7.6). To avoid this defect, concrete blocks should be used to load the stack (Figure 7.7), especially for wood species that warp significantly, such as beech and Turkey oak. The weight of the loading plates must be dimensioned such that the pressure exerted on the stack does not exceed 1 t/m^2. At higher pressures there is a risk of crushing the material and the stacking stickers.

Coating agents. To prevent premature drying of the sawn edges and thereby avoid end checks, paraffin-containing coatings, wax dispersions, or a mixture of sawdust and PVAC adhesive (white glue) can be used for this purpose, especially for thick wood (> 40 mm) and for wood species that are highly prone to cracking, such as oak.

Figure 7.6 Cupping of the wood pieces in the upper part of the stack (source: [2])

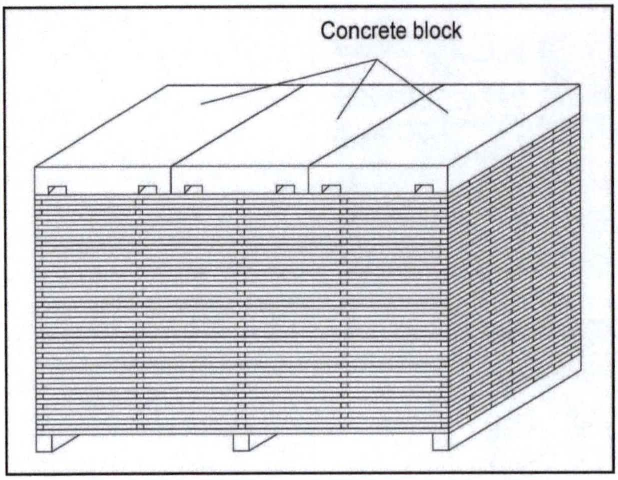

Figure 7.7 Stack loading with concrete blocks to prevent cross-curvatures (source: [1])

Stack dimensions and calculation of the wood volume in a stack. The dimensions of the stacks are generally chosen depending on the dimensions of the boards and the available means of transport. Thus, the length of the stack corresponds to the maximum length of the wood. When wood pieces of different lengths must be placed in the same stack, a left-right offset arrangement is attempted so that the stack as a whole retains the appearance of a parallelepiped (Figure 7.8).

The width of the stack is normally between 1 and 2 m, typically 1.2 m.

The height of the stack must maintain a ratio of max. 3:1 to the width to ensure stability, but is also limited by the means of transport used and is generally about 1 m.

Figure 7.8 Stacking of boards with different lengths (source: [2])

The wood volume in a stack (V_{st}) can be calculated using the following equation:

$$V_{st} = L_{st} \cdot B_{st} \cdot H_{st} \cdot f_{st} \, [m^3] \tag{7.1}$$

L_{st} stack length [m]
B_{st} stack width [m]
H_{st} stack height [m]
f_{st} stacking factor

$$f_{st} = f_L \cdot f_B \cdot f_H \tag{7.2}$$

f_L length factor, usually $f_L = 0.85$ is assumed, because not all boards are of equal length in most cases
f_B width factor, $f_B = 0.8$ (for edged timber); $f_B = 0.5$ (for unedged timber)
f_H thickness factor

$$f_H = \frac{d}{d_{st} + 1.08 \cdot d} \tag{7.3}$$

d wood thickness [mm]
d_{st} thickness of stacking sticks [mm]
1.08 coefficient for wood shrinkage

The wood volume in a standard stack (B_{st} = 1.2 m and H_{st} = 1 m) of 4 m long and 48 mm thick spruce boards is:

$$V_{st} = 4 \cdot 1.2 \cdot 1 \cdot f_{st} \, [m^3] \tag{7.4}$$

where:

$$f_{st} = 0.85 \cdot 0.8 \cdot f_H \tag{7.5}$$

where:

f_B = 0.8 because softwoods are edged and dried and

$$f_H = \frac{d}{d_{st} + 1.08 \cdot d} = \frac{48}{20 + 1.08 \cdot 48} = 0.67 \tag{7.6}$$

Thus:

f_{st} = 0.45 and V_{st} = 2.16 m^3 (much smaller than the volume of the parallelepiped: 4.8 m^3).

7.2 Placement of Wood Moisture Measurement Points

Wood moisture serves as a reference parameter in fresh air/exhaust air drying. Therefore, it must be measured continuously. All measurement methods already mentioned in Section 2.1.3 can be used, but the resistive measurement method is most commonly preferred, even though its accuracy is limited, especially at wood moisture levels above the fiber saturation range.

A measurement point consists of two electrodes (Figure 7.9). These are driven into pre-drilled holes on the wood surface (not on the edge), perpendicular to the grain, at a distance of at least 50 cm from the end of the board and approximately 30 mm apart from each other (Figure 7.10). Attaching the electrodes to the underside of the board can be advantageous to avoid the accumulation of water in the measurement hole.

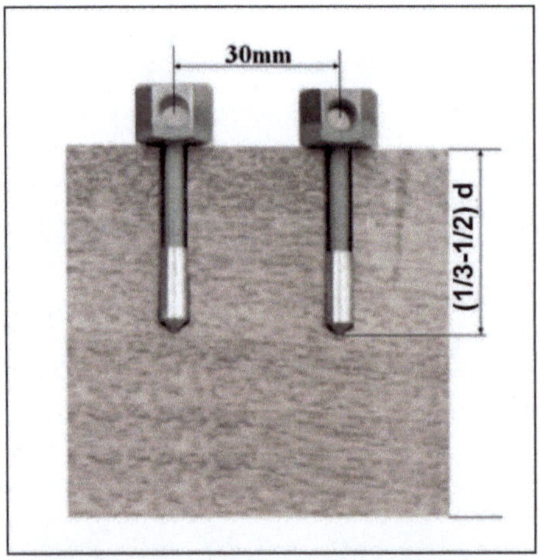

Figure 7.9 Electrodes for measuring wood moisture during drying (source: [1])

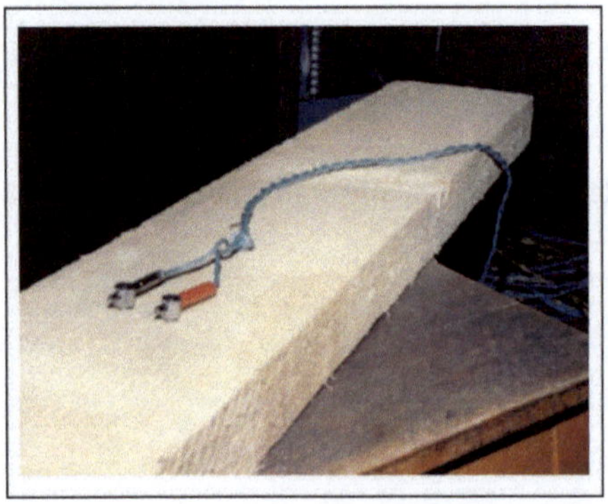

Figure 7.10 Resistive measurement of wood moisture during drying

The penetration depth should be 1/3 – 1/2 of the board thickness. The electrodes must penetrate the holes by friction (not loosely). Electrodes with silicone-insulated sheathing are preferable.

The measured values must be representative of the entire charge. Therefore, it is extremely important how many measurement points are used and how they are positioned within the stack. As a first rule, the longest and wettest boards should be selected for the insertion of the electrodes.

The number of measurement points should be as high as possible. At least four measurement points are necessary to obtain a reliable mean value. The measurement points should be distributed throughout the material being dried, preferably in those positions from which wood moisture escapes most difficultly, e.g. in the center of the stack, in the lower third of the stack height, on the air outlet side, etc.

The connection of the electrodes to the control and regulation system is made via special cables after the stacks have been introduced into the drying plant.

7.3 Preparation of the Drying Plant

Before loading the drying chamber, a number of inspection operations must still be carried out, such as:

- whether the door seal is intact;
- check the spray device for lime deposits; if necessary, replace nozzles;
- check fan centering;
- verify normal opening and closing of the solenoid valves and the dampers in the air ducts;
- check measuring cables for breaks;
- replace the hygroscopic disc of the EMC sensor.

During loading of the dryer, care must be taken to fill the space as completely as possible. The maximum distance between the top stack and the false ceiling should be 10 cm; otherwise the airflow will preferentially take the empty gap and bypass the stack, which significantly slows down the drying process. Any gaps present should be masked with a rubber canvas or an articulated metal plate (Figure 7.11), so that the air does not flow past the stacks.

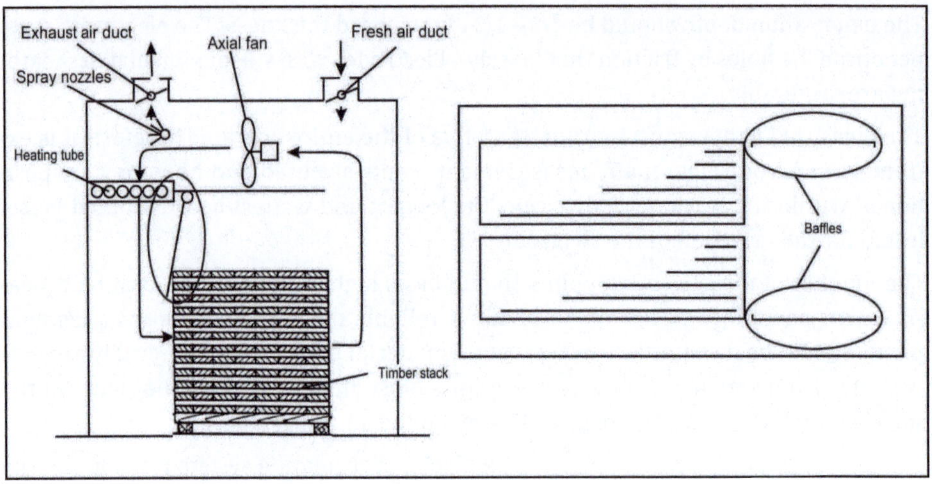

Figure 7.11 Baffling of empty spaces in the dryer (source: [4])

7.4 Creation/Verification of the Drying Schedule

Since nowadays the drying process is in most cases controlled by computer, the drying schedule is specified by the software of the control system. The kiln operator only needs to enter the initial data for the charge: wood species, thickness and desired final moisture content. Once these have been entered, the software selects the appropriate drying program from its database. The operator should query and visualize the intended drying schedule in order to review it. The recommendations explained in Section 4.2 should be taken into account for this purpose. These correspond to mild drying conditions, which should in any case be observed during the first drying run.

In the event of a successful drying run, the drying conditions may be intensified in order to shorten the drying time. However, changes to the specified drying schedules should be made cautiously (± 1 – 2 °C or ± 1 – 2 % U_{gl}) and in a stepwise manner, i.e. under no circumstances should the temperature be increased and the air humidity reduced simultaneously.

7.5 Kiln Control

The control of the drying process involves the continuous measurement of the main parameters (*t* and *EMC*), comparison with the values specified in the drying schedule, and the corresponding regulation of the controlled variables. In this process, the value of the controlled variable (actual value) is continuously recorded and compared with the setpoint value. In the event of deviations, the command is triggered that leads to a change in the controlled variable.

In wood drying, the wood moisture content serves as the reference variable and must therefore be measured continuously.

Air temperature and air humidity are the controlled variables. Accordingly, there are two control loopsin the drying process:

- the temperature control loop;
- the humidity control loop.

In order to carry out proper regulation, the plant must be equipped on the one hand with elements for measuring the actual values of the air parameters in the dryer and on the other hand with actuators for changing these parameters.

Temperature control loop. The actual temperature in the dryer is measured using a thermoresistive sensor (e.g. Pt100) (Figure 7.12).

The actuator for temperature regulation is a solenoid valve (Figure 7.13), which allows or stops the supply of the heating medium to the heat exchanger.

Temperature regulation is based on the comparison of the measured temperature with the prescribed value:

- ff the two values are equal, the solenoid valve remains in its current position;
- ff the actual value < setpoint value, the solenoid valve opens to increase the air temperature in the dryer;
- ff the actual value > setpoint value, the solenoid valve closes.

Modern control systems enable stepless opening/closing of the solenoid valve by means of proportional controllers, thereby achieving a "finer" regulation than with two-position controllers.

Humidity control loop. The measurement of the actual air humidity in the dryer is most commonly performed using a sensor with a hygroscopic veneer or cellulose wafer (Figure 7.12). It is also possible to measure and control the relative humidity (φ) instead of the equilibrium moisture content (*EMC*). In this case, the psychrometer (see Section 2.2.2.4) is used as the measuring element.

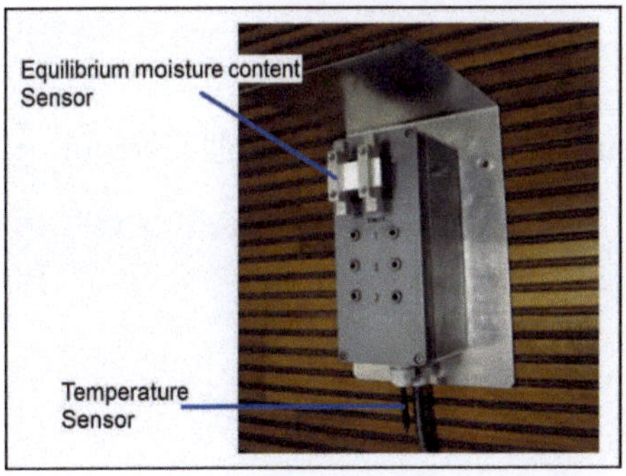

Figure 7.12 Measurement of temperature and equilibrium moisture content during wood drying

Two actuators are required for humidity control: on the one hand, an electric valve on the spray pipe of the dryer (Figure 7.13) and, on the other hand, the motors that open and close the fresh air/exhaust air dampers (Figure 7.14). The two actuators always operate in opposition: when one opens, the other must be closed.

Figure 7.13 Electric valves for controlling temperature and humidity during wood drying

Figure 7.14 Drive motor of the fresh air/exhaust air dampers for controlling (reducing) humidity during wood drying

The humidity control is based on the comparison of the measured humidity with the prescribed value:

- if the two values are equal, both actuators remain in their current position;
- if the actual value < setpoint value, the air dampers are first closed and then the spray valve is opened;
- if the actual value > setpoint value, the spray valve closes and the dampers are opened. In temperate climate zones, the humidity in the dryer decreases by the intake of fresh air from the outside atmosphere.

Mutual influence of the two control loops. It should be noted that there is a close interdependence between the two air parameters, so that the two control loops should not only be considered individually, but also with regard to their mutual influence.

For example, it should be noted that an increase in temperature immediately causes the air humidity to decrease. Therefore, especially during the heating phase and when drying above the fiber saturation point, where achieving a high humidity plays a very important role, the temperature increase should only occur in small steps so that the humidity can keep pace.

It should also be taken into account that, depending on the spray medium used for humidification, the temperature may also be slightly changed during humidification. For example, when humidifying with steam, the air temperature is slightly increased, while humidification with cold water can cause a cooling of the air. This effect should be taken into account when setting the temperature during the conditioning phase.

Control of air velocity. In addition to the two main parameters, modern drying plants also offer the possibility of adjusting the air velocity in order to save energy. Continuous or stepped speed control is possible. Continuous speed control in the interval 1430 – 760 rpm (from 100% down to 68%) (Variant I in Table 7.3) leads to a saving of approximately 50% of electrical energy consumption, but also to an extension of drying time of up to 21% when drying softwood and up to 13% when drying hardwood. The stepped control (Variant II in Table 7.3) results in a much smaller time extension while still achieving an energy saving of approximately 40%.

Table 7.3 Influence of Speed Control on the Duration and Electrical Energy Consumption in Wood Drying [5]

Var	Speed* [rpm]	Wood moisture content [%]	Duration [h]	Electrical energy consumption [kWh/fan with P = 3 kW]		
Spruce, 25 mm		**Initial moisture content 80%, final moisture content 12%**				
I	1,430 – 980	80 – 35	25	40.5	–	–
II	1,430		22	–	63.8	–
III	1,430		22	–	–	63.8
I	980 – 760	35 – 20	18	15.4	–	–
II	980		16	–	15.2	–
III	1,430		14	–	–	40.6
I	760	20 – 12	15	9.5	–	–
II	760		15	–	7.5	–
III	1,430		12	–	–	34.8
I	Total var I	58	65.4	–	–	
II	Total var II	53	–	86.5	–	
III	Total var III	48	–	–	139.2	
I	Change compared to variant III	21 %	53 %	–	–	
II		10 %	–	39 %	–	
Oak, 40 mm		**Initial moisture content 60 %, final moisture content 10 %**				
I	1,430 – 980	60 – 35	205	439	–	–
II	1,430		182	–	546	–
III	1,430		182	–	–	546
I	980 – 760	35 – 20	160	147	–	–
II	980		150	–	150	–
III	1,430		138	–	–	474
I	760	20 – 10	165	114	–	–
II	760		165	–	90	–
III	1,430		148	–	–	444
I	Total var I	530	700	–	–	
II	Total var II	470	–	786	–	
III	Total var III	468	–	–	1404	

Var	Speed* [rpm]	Wood moisture content [%]	Duration [h]	Electrical energy consumption [kWh/fan with P = 3 kW]		
I	Change compared to variant III	13 %	50 %	–	–	
II		0.4 %	–	44 %	–	
* Variant I = continuous speed control from 1430 rpm to 760 rpm Variant II = stepped speed control: 1430/980/760 rpm Variant III = without speed control (constant speed: 1430 rpm)						

It is established that no drying occurs during the heating phase, so little water is removed from the wood and the air velocity can be lower. However, this can promote fungal infestation.

As soon as drying begins, the role of air velocity becomes essential in order to carry the free water removed from the wood away from the stack as quickly as possible. In this phase, the maximum fan speed is required.

As drying progresses and the amount of water that must be removed per unit of time decreases, the role of air velocity diminishes again. This is based on the theory that during the removal of bound water until the end of drying and during the heating phase, the speed can be reduced (e.g., 760 rpm instead of 1430 rpm). However, considering the other role of speed, namely ensuring heat exchange between the environment and the wood, it is questionable whether the economic savings achieved by reducing the speed offset the increase in costs caused by the extension of the drying time.

In contrast to the other two air parameters, speed control is not achieved by comparing a measured value with the prescribed value, but by the simple assignment (via programming) of a specific percentage of the maximum speed to each phase of the drying process. In Table 7.4 an example is presented.

Table 7.4 Sample Schedule for Speed Control in Wood Drying

Phase	Fan speed (as a percentage of the maximum value)
Heating phase	70 – 77 %
Drying in the region above the fiber saturation point	100 %
Drying in the region below the fiber saturation point	80 – 74 %
Conditioning	58 %
Cooling	5 %

No measuring element is required for this type of control. The actual speed change is carried out using the frequency converter integrated in the electric motor of each fan.

Control systems. Currently, the management of the drying process in most companies is carried out automatically using a computer. Fully automatic systems can independently control the entire drying process, so that after entering the initial data (wood species, thickness, desired final moisture content) no operator intervention is required. Despite the advanced level of automation, practice has shown that the operator's experience is indispensable for optimizing the process (relationship between quality and cost of drying).

An automated system for controlling the drying process generally comprises the following components:

- central processing unit;
- control cabinet;
- signal amplifiers;
- measuring elements;
- actuators.

The placement of the individual components relative to the drying plant and the connections between them are illustrated in Figure 7.15.

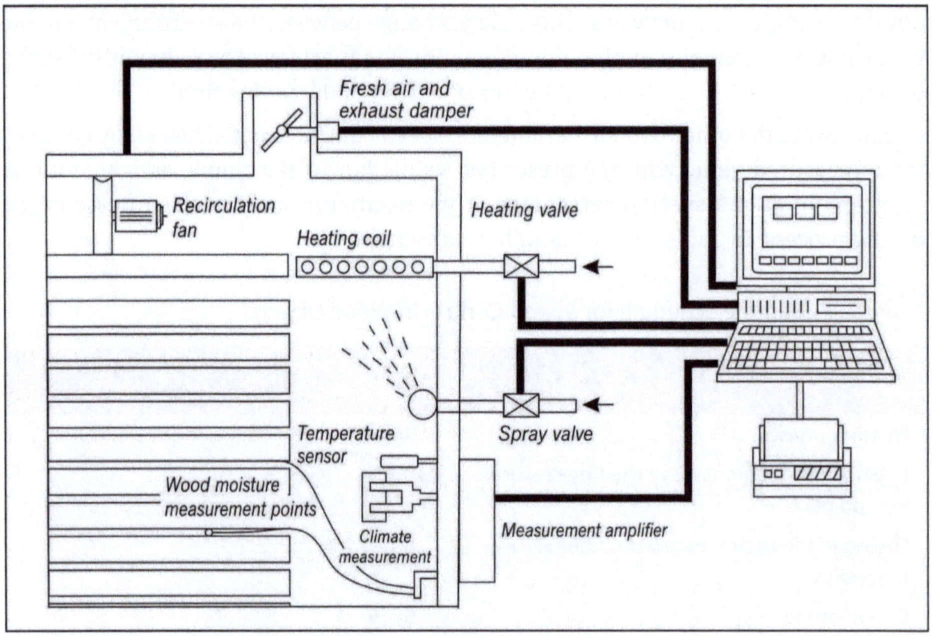

Figure 7.15 Diagram of an automatic control system (based on [2])

The **central unit** is the element that enables the centralization of drying data and the remote management of the process from several plants simultaneously. Thus, one central unit can serve multiple dryers (typically up to 16). The central unit is not a mandatory component of the control system, as its functions can be ensured individually for each dryer from the control cabinet. The acquisition of a central unit is considered economically viable when the company has a minimum of three drying plants.

The **control cabinet** is the basic component of the control system. One control cabinet is required for each drying plant. It must be placed in a room in close proximity to the dryer. As a rule, the control cabinet is equipped with a monitor (sometimes of the touchscreen type), through which the operator enters the initial data for the batch and through which he can retrieve the planned drying schedule, the measured values, or the current status of the control elements at any time.

The microprocessor system built into modern systems enables the definition of the drying schedule, determines when the transition from one phase to another should be made, among other things. In some systems, it is also possible to measure energy consumption, store drying diagrams ("archive"), etc.

At the same time, the control cabinet receives information about the values of the air parameters and wood moisture content measured by the sensors in the chamber, and issues the corresponding commands to the actuators when control actions are required.

The **signal amplifier** has the task of receiving the signals transmitted by the sensors in the drying chamber, amplifying them, and subsequently forwarding them to the control cabinet. The amplifier is usually placed directly on the outer wall of the drying chamber next to the cable centralization box of the measuring elements (see Figure 7.15).

The **measuring elements (sensors)** are located inside the drying chamber and have the task of continuously measuring the actual values of the drying parameters (wood moisture content, air temperature, and equilibrium moisture content of the air). For wood moisture content, a minimum of four (preferably eight or even 16) probes per batch are used. For temperature and equilibrium moisture content measurement, one probe per parameter is sufficient for plants with non-reversible air movement. These are always mounted on the chamber wall on the side where the air enters the stack. For plants that allow reversible air circulation through the stacks, two probes per air parameter should be used, so that the pair located on the air inlet side at any given time is the one measuring.

The **actuators** are the heating valve, the spray valve, the servomotors of the air dampers, and the frequency converters of the motors driving the fans. The actuators receive commands from the control cabinet and have the task of adjusting the air parameters so that they reach the values specified in the drying schedule.

Automatic systems for controlling the drying process offer safety and convenience in the operation of drying plants. In this case, the role of the operator is limited to entering the initial data. Regular monitoring (daily readings) is nevertheless recommended in order to address in time any potential damage situations that may occur during the process, e.g., blockage of a solenoid valve, etc.

7.6 Drying Record and Drying Diagram

The document certifying how the drying of a batch proceeded can be presented in tabular form (as a drying record – see Table 7.5) or graphically, as a drying diagram (see example in Figure 7.16). In both forms, it shows the time course of all drying parameters (wood moisture content, air temperature, air humidity) over the total duration of the drying process.

The **drying record** (Table 7.5) is most often created by hand based on daily readings. It enables the verification of the drying parameters during the process, not only at the end (as in the case of the drying diagram), so that if problems arise, they can be identified and resolved in time. Furthermore, it is not particularly useful when the control system is not equipped with a printer and also has no storage function for the drying diagram. In such cases, the drying diagrams of completed batches may be lost. Handwritten drying records help to ensure an archive of the dried batches.

Table 7.5 Drying Log

Customer:					Wood species:						
Order number:					Dimensions:						
Batch number:					Wood thickness:						
Date	Time	Wood moisture						Temperature		EMC	
		M1	M2	M3	M4	...	Average	Actual	Target	Actual	Target

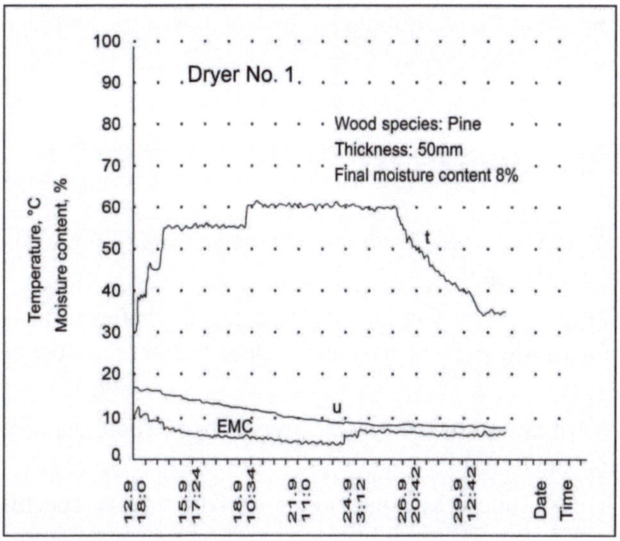

Figure 7.16 Drying diagram: *t* = temperature; *u* = wood moisture; *EMC* = equilibrium moisture content (source: [6])

The most frequently occurring problems that can be identified using both instruments (drying log and drying diagram) are listed in Table 7.6.

Table 7.6 Problems That Can Be Identified from the Drying Log and Their Causes

Problems	Possible causes
The intended air temperature was not reached	▪ Insufficient boiler capacity ▪ Insufficient sizing of the radiator ▪ Malfunction of the heating valve ▪ Dirty, corroded heating registers ▪ Leaks (particularly at the doors)
The intended air humidity was not reached	▪ Faulty function of the spray valve ▪ Calcification of the spray nozzles ▪ Incorrect pressure of the spray medium (too high when using steam or too low when using cold water - no "mist" is produced)
The air humidity does not drop quickly enough	▪ Malfunction of the air dampers ▪ Insufficient sizing of the air ducts

After the drying process is complete, the stacks are moved out and stored in a closed, air-conditioned room for at least two days for softwood and at least seven days for hardwood, in order to cool down completely and relieve drying stresses. Under no circumstances is further processing or film packaging of the wood permitted while it is still warm.

7.7 Quality Control: Drying Defects

Optimization of the drying process is not possible without knowing the most important drying defects and, in particular, their causes.

Any deterioration in material quality caused by the drying process is referred to as a drying defect. When assessing drying defects, however, a clear distinction must be made between:

- unavoidable defects, which are caused by the specific properties of wood as a non-homogeneous and anisotropic material;
- actual drying defects, which are caused by non-compliance with process-specific factors.

The following are considered unavoidable drying defects:

- defects caused by certain structural anomalies in the wood, such as knots, diagonal grain, spiral grain, reaction wood, etc.;
- defects caused by the uneven shrinkage of wood along the three structural directions (assuming that stacking was carried out correctly);
- defects caused by biological factors (insects, fungi) and atmospheric factors (wind, solar radiation) during storage of the material prior to drying.

In order to highlight these defects, it is necessary to check the quality of the material before drying and to mark the defects (especially cracks) on the wood with chalk prior to drying, as otherwise they cannot be distinguished from defects caused by the drying process.

The causes (Figure 7.17) of the actual drying defects (Table 7.7) can be found in non-compliance with the technological rules throughout the entire process, e.g.:

- excessively long storage of the wood outdoors; this promotes case hardening, fungal and insect infestation;
- non-compliance with stacking rules; this leads to distortion and uneven final moisture content;
- application of an unsuitable drying schedule; this most commonly leads to crack formation and sometimes to discoloration;
- incorrect operation of measuring and/or actuating elements; this leads to non-compliance with the drying schedule and most commonly to an extension of the drying time;
- faulty design or improper maintenance of the drying plant; this can lead to uneven distribution of air parameters, heat losses, or the formation of condensate;

- storage of the wood under inadequate conditions after drying; this leads to an undesirable change, most commonly an increase in wood moisture content.

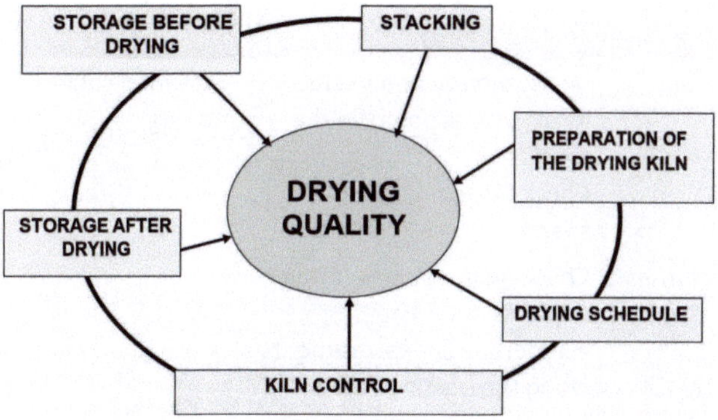

Figure 7.17 Factors influencing drying quality influencing

The **quality control** is carried out during destacking. In principle, the European standard EN 14298 [7] applies for the determination of drying quality. According to this standard, the accuracy and variation of the final moisture content and, where applicable, case hardening must be checked. Even if the standard does not require it, a visual inspection of the boards is also recommended as a basis for process optimization, in order to determine the proportion and severity of any cracks, distortions, and discolorations.

Table 7.7 Actual Drying Defects – Causes, Prevention Options and, Where Applicable, Remedial Measures.

Drying defects	Causes	Preventive measures	Corrective measures
1. Uneven final moisture content	Uneven or insufficient air velocity	Measurement of air velocity at various points in the dryer Check blanking of empty spaces	Extended conditioning
	Unevenly distributed temperature	Check heating pipes for dirt deposits or damage. Temperature measurements along the heating register to detect possible blockages	
	Unevenly distributed air humidity	Check spray pipes for limescale deposits. Check for possible condensation on the suspended ceiling or at the fan openings	
	Too many stacks in the direction of airflow	Purchase dryers with fewer than four stacks across the dryer width	Install more powerful fans
2. Case hardening	Long storage of the wood outdoors Too severe drying conditions (too low air humidity) during the heating phase and at the beginning of drying	High air humidity (*EMC* = 18 % or φ = 90 %) during the heating phase	Humidification (only effective if the defect is detected in time, before cracks develop)
3. Collapse	Too severe drying conditions (too high temperature) at the beginning of drying	Special drying schedule for sensitive wood species	4 – 8-hour steaming at 100 °C and φ = 100 % (*EMC* = 19 %) (only effective if the defect is detected in time)

Drying defects	Causes	Preventive measures	Corrective measures
4. Surface checks	Internal stresses that develop at the beginning of drying when the moisture content at the surface is already below the fiber saturation point while free water still exists in the interior	Adherence to the drying schedule at the beginning of the actual drying process: moderate temperature and high air humidity until the wood reaches the critical moisture content	None
5. Honey-combing	Internal stresses towards the end of drying	Adherence to the drying schedule: wait until the critical moisture content is reached and only then increase temperature and reduce air humidity	None
6. End checks	Higher diffusion rate of water along the grain Too high air velocity	Protect end grain with a coating or nailed-on protective strips or stack stickers protruding in front of the end face (especially recommended for boards thicker than 40 mm)	None
7. Deformations	Stacking defects Absence of load on top of the stack.	Respect stacking rules Stack load	Does not exist
8. Blue stain	Fungal attack, especially when the initial moisture content is very high	Light steaming (2 – 3 hours at 70 – 75 °C), then let the wood cool down and then apply the normal drying schedule	Surface blue stain can be removed by planing
9. Sticker stain	Use of wet or dirty stickers	Dry stickers to $u \leq 15$ % before use Use of planed stickers (see Figure 7.5)	Can be removed by planing
10. Condensation stains	Direct contact between warm wood and cold water droplets (or spray steam)	Alignment of spray nozzles must be checked False ceiling must be checked (for condensate forming there, especially common in winter)	Can be removed by planing

Drying defects	Causes	Preventive measures	Corrective measures
11. Chemical discolorations	Contact between iron-containing parts and tannin-rich wood species (e.g. oak), at high moisture content Oxidative reaction of tannin-rich wood species at too high a temperature (> 30 °C) during the heating phase Oxidative reaction of starch-containing wood species (e.g. alder) at too low a temperature (< 45 °C) during the heating phase	Specially adapted drying schedules	Surface discolorations can be removed by planing Internal discolorations can no longer be remedied.

a) Determination of the uniformity of the final moisture content within a batch is, according to EN 14298, the main criterion for assessing drying quality. This defect refers to the difference between the final moisture content of some empirically selected boards from the respective batch and the target final moisture content.

The sampling is governed by ENV 12169 [8]. Depending on the batch size (see Table 7.8 and Table 7.9), a certain number of boards is randomly selected from a certain number of packages and their average moisture content is determined using a resistance meter.

To assess the conformity of the batch, the arithmetic mean of the individual average values is calculated and the number of non-conforming values is determined.

The batch is considered conforming (standard quality) when two conditions are met:

- the average moisture content (the arithmetic mean of the individual measured values) lies within the limit values specified in EN 14298 (Table 7.10);
- 93.5 % of the individual moisture measurements must be conforming. Boards are considered conforming if their moisture content is within the range of (0.7 – 1.3) of the target moisture content.

b) The outer formwork consists in a hardening of the wood surface, which occurs as a result of excessively rapid moisture removal from the surface layer of the wood (plastic deformation). The case-hardened surface acts as a barrier to moisture removal from the interior of the wood. This creates a large moisture difference between the core and the surface layer. If this is not detected in time, the accumulation of water vapor inside the wood can lead to internal cracks.

Table 7.8 Number of Packages to Be Opened [8]

Number of packages in the batch	Number of packages to be opened
1	1
2 - 5	2
6 - 11	3
≥ 12	4

Table 7.9 Single Sampling According to AQL 6.5 [8]

Number of boards in the batch	Number of samples	Max. number of non-conforming boards
100 - 150	20	3
151 - 200	32	5
201 - 500	50	7
501 - 1200	80	10
1,201 - 3,200	125	14
> 3,200	200	21

Table 7.10 Permissible Range of Average Moisture Content of a Batch Depending on the Target Final Moisture Content [7]

Target final moisture content [%]	Limit deviations [%]
7 - 9	± 1
10 - 12	± 1.5
13 - 15	± 2
16 - 18	± 2.5

The determination method (Figure 7.18) is regulated by ENV 14464 [9] and consists of cutting a 15 mm wide test piece at a distance of at least 300 mm from the end of the board. Its ends are removed so that a test piece with dimensions of 100 mm × 15 mm × wood thickness (*d*) is obtained. This is split in the middle. The two resulting halves are conditioned at a temperature of 20 °C and a relative humidity of 55 % for 24 h for softwoods and 48 h for hardwoods. During this process, the two halves bend inward. The more case-hardened the wood is, the greater the curvature. The maximum value of the opening in mm indicates the degree of case hardening.

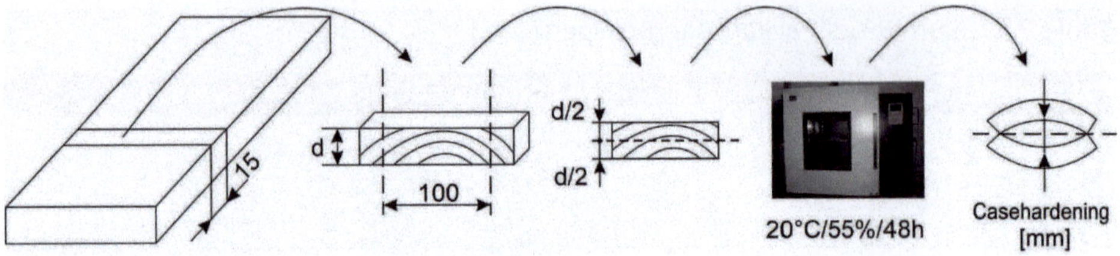

Figure 7.18 Measurement of case hardening

According to EN 14298 [7], the determination of case hardening is optional. The determination is carried out on every second piece of the samples used for final moisture content determination (i.e., half of the samples to be measured according to Table 7.9). The batch is compliant with regard to case hardening if 80 % of the tested pieces are below the permissible limit value (usually 3 mm).

As already indicated in Table 7.7, there are also other drying defects that should be identified during quality control.

c) Cell collapse is the most severe drying defect, but it occurs only in some wood species, e.g., eucalyptus, walnut, oak, aspen, especially when the wood has a high initial moisture content. This drying defect consists in the collapse of the cell walls of the wood as a result of the excess pressure that develops in the cell cavities when drying is carried out at too high a temperature with a high initial moisture content of the wood. In this case, the free water evaporates rapidly and in large quantities, without the diffusion of the resulting vapor being able to occur to the same extent. Cell collapse can be recognized by a waviness of the surface and an uneven reduction in thickness. It occurs at the beginning of drying at moisture contents above the fiber saturation point.

The determination method consists of measuring the minimum thickness (d_{min}) and the maximum thickness (d_{max}) (Figure 7.19). The difference between the two dimensions (d_c) provides information about the severity of this defect [10]: If $d_c > 5$ mm, the situation is critical.

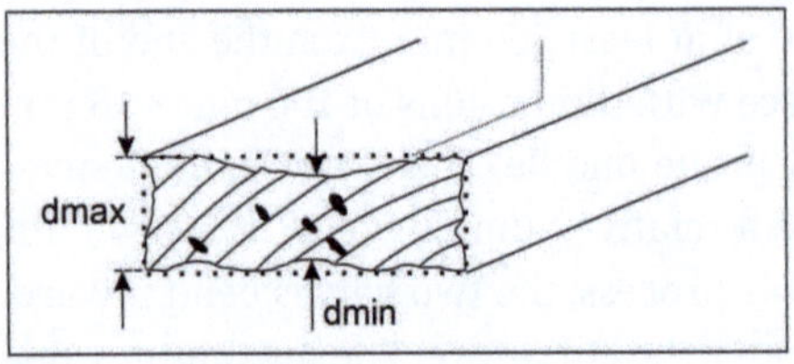

Figure 7.19 Collapse determination (based on [10])

d) Cracks, which can appear during wood drying, are divided into three categories (Figure 7.20): **surface checks, end checks** and **internal cracks.**

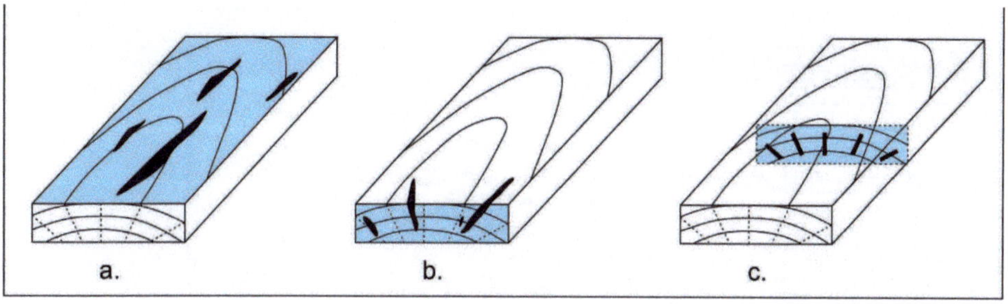

Figure 7.20 Drying cracks: a – surface checks; b – end checks; c – internal cracks (based on [11])

Surface checks (Figure 7.21) – these cracks occur during the first drying phase, when the surface layer of the wood pieces dries too quickly due to excessively low relative humidity. Due to this phenomenon, the surface layer tends to shrink, but is restrained by the inner layers, which are still very moist and have not yet begun to shrink. As a result of this restraint, drying stresses develop in the surface layer, and when their value exceeds the tensile resistance of the wood perpendicular to the grain, surface checks occur.

Surface checks generally occur in thick, wide wood pieces that run tangentially to the annual rings [12]. Sometimes these checks also continue along the edges.

The severity of surface checks as a defect is assessed by their number and depth of penetration. The maximum depth of penetration is determined by successive planing passes of 1 mm thickness until the peeled shaving remains continuous (Figure 7.22). For checks with a penetration depth exceeding 5 mm on either side, optimization is susceptible.

Figure 7.21 Surface checks in a turkey oak board

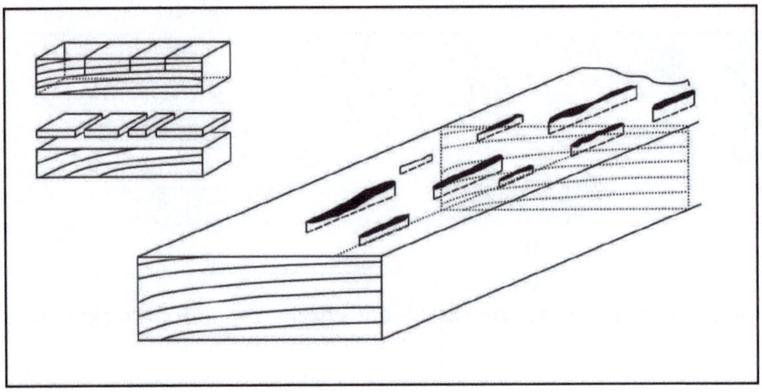

Figure 7.22 Determination of the depth of surface checks (based on [10])

End checks (Figure 7.23) – this type of check occurs at the ends of wood pieces at the beginning of the drying process. End checks develop because the rate of water diffusion in the longitudinal direction is higher than perpendicular to the grain. Due to this property, the end grain of the boards dries faster than the center, leading to drying stresses.

Wood pieces with large width and thickness (> 40 mm) are more susceptible to end checks. To avoid them, it is recommended to use high relative humidity at the beginning of the drying process and to apply end coatings. To achieve effective protection, it is recommended to apply the coating to the board ends immediately after sawing the log, or even directly to the ends of the log before sawing [12,13].

End checks can extend along the length of the wood. This disadvantage can be counteracted by placing the sticker boards precisely under the ends of the wood pieces [12].

The severity of end checks as a defect is assessed by their number and maximum length. The maximum length of the check is determined by successive cuts of 50 mm from the end until a clean end without checks or scratches is reached. For lengths exceeding 200 mm in more than 10% of the batch, optimization is susceptible.

Figure 7.23 End checks in a 50 mm thick turkey oak board

Internal checks – these are not visible on the surface (see Figure 7.20 c) and, to determine them, the wood pieces must be cut at a distance of at least 300 mm from the end. However, this assessment is normally carried out on wood pieces of large thickness when the casehardening test has indicated the presence of internal stresses. Internal checks can also be detected using a non-destructive method based on determining the speed of sound through the wood [13].

To assess the quality of drying with regard to internal checks, the percentage of pieces exhibiting this type of defect must be determined. If this is higher than 10% of the batch, optimization is susceptible.

Deformations – are rarely an actual drying defect, as their main cause is the anisotropy of wood. They are recognized as drying defects only when caused by improper stacking (see Figure 7.2 and Figure 7.6).

Discolorations – as with deformations, discolorations are only considered drying defects when a factor related to the drying process causes them. Figure 7.4 and Figure 7.24 to Figure 7.27 illustrate the discolorations that most commonly occur both during air drying and conventional drying.

Figure 7.24 Blue stainin a spruce board

Figure 7.25 Condensation stains

Figure 7.26 Iron-tannin discoloration in oak wood

Figure 7.27 Oxidative reddish discoloration in oak wood

In practice, quality control of drying should always include three procedures:

- visual inspection of the parts to detect defects visible to the naked eye (discoloration, surface checks, distortions). The greatest possible efficiency in distinguishing actual drying defects from growth or storage defects is achieved when this inspection is also carried out before drying during stacking and the visible defects are marked with chalk;
- measurements to verify the final moisture content on a number of boards, depending on the size of the batch, selected at random;
- cutting a test sample at a distance of 300 mm from the end to check for the possible presence of internal checks and/or internal discoloration.

If necessary – for hard hardwood species, thick sizes with high initial moisture content, etc. – it is also recommended to carry out the casehardening test.

Table 7.11 shows a sample form that should be filled in during drying quality control to record the results and thus provide a basis for process optimization.

Table 7.11 Inspection Report [6]

Part No.	Inspection criteria	Before drying	After drying
Visual inspection			
	Surface checks	yes/no+ Depth of penetration	yes/no+ Depth of penetration
	End checks	yes/no+ Length	yes/no+ Length
	Collapsed surface (collapse)	yes/no+d_c	yes/no+d_c
	Spiral grain	yes/no	yes/no
	Lengthwise distortions	yes/no	yes/no
	Other distortions	yes/no (if yes, which)	yes/no (if yes, which)

Part No.	Inspection criteria	Before drying	After drying
	Discolorations	yes/no (if yes, which)	yes/no (if yes, which)
Inspection			
	Number of packages/ batch		
	Number of opened packages		
	Number of wood pieces/ batch		
	Number of inspected samples		
	Mean value of final moisture content, %	–	
	Deviation of the mean value from the target moisture content, %	–	
	Maximum ±-deviation of individual final moisture contents from the target moisture content, %	–	
	Number of non-conforming wood pieces with respect to final moisture content uniformity		
	Maximum casehardening, mm	–	

References

[1] Campean, M. (2003): *Heat Treatments of Wood*. Transilvania University, Brasov, Romania

[2] Trübswetter, T. (2009): *Holztrocknung. Verfahren zur Trocknung von Schnittholz – Planung von Trocknungsanlagen*. 2nd ed., Carl Hanser Verlag, Munich, Germany

[3] Frühwald, A.; Ressel, J. (1982): *Holztrocknung. Teil 1. Grundlagen der Holztrocknung*. Universität Hamburg, Germany

[4] Câmpean, M. (1997): *Tratamente termice ale lemnului. Uscarea cherestelei*. Transilvania University, Brasov, Romania

[5] Schmidt, R. D. (1984): *Energie sparen durch veränderbare Luftgeschwindigkeit in Schnittholztrocknern – Sinn oder Unsinn?* Worblingen, Germany

[6] Bedelean, B.; Campean, M. (2013): *Tratamente termice ale lemnului. Îndrumar pentru lucrări practice*. Transilvania University, Brasov, Romania

[7] EN 14298 (2001): Schnittholz – Ermittlung der Trocknungsqualität

[8] ENV 12169 (2000): Kriterien zur Konformitätsprüfung eines Loses Schnittholz

[9] ENV 14464 (2003): Schnittholz – Verfahren zur Ermittlung der Verschalung

[10] Welling, J. (1994): *EDG-Recommendation: Assessment of Drying Quality of Timber*

[11] Yin, Q.; Liu, H. H. (2021): *Drying Stress and Strain of Wood: A Review*. In: Applied Sciences, 11(11), 5023. *https://doi.org/10.3390/app11115023*

[12] Ward, J.; Simpson, W. (2001): *Drying Defects*. In: Dry Kiln Operator's Manual. Forest Products Laboratory, Madison

[13] Bergman, R. (2021): *Drying and Control of Moisture Content and Dimensional Changes*. In: Wood Handbook, Kapitel 13, S. 1 – 21. U.S. Department of Agriculture, Forest Service, Forest Products Laboratory

8 Veneer Drying

8.1 Objectives and Special Features

Veneers are thin wood layers with a thickness of 0.08 to 6 mm. They are produced by peeling or slicing. Peeling consists of the rotating movement of the log around its longitudinal axis (Figure 8.1 a). Slicing consists of a rectilinear reciprocating movement of a flat knife relative to the log (Figure 8.1 b).

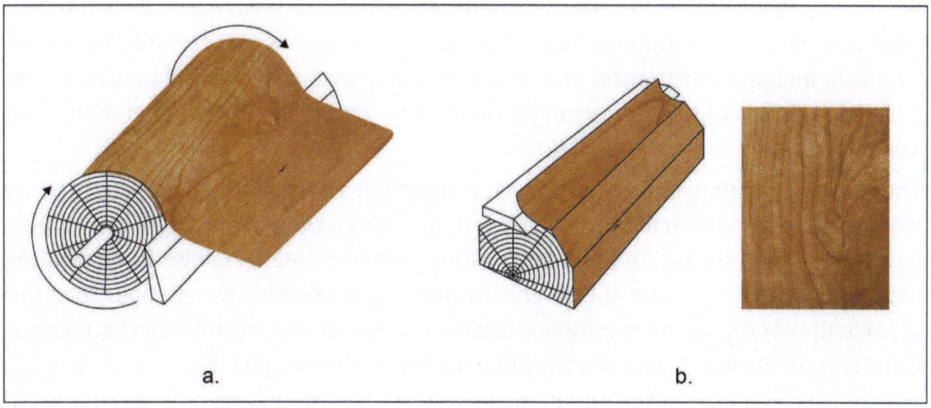

Figure 8.1 Veneer production by a - peeling (peeled veneers); b - slicing (sliced veneers)

To facilitate cutting and to obtain good surface quality of the veneers, the logs are heat-treated before cutting by steaming or hot-water soaking. Therefore, the veneer moisture content immediately after cutting is high, e.g. 80 – 100 % for pine sapwood, 80 – 120 % for alder, 70 – 90 % for linden, 60 – 90 % for birch, 60 – 100 % for beech, 60 – 80 % for oak.

The purpose of veneer drying is to reduce its moisture content as quickly as possible in order to prevent fungal attack (blue stain, mold) and to avoid certain chemical reactions that only occur in the presence of free water and lead to undesirable discoloration.

The final moisture content that the veneers must reach after drying depends on their area of application and on the type of adhesive used. A distinction is made between:

- decorative veneers – these are produced from high-quality, often exotic wood species, since aesthetics are the decisive factor in this case. They are mostly produced by slicing and are thinner than peeled veneers. They are used in the furniture industry as face veneers applied to low-cost substrate materials such as particleboard and MDF panels;
- technical veneers – these are mostly produced as peeled veneers from lower-cost wood species such as softwood, birch, poplar, and beech, because in this case it is not aesthetics but strength and stability that play a much more important role. These veneers are used, by bonding several layers together, to produce plywood or laminated veneer lumber (LVL).

In general, the moisture content of dried veneers should be below 7 % [1]. Depending on the type of adhesive used for further processing, a final moisture content of 5 – 6 % is recommended for technical core veneers, 6 – 8 % for technical face veneers, and 7 – 14 % is recommended for decorative veneers [2].

Insufficient drying leads to defective bonding, e.g. blistering. An excessively low final moisture content is also undesirable, because over-dried veneers tend to break easily due to increased brittleness and exhibit rougher surfaces, which lead to increased adhesion [3]. In addition, longer drying increases energy consumption and thus production costs.

The complex relationships between heat conduction and moisture transport in wood, which essentially characterize the drying of sawn timber, are of only secondary importance for veneer drying. The smaller thickness of the veneers enables faster heating. Furthermore, the structure of the veneer is more permeable to water vapor than that of sawn timber due to the loosening caused during production (knife checks, mechanical stress on the wood between the knife and the pressure bar) [4].

The drying of veneers differs from the drying of sawn timber. The reason lies in the much larger ratio between surface area and volume (Table 8.1).

Therefore, the drying time of the veneers is much shorter and is on the order of minutes.

Due to the smaller thickness of the veneers, the moisture gradient across the thickness of the material is practically non-existent, so that in this case drying does not cause internal stresses. Thus, higher drying temperatures can be used (above 100 °C, often 150 – 210 °C).

Table 8.1 Specific Evaporation Surface Area for Veneers vs. sawn Timber (source: [5])

Wood assortment	Thickness [mm]	Specific surface area [m^2/m^3]
Boards	30	71
	24	88
	18	115
	12	171
Veneers	3	671
	2.5	804
	2	1,004
	1	2,004

The type of airflow and the air velocity play a very important role in veneer drying. In order to remove the moisture extracted from the veneer from the surrounding area as quickly as possible, turbulent flow at a velocity of more than 10 m/s is required.

Due to their small thickness and shrinkage anisotropy, veneers tend to warp during drying. It is therefore extremely important to press the veneers during drying so that they retain a flat (planar) shape.

Cracks at the ends of the veneers are also a common defect in veneer drying. It is recommended to overlap the ends of the veneers at intervals of 5 – 10 cm.

Figure 8.2 Veneer moisture measurement with brush sensors (source: © Raute Corporation, used with permission)

The highly variable initial moisture content, can lead to an unevenly distributed final moisture content, especially in thicker veneers. In order to achieve precise and consistent control of the final moisture content, an increasing number of industrial veneer dryers are being equipped with sensors for extensive data acquisition [6, 7]. Measuring the initial moisture content, e.g. with brush sensors, (Figure 8.2) creates the prerequisite for sorting the veneers according to their initial moisture content and forming two or more moisture groups (e.g. initial moisture contents above and below 40 %), which are then dried in separate batches. This allows the drying time of each batch to be optimized and a largely more uniform final moisture content of the veneers to be achieved [6].

8.2 Drying Time: Influencing Factors

The drying time of the veneers is, as with sawn timber, influenced by many factors: wood species, thickness, initial moisture content and final moisture content, temperature, air velocity, direction of airflow relative to the veneer and degree of turbulence, drying method, and others.

In the case of convective hot-air drying, Equation 8.1 and the nomograms in Figure 8.2 and Figure 8.3 can be used to determine the drying time:

$$D = \left(\frac{U_i - U_{FS}}{V_u} + \frac{2{,}3}{K_z} \cdot \ln \frac{U_{FS}}{U_f} \right) \cdot K_{wood} [\text{minutes}] \tag{8.1}$$

U_i initial moisture content [%]
U_{FS} fiber saturation moisture content [%]
U_f final moisture content [%]
V_u coefficient accounting for the drying rate above the fiber saturation point (taken from Figure 8.3)
K_z coefficient accounting for the drying rate below the fiber saturation point (taken from Figure 8.4)
K_{wood} coefficient depending on the wood species: K_{wood} = 1 for beech, K_{wood} = 1.1 for oak and K_{wood} = 0.9 for linden

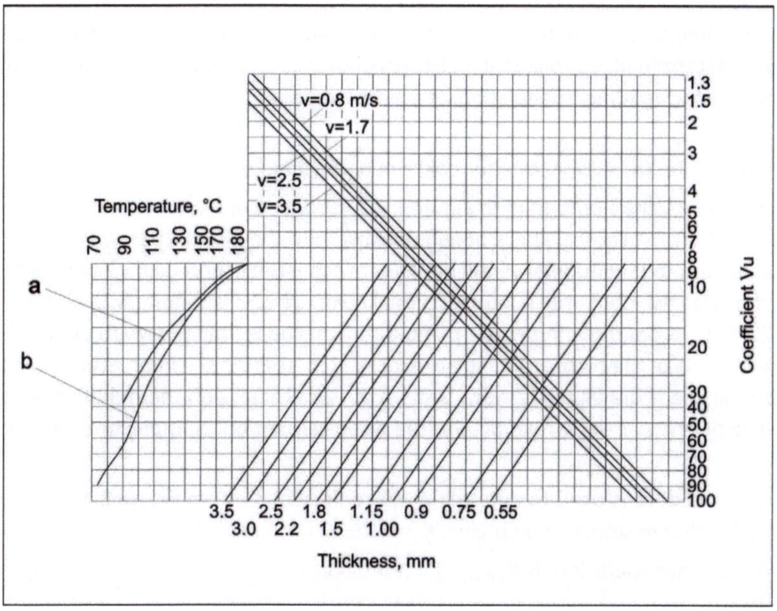

Figure 8.3 Nomogram for determining the coefficient V_u (v = air velocity): a – cross ventilation; b – longitudinal ventilation (based on [2])

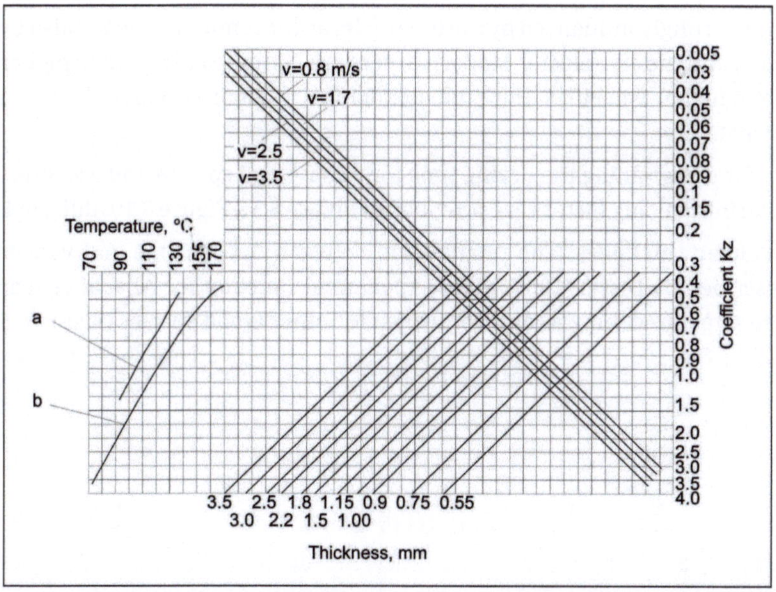

Figure 8.4 Nomogram for determining the coefficient K_z (v = air velocity): a – cross ventilation; b – longitudinal ventilation (based on [2])

The drying time for 2.2 mm thick oak veneers with an initial moisture content U_i = 60 % is to be determined, up to a final moisture content U_f = 10 %. Drying is carried out in a dryer with longitudinal ventilation, at a temperature of 130 °C, with an air velocity of 3 m/s.

One knows U_i, U_f, U_{FS} = 22 – 24 % (for oak). Let us assume: U_{FS} = 22 %.

To determine V_u, the nomogram in Figure 8.3is used. The entry point is at the temperature (130 °C). A vertical line is drawn downward to the intersection with curve *b* (corresponding to longitudinal ventilation). From the intersection point, a horizontal line is drawn to the right until it intersects the line of the corresponding veneer thickness (2.2 mm). From the intersection point, a vertical line is drawn upward until it intersects the line of the corresponding air velocity (3 m/s). From the intersection point, a horizontal line is drawn to the right, and on the vertical axis on the right the value of the coefficient V_u can be read off. For the given calculation example, one obtains: $V_u = 7$.

A similar procedure is used to determine the coefficient K_z from Figure 8.4. For the given calculation example, one obtains: $K_z = 0.26$.

The wood species coefficient is $K_{wood} = 1.1$ (for oak).

Thus:

$$D = \left(\frac{60 - 22}{7} + \frac{2.3}{0.26} \cdot \ln \frac{22}{10}\right) \cdot 1.1 = 13.6 \text{ [minutes]} \tag{8.2}$$

The drying time is strongly influenced by the type of dryer, the temperature, but also by the air circulation and the air velocity. Modern dryers are equipped to generate higher temperatures and higher velocities. Here the nomograms in Figure 8.3 and Figure 8.4 may be insufficient.

Further graphs for determining the drying time for other wood species and as a function of additional influencing factors are provided in Figure 8.5 – Figure 8.10 indicated.

Influence of veneer thickness. The relationship between drying time and veneer thickness is described by Fessel (1964) as an exponential function for Abachi veneer (Figure 8.5). The same function was also confirmed for other wood species (Figure 8.6 – Figure 8.8).

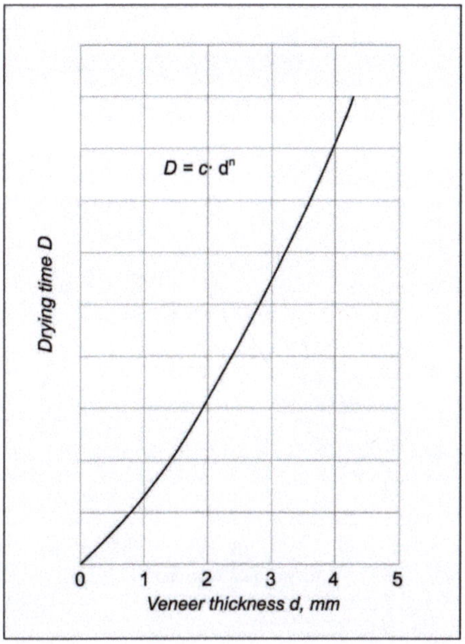

Figure 8.5 Dependence of drying time on veneer thickness in the drying of Abachi peeled veneers: *c* – coefficient; *n* – exponent of the parabolic function (adapted from [8])

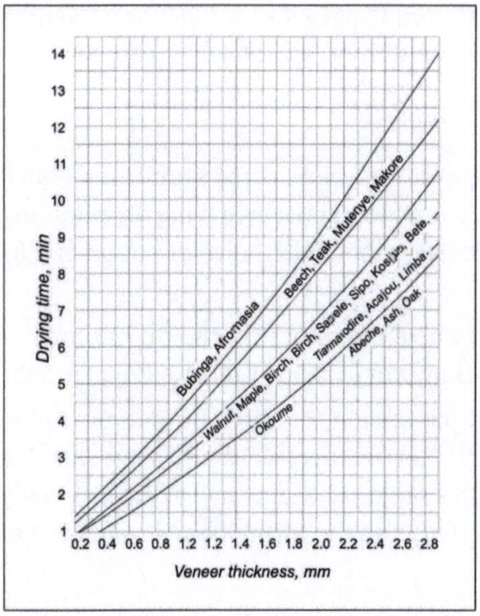

Figure 8.6 Dependence of drying time of various fancy veneers on veneer thickness in a jet-nozzle belt dryer by Cremona (adapted from [2])

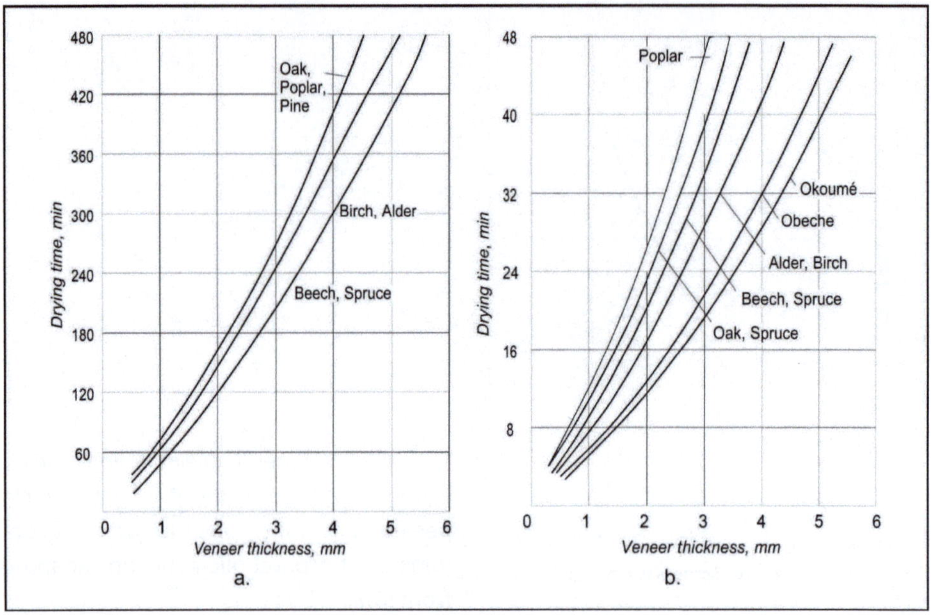

Figure 8.7 Drying time of veneers of various wood species as a function of veneer thickness in a roller dryer (a) and a belt dryer (b) by Büttner-Schilde-Haas (adapted from [8])

Influence of initial and final moisture content. Experiments by Marinescu (1980) with beech veneers led to a linear dependency function between drying time and initial moisture content for various veneer thicknesses and final moisture contents (Figure 8.8).

Influence of temperature. The influence of air temperature on the drying time of veneers is described according to Fessel (1964) by a hyperbolic function (Figure 8.9).

From an economic point of view, the temperature should therefore be chosen as high as possible in order to reduce drying time. Higher temperatures also have positive effects on the equilibrium moisture content (the veneers become more dimensionally stable) [10], on the surface roughness [11], and also on the formaldehyde emission of the plywood produced from them [12].

Excessively high temperatures, on the other hand, negatively affect the mechanical properties of the wood [13, 14]. The veneers become more brittle with increasing temperature, and thus the risk of cracking during transport and handling after drying increases. Ozsahin et al. (2019) found that the optimal drying temperature for Scots pine veneers (*Scots pine*) and alder veneers (*Alnus glutinosa*) is 170 °C and 190 °C, respectively, in order to achieve the best results regarding shear strength, bending strength, and modulus of elasticity of the plywood panels produced from these veneers [15].

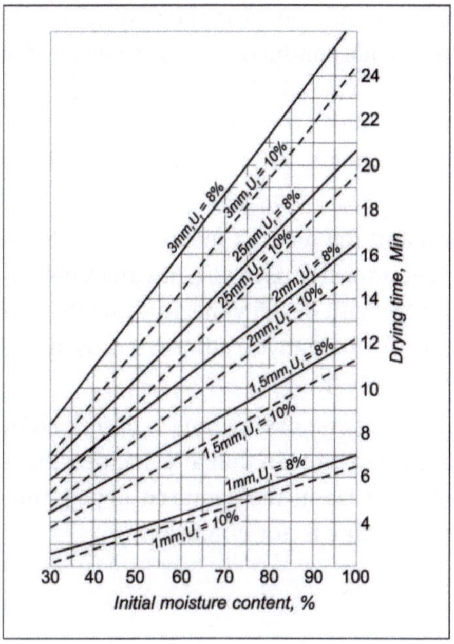

Figure 8.8 Dependence of drying time on initial moisture content and final moisture content U_f for beech veneers (source: [2], with permission)

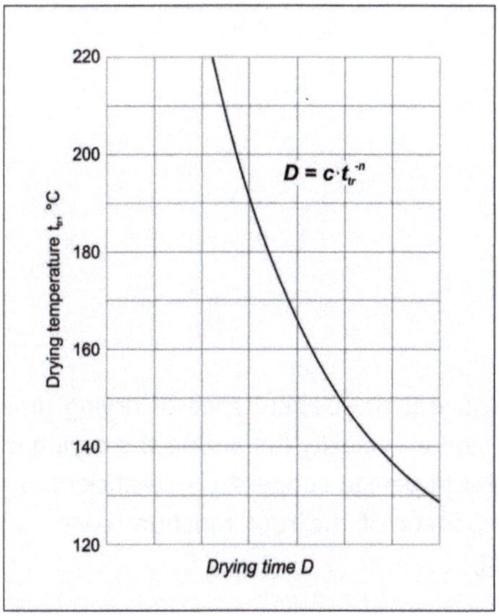

Figure 8.9 Dependence of drying time on drying temperature in the drying of Abachi peeled veneers: c – coefficient; n – exponent of the hyperbolic function (adapted from [8])

Ozturk (2023) determined, using ANN (artificial neural networks), the optimal values of air temperature, final moisture content, and thermal conductivity of the veneer that lead to the best mechanical properties of the plywood panels. For poplar veneers, the best bond line strength (1.95 N/mm^2 when drying veneers that were dried at 128 °C to a final moisture content of 5.4 %, and for spruce veneers, 110 °C and 4.6 % were the optimal drying conditions [16].

In practice, the thermal resistance of the wood and the heating medium used determine the limits of this influencing variable [4]. The adhesive used also has an influence. Thus, tests conducted by Ozsahin and Aydin (2014) with beech veneers show that the optimal drying temperature is 169 °C when urea-formaldehyde adhesive is used and 125 °C when phenol-formaldehyde adhesive is used [17].

Influence of air velocity. The influence of air velocity on drying time is described by Fessel (1964) using a root function (Figure 8.10). It can be seen that at velocities above approximately 15 m/s, only a relatively small decrease in the required drying time occurs, so that increasing the velocity beyond this value is not practical.

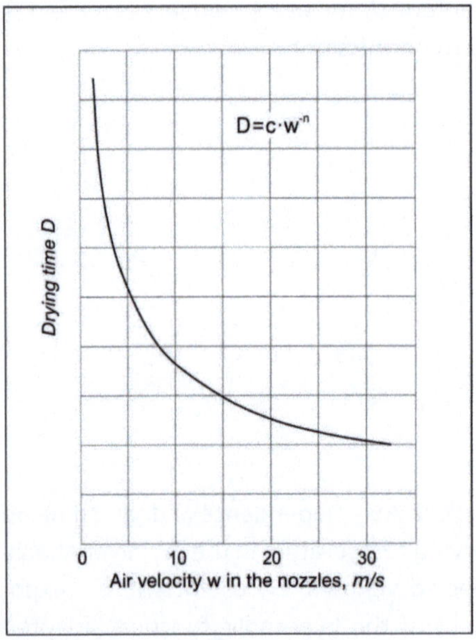

Figure 8.10 Dependence of drying time (*D*) on air velocity (*w*) during the drying of Abachi peeled veneers: *c* – coefficient; *n* – exponent of the root function (based on [8])

Modeling. Convective veneer drying is a complex process influenced by many factors. The literature includes numerous contributions to process modeling. These physics-based models often yielded accurate results, but also required detailed thermodynamic inputs that are not accessible and/or applicable under industrial conditions.

Pang et al. (1997) modified their existing 2D model for wood drying to create a model for simulating the mechanisms of drying Radiata pine (*Pinus insignis*) veneers in a jet dryer [18].

Hanhijärvi et al. (2003) started from Fick's law of diffusion across the veneer thickness and used experimentally obtained values of the diffusion coefficient. This approach enabled the development of a model [19] for predicting the drying rate of veneer, which, however, has limited applicability (only for calculations corresponding to the experimental conditions).

Qiu and Cool (2023) conducted a five-month study on an industrial veneer dryer, collecting and analyzing data. The aim was to gain insights into drying quality in order to develop an improved, data-driven quality control system. Logistic regression classifiers and a random forest classifier were set up for prediction [7]. The results showed that more accurate sorting of veneers based on initial moisture content could reduce the number of re-drying operations required.

8.3 Drying Methods and Equipment

As in the case of lumber drying, the methods of veneer drying are classified according to the type of heat transfer. The following options are available:

- convection drying with hot air – here, the climatic and flow conditions in the boundary layer are of particular importance;
- contact drying – here, the temperature of the heating plates plays the decisive role;
- vacuum drying – the reduction in pressure leads to an increase in drying rate compared to hot air drying;
- drying with infrared radiation – heat transfer by infrared radiation allows rapid heating of a thin surface layer (without penetrating into the interior of the veneer); due to high energy costs, the process is rarely used;
- drying in high-frequency alternating fields – this process allows the selective drying of so-called moisture pockets by preferential heating of the wetter spots/zones. The process is applicable to veneers of various thicknesses. For cost reasons (energy costs, high technological effort), the process is only rarely used.

8.3.1 Convection Drying

The principle. Convection drying with hot air is the most commonly used process and proceeds in four stages (Figure 8.11):

- the heating phase – this is relatively short. During this time, the temperature inside the veneers rises. It must reach the wet-bulb temperature; otherwise, moisture condenses onto the veneer and drying does not occur. The time required depends on the wood density and the veneer thickness. Once the wet-bulb temperature is reached inside the veneers (approximately 80 °C), the drying period begins;
- the first drying period – is characterized by the evaporation of capillary water. In this section, the hot air velocity in the gas–wood boundary layer is extremely important, as it influences both the heat transfer to the veneer surface and the removal of the free water evaporated from the veneer. At the beginning of this drying period, there is a risk of case hardening, especially with very wet and thick veneers. Thin veneers tend to deform [5]. Both risks can be reduced by appropriate drying conditions;
- the second drying period – is characterized by the migration of the evaporation front into deeper wood layers and the rise of the veneer temperature above the wet-bulb temperature. So-called internal evaporation of free water occurs. The veneer moisture content decreases after a distinct first inflection point. Due to the high temperatures compared to lumber drying and the increased permeability of the veneers, mass transport now occurs as vapor;
- the third drying period – occurs after the fiber saturation point is reached. It is characterized by a steeply rising temperature inside the veneers. After a weakly pronounced second inflection point in the drying rate curve, evaporation of bound water begins. The drying rate decreases. Drying is terminated when the veneer moisture content reaches approximately 6 %. Due to the varying final moisture content, it is advisable to connect a conditioning zone and then also a cooling zone to the dryer.

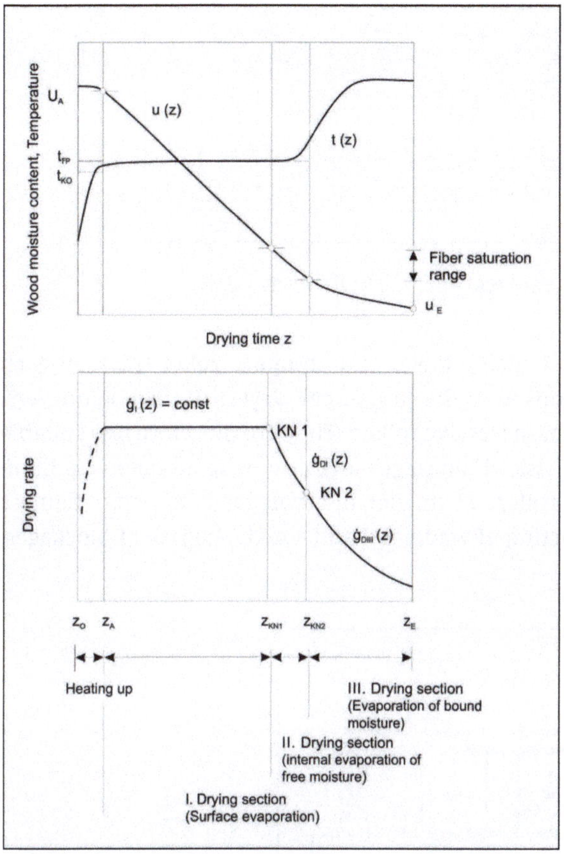

Figure 8.11 Schematic profile of the temperature inside the veneer, the wood moisture content, and the drying rate during veneer drying (based on [20, 21])

Construction types. Convective veneer dryers are mostly metallic continuous-flow systems with one, two, or three decks. Depending on how the veneers are transported from the so-called "green" end to the "dry" end of the dryer, a distinction is made between:

- Belt dryer (Figure 8.12) – with two endless belts per deck. The lower belt ensures the transport of the veneer and the upper belt ensures the pressing of the veneers and thus the reduction of waviness. The belts are made of coarse-mesh steel wire mesh to allow air to pass through. Belt dryers can be used for any veneer thickness, but are mainly used for long and thin veneers.

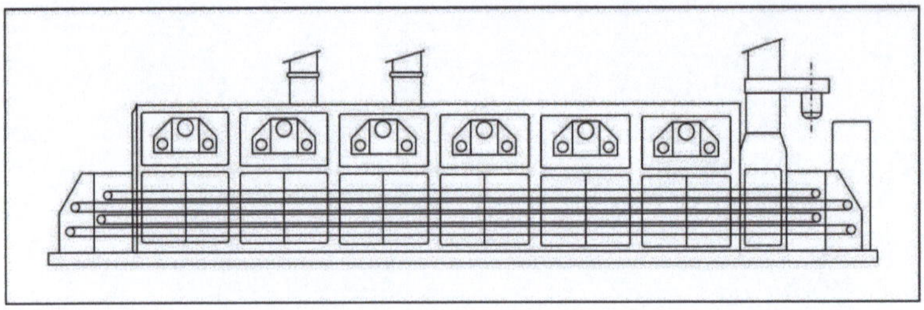

Figure 8.12 Belt dryer for veneers (source: [2], with permission)

- Roller dryer (Figure 8.13) – in which a lower and an upper roller track on each deck perform the same functions as in the case of belt dryers. Unlike belt dryers, roller dryers have gaps between successive rollers. They are therefore less suitable for thin veneers, as there is a risk of blockages and also of spontaneous ignition of the veneers. Mounting the rollers at smaller intervals leads to better flatness retention of the veneers (reduction of waviness) and avoids the risk of blockages, but it impedes air circulation.

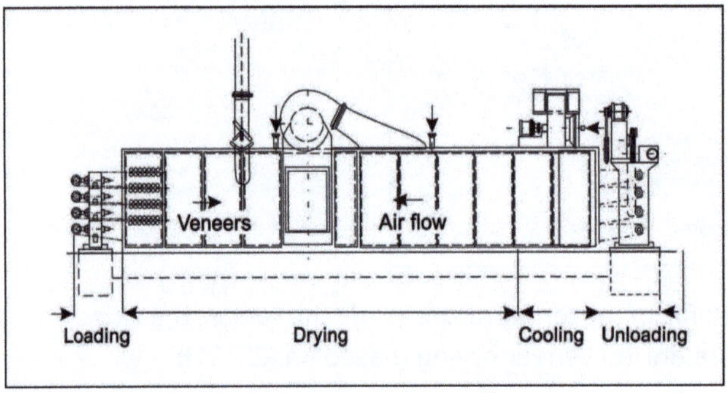

Figure 8.13 Roller dryer for veneers (based on [8])

- Drum dryer (Figure 8.14) – in which the veneers are guided over several successive suction nozzle drums so that they alternate their bending direction and are thus kept flat. The upper and lower heating registers ensure the heating of the air. This drying method is suitable for thin sliced veneer with a thickness of 0.5 – 0.8 mm. When drying thicker veneers, the energy consumption for operating the fans that generate the negative pressure increases significantly.

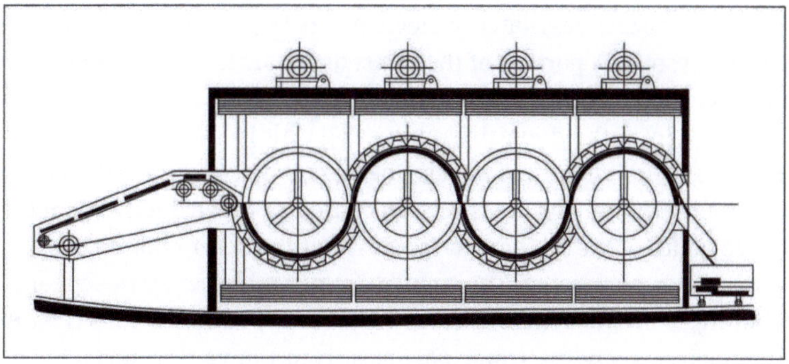

Figure 8.14 Drum dryer for veneers (based on [5])

Depending on the direction of the airflow in relation to the veneers (Figure 8.14), there are systems:

- with longitudinal ventilation (Figure 8.15 a), ensured by a centrifugal fan that moves the air in counterflow to the direction of the veneers;
- with transverse ventilation (Figure 8.15 b), ensured by individually operated axial fans;
- with perpendicular nozzle ventilation (Figure 8.15 c), in which the air jet is directed perpendicular to the veneer surface.

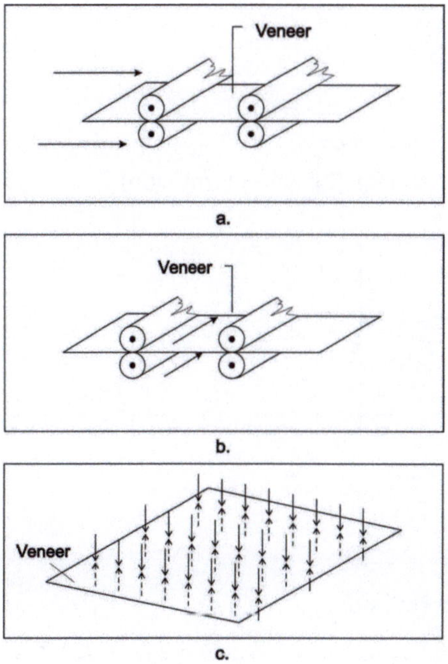

Figure 8.15 Classification of veneer dryers according to the type of ventilation: a – longitudinal ventilation; b – transverse ventilation; c – nozzle ventilation (based on: [4])

The operating principle of nozzle-ventilated veneer dryers (Figure 8.16) is based on the fresh air/exhaust air system. A portion of the moist dryer air is discharged via exhaust air ducts and replaced by fresh air. The recirculated air mixed with fresh air is drawn in by the fans, heated in the downstream heating unit located in the airflow, and supplied under pressure to the nozzle boxes. The nozzle box (Figure 8.17) has an inlet cross-section of 200 × 180 mm. The cross-section decreases along the length of the box (equal to the active width of the system) and is closed at the other end. On the side facing the veneer, each box is perforated. The small diameter of these nozzles accelerates the air, which impinges on the veneer surface at a velocity of 15 – 20 m/s (Figure 8.18). In doing so, they carry away the water particles from the veneer surface. The velocity profiles for roller dryers with and without nozzles are shown in comparison in Figure 8.19.

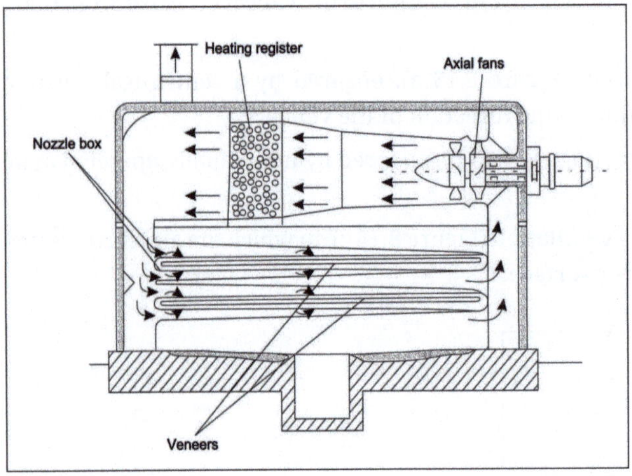

Figure 8.16 Veneer dryer with jet ventilation (source: [2], with permission)

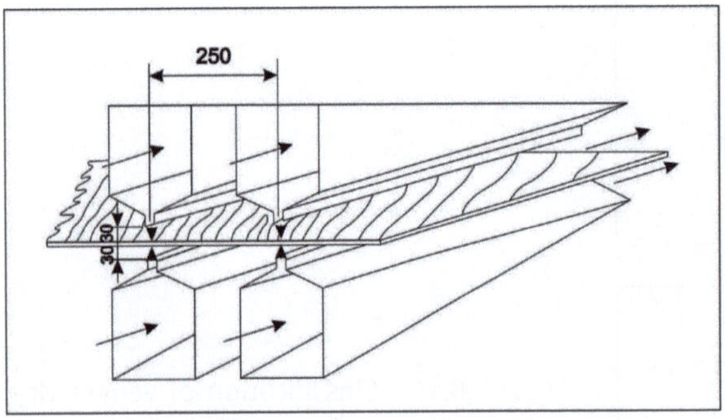

Figure 8.17 Nozzle boxes (source: [2], with permission)

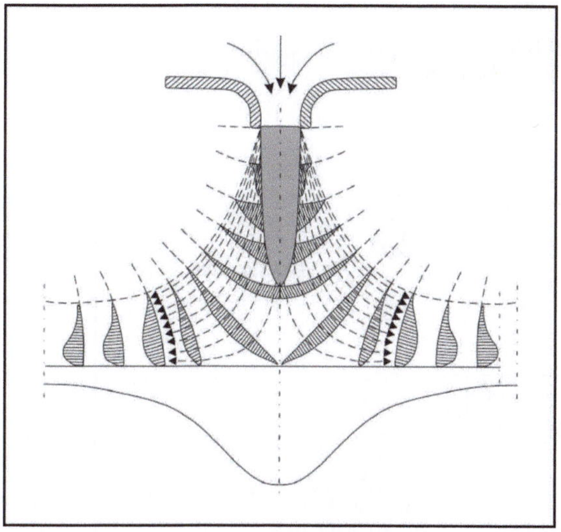

Figure 8.18 Perpendicular air jet in jet ventilation (adapted from [22])

The perpendicular jet ventilation not only results in a reduction of drying time, but also produces very uniformly dried veneers. It is particularly recommended for delicate veneers with a thickness of 0.2 – 2.2 mm.

In systems without nozzles, the drying temperature is limited. It generally varies between 80 – 120 °C at the green end of the channel and a maximum of 110 – 140 °C at the dry end, with the lower limits applying to belt dryers and the upper limits to roller dryers. The drying time for roller dryers is 7 – 28 minutes and is 25 – 100 % longer for belt dryers.

The introduction of the jet drying concept has made it possible to reduce the drying time of veneers to as little as 1 – 3 minutes. This significant reduction is attributable both to the turbulent airflow at higher velocity and to the possibility of applying higher temperatures (also above 200 °C).

For a typical single-deck belt dryer (length 8 – 12 modules), an average production output of 2000 to 2400 m^2/h (~ 1.5 m^3/h at a veneer thickness of 0.7 mm) is achieved with manual loading. The amount of water evaporated varies between 180 – 325 kg/h. The specific heat consumption is 4800 – 9400 kJ/kg of evaporated water [4].

The dryer efficiency is derived from the ratio of the total specific heat demand (required for heating the veneers and for evaporating and removing the moisture) to the total specific heat consumption (the amount of heat supplied to the veneers via the heating registers) and is in the range of 0.485 – 0.540 [23].

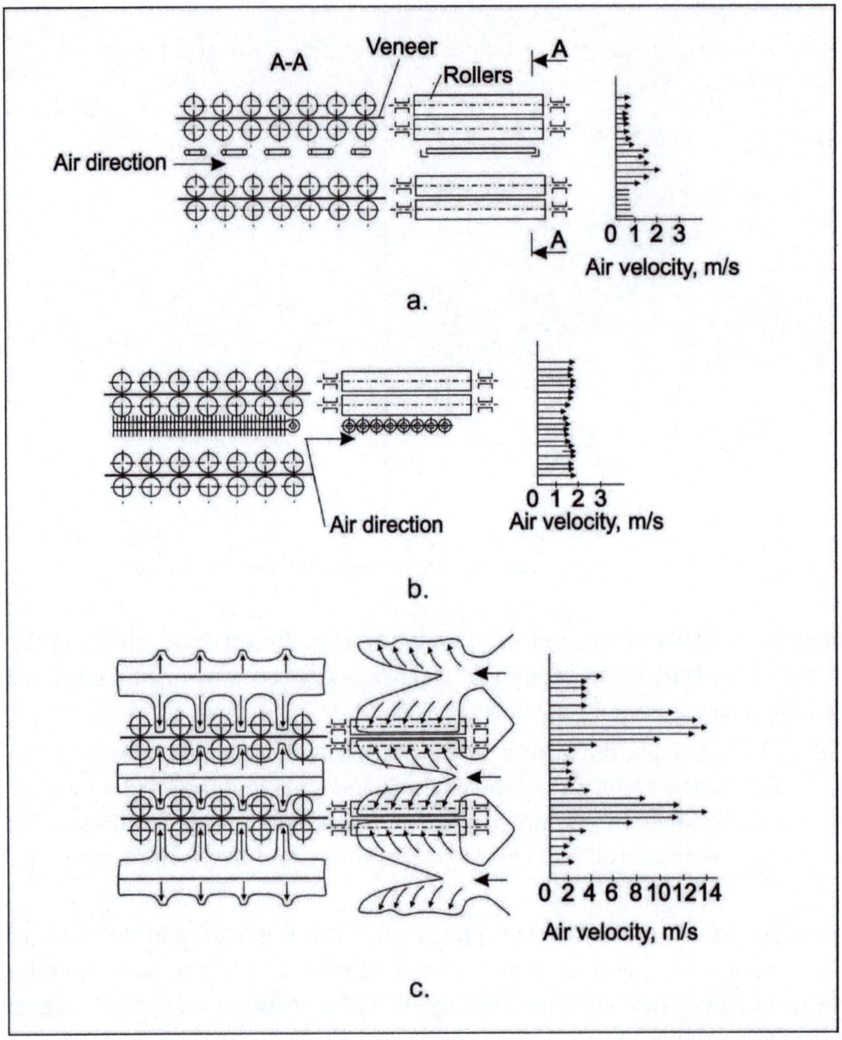

Figure 8.19 Velocity profiles in a veneer roller dryer with a – heating register in serpentine form; b – finned heating tubes; c – jet air guidance (source: [2], with permission)

Energy saving is possible through the use of heat exchangers. In the case of a plate heat exchanger (Figure 8.20), the required heating energy for warming up to the recirculated air temperature is 14 – 17 % lower than without a heat exchanger [4].

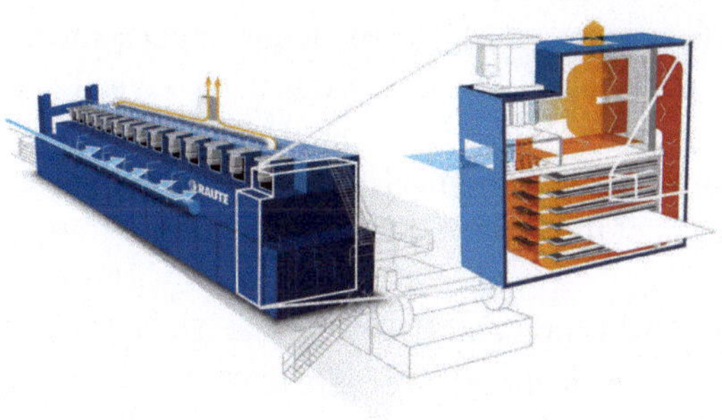

Figure 8.20 Use of a plate heat exchanger for energy saving in a veneer dryer (source: © Raute Corporation, with permission)

8.3.2 Contact Drying

In contact drying, the veneers are inserted sheet by sheet between the heated metal platens of a press. The temperature of the platens varies between 110 and 145 °C and must be correlated with the sensitivity of the wood species in order to avoid local scorching [24]. The pressing time (required for heating the wood and evaporating the water) is 6 – 10 minutes. The water removed from the veneer is converted into superheated steam, which escapes when the press is opened.

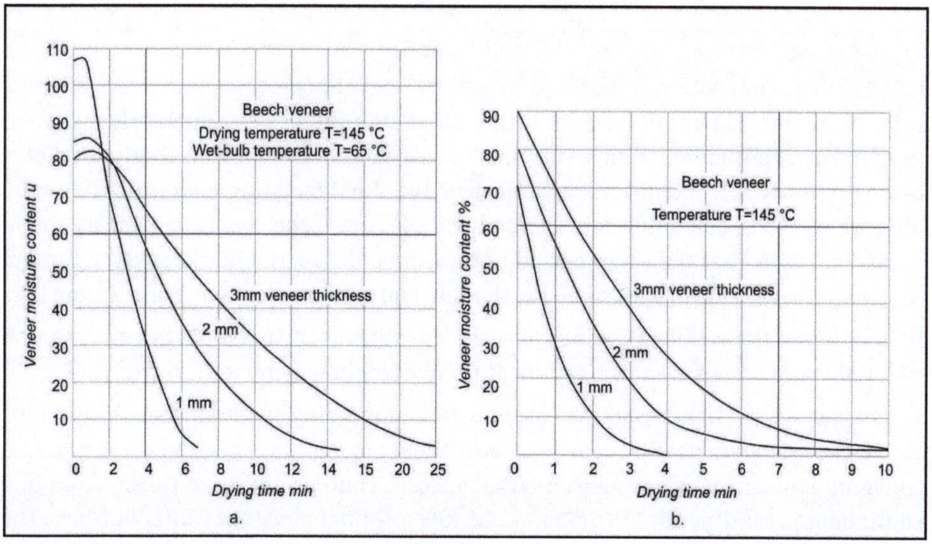

Figure 8.21 Drying times of beech veneers of various thicknesses in a - convective drying; b - contact drying (adapted from [25])

The drying time can be significantly reduced by direct contact between the veneer and the heating surface (conductive heat transfer). The reduction effect is much more pronounced than by increasing the air velocity or temperature in convective drying (Figure 8.21). Particularly at the beginning of drying, the rapid heat transfer increases the pressure of the water evaporated inside the veneer. As the steam expands, it mechanically forces the liquid water out of the veneer (Table 8.2).

Table 8.2 Comparison of Drying Times for Contact Drying and Convection Drying (without Nozzle Ventilation) of 1.5 mm Thick Plywood Veneers [26]

Method	Temperature [°C]	Pressure [bar]	Initial moisture [%]	Final moisture [%]	Drying time [s]
Contact	120	1	77	38	40
Convection			62	38	150
Contact	120	1	77	8	250
Convection			62	8	350
Contact	160	1	73	33	20
Convection			80	33	105
Contact	160	1	73	5	90
Convection			80	5	180
Contact	200	1	84	28	20
Convection			75	28	80
Contact	200	1	84	5	80
Convection			75	5	150

Lutz (1974) found that even shorter drying times can be achieved when steam at atmospheric pressure (1 bar) and temperatures above 100 °C is also introduced during press drying. His experiments with 1 mm thick red oak veneers showed that with steam press drying at 121.1 °C, the veneer reached 10% final moisture content in two minutes, whereas in a conventional (convective) roller dryer the veneer moisture content was still at 50% after the same period of time. In addition, the final moisture content in steam press drying proved to be much more stable than in simple press drying [27].

In addition to the reduced drying time, the advantages include energy consumption reduced by up to half [4] and a significant reduction in waviness during drying.

Steinhagen et al. (1999) used this process as a smoothing method after drying [28]. *Eucalyptus globulus*sliced veneer with a thickness of 0.5 mm was first pre-dried in a conventional convective system to 30% moisture content and then finish-dried in a continuous press dryer at 115.5 °C and 16.5 kPa to a final moisture content of 10%. The result was reduced waviness and a much better quality (grade A instead of grade C, which was achieved with conventional drying without the press process).

The main disadvantage of contact drying is the low productivity and the risk of veneer adhesion to the press plates. To prevent sticking, it is necessary to spray the plates with water or other release agents before the veneer is fed in.

8.3.3 Vacuum Drying

In order to shorten the duration of veneer drying and improve quality, special drying processes have also been tested.

For example, experiments with a combined vacuum-contact drying process were conducted at the University of Helsinki [26, 29]. The drying system consists of a heated plate and a cold plate, between which the pressure is reduced below atmospheric level. The veneer is placed on the cold plate via a layer of porous material and a wire mesh (Figure 8.22). The temperature difference between the two surfaces of the veneer and the press effect lead to an accelerated movement of water in the veneer downward. The water is absorbed by the porous material and condenses upon contact with the cold plate.

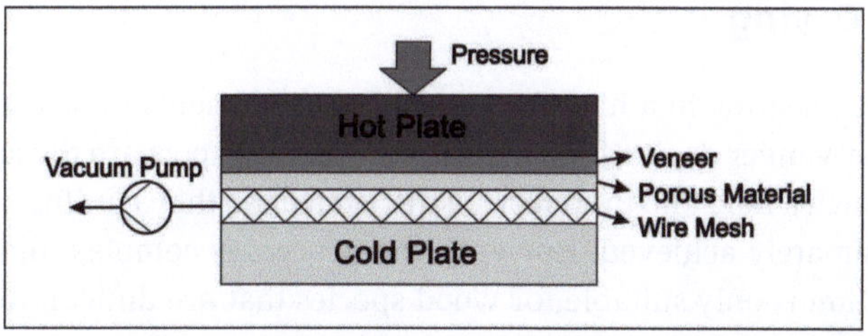

Figure 8.22 Combined contact-vacuum drying of veneers (adapted from: [26])

The experiments were carried out at temperatures of 120 °C, 160 °C and 200 °C and a negative pressure of 0.25 bar. The shortest time achieved at a temperature of 200 °C, which enabled a moisture reduction of 1.5 mm thick veneers from 30% to 5%, was 70 s (Figure 8.23). Although the process offers good application prospects for the plywood sector, it has not yet found industrial adoption.

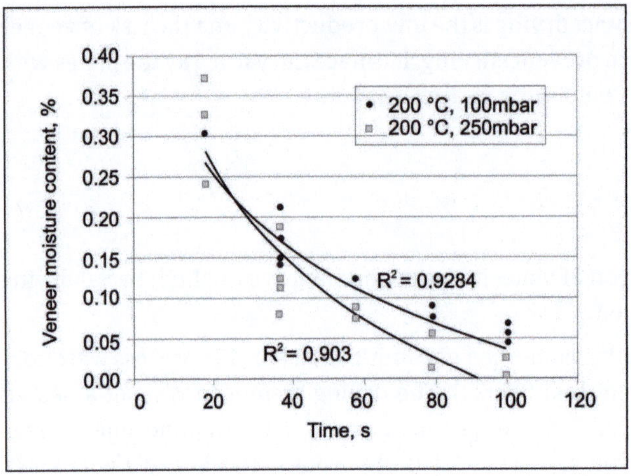

Figure 8.23 Drying time in combined contact-vacuum drying of plywood veneers (adapted from [26])

8.3.4 High-Frequency Drying

From a physical point of view, heating in a high-frequency field represents the ideal method of heat generation for veneer drying [30], because the wettest spots are those that heat up most rapidly in such a field and therefore also dry faster, so that a uniform final moisture content is ultimately achieved. However, the process is complex and energy-intensive. Its application is only suitable for wood species that are difficult to dry convectively and for wood species with wet pockets (e.g. western hemlock).

Studies have shown that microwaves, compared to radio waves, enable a higher heating intensity thanks to the higher frequency (see Section 6.7). They are therefore particularly suitable for equalizing the final moisture content in cellulosic materials, including veneers [31].

Good results regarding the uniformity of drying were achieved with the hot-air microwave dryer from Cryodry Inc. (Figure 8.24). In this system, the veneers are transported via rollers. Two 25 kW magnetrons generate microwaves at a frequency of 915 MHz. Hot air jets positioned above the veneer ensure the evaporation of moisture. To achieve different final moisture content levels in the veneer, the system is equipped with a feed speed control system.

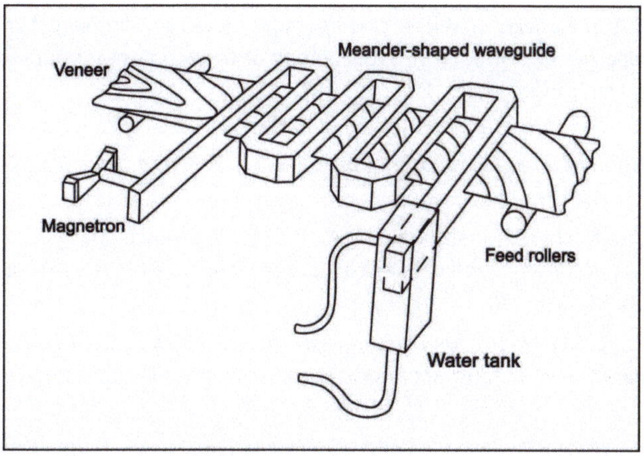

Figure 8.24 Meandering waveguides in the microwaveveneer dryer from Cryodry Inc. (adapted from [31])

References

[1] Lutz, J. F. (1978): *Wood Veneer: Log Selection, Cutting, and Drying*. In: Technical Bulletin Nr. 1577, U.S. Department of Agriculture, Washington, DC, USA

[2] Marinescu, I. (1980): *Uscarea și tratarea termică a lemnului. Vol. 2*. Editura Tehnică, Bucharest, Romania

[3] Wagenführ, A.; Buchelt, B.; Kairi, M.; Weber, A. (2023): *Veneers and Veneer-Based Materials*. Chapter 26 in *Springer Handbook of Wood Science and Technology*, Springer Nature Switzerland, p. 1365

[4] Frühwald, A.; Ressel, J. (1982): *Holztrocknung. Teil 3. Furniertrocknung*. Universität Hamburg, Germany

[5] Janik, W. (1965): *Handbuch der Holztrocknung*. 2nd ed, VEB Fachbuchverlag, Leipzig, Germany

[6] Kaislaaja, J.; Antikainen, T.; Sihvonen, P. (2025): *Veneer Drying Process and Benefits*. Raute Corporation, Finnland. Online: *https://www.raute.com/knowledge/veneer-drying-process-and-benefits/*, retrieved 05. 08. 2025

[7] Qiu, Q.; Cool, J. (2023): *Transitioning to Data-Driven Quality Control in Industrial Veneer Drying: A Case Study*. In: European Journal of Wood and Wood Products, 81, p. 1033 – 1044

[8] Fessel, F. (1964): *Furnier-Durchlauftrocknung mit Düsenbelüftung – Untersuchungen und Versuchsergebnisse*. In: European Journal of Wood and Wood Products, 22(4), p. 129 – 139

[9] Kollmann, F. F. P.; Kuenzi, E. W.; Stamm, A. J. (1975): *Principles of Wood Science and Technology. Vol. II. Wood Based Materials*. Springer Verlag, Berlin/Heidelberg/New York

[10] Aydin, I. (2014): *Effects of Veneer Drying at High Temperature and Chemical Treatments on Equilibrium Moisture Content of Plywood. Maderas*. In: Ciencia y Tecnología, 16(4), p. 445 – 452

[11] Aydin, I.; Colakoglu, G. (2002): *The Effects of Veneer Drying Temperature on Wettability, Surface Roughness and Some Properties of Plywood*. In: Proceedings of the 6th Panel Products Symposium, 9. – 11. Oktober, Llandudno, Wales, UK, p. 60 – 70

[12] Bekhta, P.; Sedliačik, J.; Bekhta, N. (2020): *Effect of Veneer-Drying Temperature on Selected Properties and Formaldehyde Emission of Birch Plywood*. In: Polymers, 12, 593. *https://doi.org/10.3390/polym12030593*

[13] Jamalirad, L.; Doosthoseini, K.; Koch, G.; Mirshokraie, S. A. (2011): *Physical and Mechanical Properties of Plywood Manufactured from Treated Red-Heart Beech (Fagus orientalis L.) Wood Veneers*. In: BioResources, 6, p. 3973 – 3986

[14] Nazerian, M.; Ghalehno, M. D. (2011): *Physical and Mechanical Properties of Laminated Veneer Lumber Manufactured by Poplar Veneer*. In: Journal of Agricultural Science and Technology, 1, p. 1040 – 1045

[15] Ozsahin, S.; Demir, A.; Aydin, I. (2019): *Optimization of Veneer Drying Temperature for the Best Mechanical Properties of Plywood via Artificial Neural Network*. In: Journal of Anatolian Environmental and Animal Sciences, 4(4), p. 589 – 597. *https://doi.org/10.35229/jaes.635302*

[16] Ozturk, H. (2023): *Prediction of Optimum Veneer Drying Parameters with Artificial Neural Networks for Production of Plywood with High Mechanical Properties*. In: Drvna Industrija, 74(3), p. 297 – 308. *https://doi.org/10.5552/drvind.2023.0074*

[17] Ozsahin, S.; Aydin, I. (2014): *Prediction of the Optimum Veneer Drying Temperature for Good Bonding in Plywood Manufacturing by Means of Artificial Neural Network*. In: Wood Science and Technology, 48(1), p. 59 – 70. *https://doi.org/10.1007/s00226-013-0583-2*

[18] Pang, S.; Riley, S. G.; Haslett, A. N. (1997): *Simulation of Pinus Radiata Veneer Drying: Moisture Content and Temperature Profiles*. In: Forest Products Journal, 47(7–8), p. 51 – 58

[19] Hanhijärvi, A.; Hukka, T.; Paajanen, P.; Pulkkinen, S.; Sundman, A. (2003): *Experimental Investigation of Jet Drying of Birch and Spruce Veneers and Modelling with a Simplified Approach*. In: Holz als Roh- und Werkstoff, 61, p. 83 – 88. *https://doi.org/10.1007/s00107-003-0379-4*

[20] Keylwerth, R. (1952): *Der Verlauf der Holztemperatur während der Furnier- und Schnittholztrocknung*. In: Holz als Roh- und Werkstoff, 10, p. 87–91

[21] Krischer, O. (1978): *Trocknungstechnik. Die wissenschaftlichen Grundlagen der Trocknungstechnik*. 3rd ed., Springer Verlag, Berlin/Heidelberg/New York

[22] Kröll, K. (1959): *Trockner und Trocknungsverfahren. Band 2*. Springer Verlag, Berlin/Heidelberg/New York

[23] Ressel, J. B. (1981): *Vergleichende Energieverbrauchsmessungen an Furniertrocknern*. Diploma thesis, Universität Hamburg, Germany

[24] Trabalon, P. L. (2020): *Experimental Study on the Drying Kinetics of Wood Veneer in a Contact Drying*. Master thesis, Aalto University, Finland

[25] Keylwerth, R. (1953): *Furnier-Trocknungsversuche*. In: Holz als Roh- und Werkstoff, 11, p. 11 – 17

[26] Holmberg, H.; Lahti, P.; Paajanen, O.; Ahtila, P. (2009): *An Experimental Study on Drying Times in a Contact Drying of Veneer*. In: Proceedings of the 8th World Congress of Chemical Engineering, 23. – 27. August 2009, Montreal, Canada

[27] Lutz, J. F. (1974): *Drying Veneer to a Controlled Final Moisture Content by Hot Pressing and Steaming*. USDA Forest Service, Forest Products Laboratory, Madison, Wisconsin

[28] Steinhagen, H. P.; Rozas, C.; Khattabi, A.; Loehnertz, S. P.; Poblete, M. J. (1999): *Flattening Wavy Eucalyptus Veneer During Finish-Drying*. In: Forest Products Journal, 49(1), p. 63 – 66. Online: *https://www.fpl.fs.usda.gov/documnts/pdf1999/stein99a.pdf*, retrieved March 2025

[29] Lahti, P.; Paajanen, O.; Holmberg, H.; Kairi, M. (2010): *Introducing a New Method of Veneer Drying*. In: Proceedings of the 11th International IUFRO Wood Drying Conference, 18.–22. Januar 2010, Skellefteå, Sweden

[30] Gefahrt, J. (1967): *Die Verwendung der Hochfrequenzenergie in der Holzindustrie*. In: Holz als Roh- und Werkstoff, 25, p. 125 – 129

[31] Resch, H. (2009): *Drying Wood with High-Frequency Electric Current*. Society of Wood Science and Technology

9 Steaming of Wood and Wood Products

9.1 Objectives and Effects

During steaming – also known as **hygrothermal heat treatment** – the wood is exposed to the action of a saturated steam atmosphere. This can involve saturated steam at atmospheric pressure (1 bar) and a temperature of 99 – 100 °C up to a maximum of 110 °C, or low-pressure steam (0.1 – 0.7 bar) and a temperature of 45 – 90 °C [1]. Steaming can also take place at overpressure (1.6 – 1.8 bar) and temperatures of 120 – 160 °C; however, autoclaves are required for this process.

The saturated state of the steam is very important in this heat treatment, so that during steaming only heating of the wood occurs and no drying takes place. An unfavorable combination of temperature and pressure can result in an unsaturated steam atmosphere. This then acts like a drying agent, causing the wood moisture content to decrease. The physical parameters of saturated steam are given in Table 9.1.

The higher the temperature, pressure, and duration of steaming, the more degradation processes occur [3], through the unavoidable partial hydrolysis of the pentosans (hemicellulose), but also of the lignin and cellulose, albeit to a lesser extent (Figure 9.1).

Studies by Geffert et al. (2020) with birchwood showed that steaming at 105 °C and 125 °C barely affects the chemical composition of the wood. Only at 135 °C was degradation detected, specifically in the polysaccharides by 24.2% compared to the initial state [4]. With oakwood [5], it was also only the polysaccharides that were degraded by steaming. The reduction amounted to 10% at a steaming temperature of 130 °C and 36% at 140 °C.

Table 9.1 Physical Parameters of Saturated Steam as a Function of Pressure [2]

Pressure [atm]	Saturation temperature [°C]	Specific volume water [m³/kg]	Specific volume steam [m³/kg]	Enthalpy [kJ/kg]		Heat of vaporization [kJ/kg]
				Water	Steam	
0.10	45.5	0.001010	14.95	189.9954	2581.528	2391.574
0.15	53.6	0.001014	10.21	224.0114	2596.172	2372.328
0.20	59.7	0.001017	7.795	249.4082	2607.05	2357.684
0.25	64.6	0.001020	6.322	257.2742	2615.418	2345.55
0.30	68.7	0.001022	5.328	287.0642	2204.131	2335.509
0.40	75.4	0.001026	4.069	315.3062	2215.428	2318.354
0.50	80.9	0.001030	3.301	338.109	2642.614	2304.547
0.60	86.5	0.001033	2.783	357.3554	2650.146	2292.832
0.70	89.5	0.001036	2.409	374.1751	2656.422	2282.372
0.80	93.0	0.001038	2.125	389.0702	2661,861	2272.749
0.90	96.2	0.001041	1.904	406.643	2666.882	2264.381
1.00	99.1	0.001043	1.725	414.6344	2671.484	2252.666
2.00	119.6	0.001060	0.9016	501.6616	2714.579	2200.366
3.00	132.9	0.001073	0.6166	558.1456	2720.855	2162.710
4.00	142.9	0.001083	0.4706	600.8224	2733.826	2133.003
5	151.1	0.001092	0.3816	636.3864	2743.867	2107.481

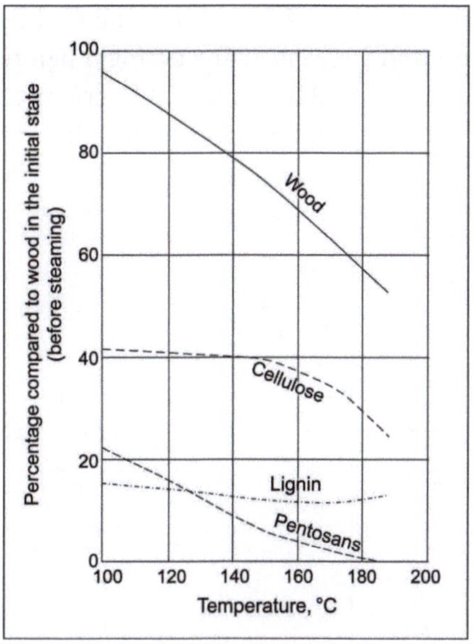

Figure 9.1 Change in the chemical constituents of trembling aspenwood, as a function of temperature (adapted from [3])

As a result of this chemical decomposition, the wood density decreases and thus so do the mechanical properties. The pH value of the wood decreases (Figure 9.2) and the wood color becomes darker. At the same time, the plasticity of the wood increases. This is a positive effect of steaming, because it allows wood to be bent without cracks.

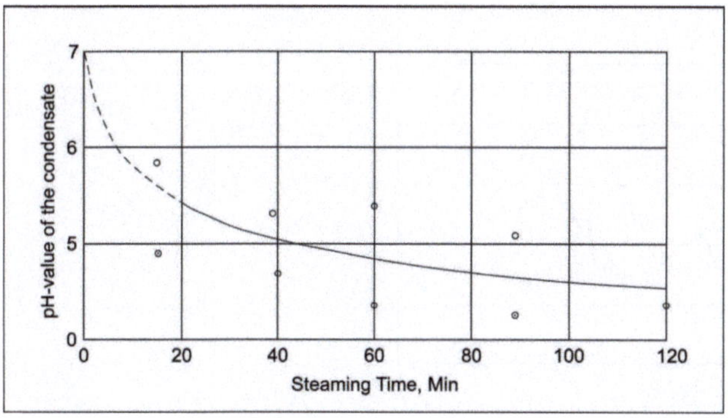

Figure 9.2 Influence of steaming time on the pH value of the condensate after steaming of beechwood (adapted from [6])

Measurements using FTIR spectroscopy have shown that steaming of various wood species [7, 8] at 80 °C causes only minor changes (Figure 9.3 – Figure 9.5), while steaming at 120 °C causes clearly visible changes in the IR spectra (Figure 9.4, Figure 9.5). In black locust, a slight decrease in absorption of guaiacyl lignin at 1512 cm^{-1} can be observed (Figure 9.4). In spruce (Figure 9.5), the decrease in absorption is much more pronounced and can be noticed at 1507 cm^{-1}, whereby the characteristic band for this chemical component extends between 1505 cm^{-1} and 1510 cm^{-1}. The syringyl lignin, which absorbs at approximately 1600 cm^{-1}, showed no visible changes. A decrease at 1175 cm^{-1} indicated the cleavage of the ether bond in cellulose and hemicelluloses already at a steaming temperature of 80 °C [8].

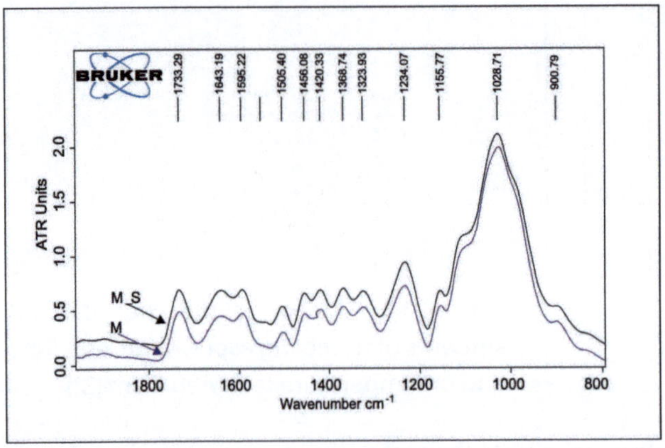

Figure 9.3 FTIR spectroscopy of a beechsample: M – unsteamed; M_S – lightly steamed at 80 °C/12 h (source: [7], with permission)

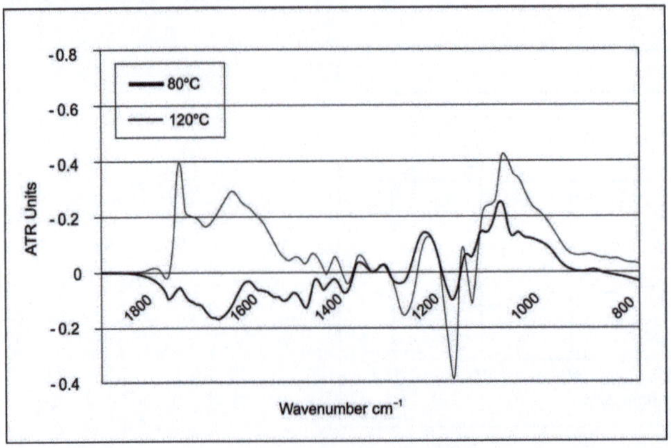

Figure 9.4 Differences in the FTIR spectra of black locust, steamed at 80 °C and 120 °C (source: [8], with permission)

The color of wood is caused by chromophores found in lignin and extractive substances [9]. The chemical reactions that take place in wood during steaming (e.g., extraction of water-soluble substances, hydrolysis of hemicelluloses, depolymerization of polysaccharides, etc.) alter the chromophoric system of wood and lead to color changes (color darkening).

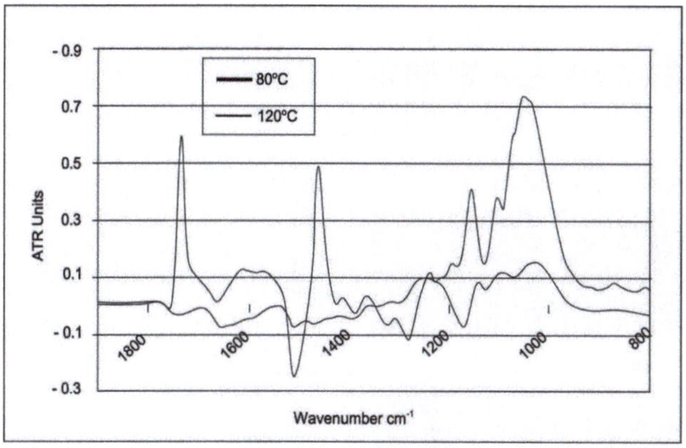

Figure 9.5 Differences in the FTIR spectra of spruce, steamed at 80 °C and 120 °C (source: [8], with permission)

Due to the high temperature, steaming not only changes the color of wood, its formability, and its strength, but the wood is also sterilized. In addition, dimensional stability improves. Swelling and shrinkage are significantly reduced by steaming. However, this only applies to pressure steaming in autoclaves. Studies by Volkmer et al. (2003) showed that after steaming oak wood (*Quercus robur* L.) at 1.7 – 4.0 bar and a temperature of 115 – 143 °C, its radial swelling was reduced by 17% and tangential swelling by 10% [10].

Steaming is used industrially, e.g., for sawn timber to change or homogenize color. Steaming is also used in veneer production (sliced and peeled veneer) to plasticize veneer logs and prisms. Most wood species require prior steaming. Only a few (mostly with low density) can be processed without steaming.

9.2 Steaming of Sawn Timber

9.2.1 Objective

The main objective of steaming in the sawn timber industry is color equalization in boards that contain both sapwood and heartwood. The slightly darker color compensates for the color differences. The most well-known application is the reduction of color differences in red-hearted beechwood. But ash, oak, black locust, cherry, and other wood species are also steamed for this purpose (Figure 9.6).

For some wood species, steaming is used to imitate the color tone of exotic wood species. For example, pear or cherry wood takes on a dark red hue through steaming, similar to mahogany wood, while steaming oak and walnut wood results in a dark chocolate tone.

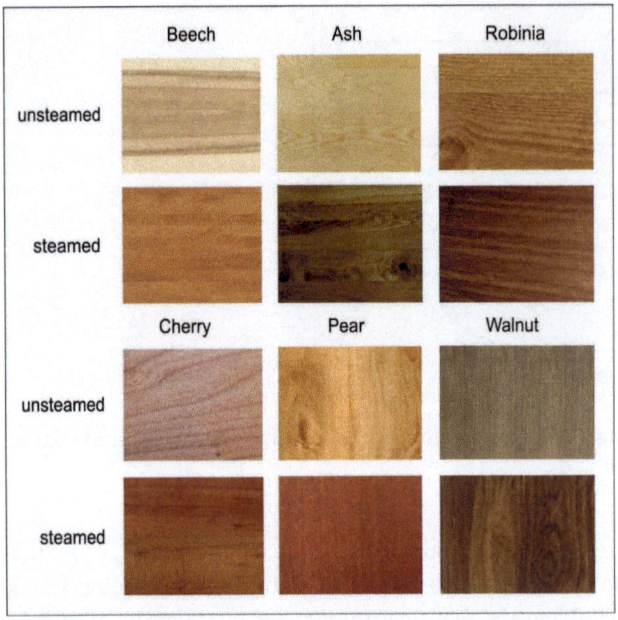

Figure 9.6 Color change of various wood species through steaming

Dzurenda and Babiak (2024) investigated the color differences in beech with and without red heart in the unsteamed state and after steaming at three different temperature/time combinations (95 °C/24 h, 105 °C/18 h, 120 °C/9 h) (Figure 9.7 and Table 9.2). The study showed that already at the second combination (105 °C/18 h) the color difference was substantially equalized (ΔE^* = 0.4) [11].

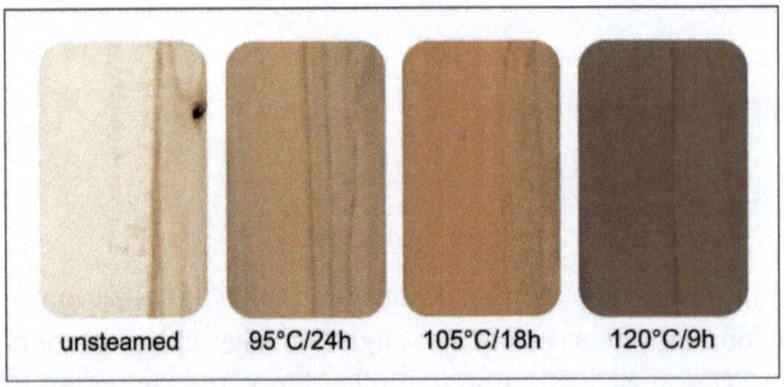

Figure 9.7 Color change of beechwood through steaming at various temperatures (based on [11])

Table 9.2 Homogenization of the Color of Beech Wood with Red Heartwood as a Result of Steaming [11]

Wood		CIE-$L^*a^*b^*$color coordinates			Color difference between sapwood and red heart ΔE^*
		L^*	a^*	b^*	
Unsteamed	Sapwood	78.5 ± 2.5	9.2 ± 1.6	19.5 ± 1.8	14.8
	Red heartwood	63.8 ± 3.6	11.5 ± 1.8	19.9 ± 1.6	
Steamed at 95 °C/24 h	Sapwood	71.0 ± 2.1	12.1 ± 1.2	19.3 ± 1.5	5.5
	Red heartwood	65.5 ± 2.8	12.5 ± 1.6	19.5 ± 1.4	
Steamed at 105 °C/18 h	Sapwood	61.5 ± 1.9	12.5 ± 1.2	19.4 ± 1.4	0.4
	Red heartwood	61.2 ± 2.3	12.3 ± 1.4	19.6 ± 1.4	
Steamed at 120 °C/9 h	Sapwood	56.0 ± 1.8	12.3 ± 1.1	19.5 ± 1.2	0.3
	Red heartwood	55.8 ± 2.1	12.4 ± 1.3	19.7 ± 1.4	

Color measurements on robinia wood [8] also showed that the lightness (L^*) decreases due to steaming and the degree of redness (a^*) increases. A difference compared to other wood species appears in the b^*coordinate. Robinia wood is one of the few wood species whose natural color is yellowish (due to its high robinetin content). The rapid degradation of this substance during steaming leads to a negative Δb (the b^*color coordinate decreases with increasing steaming temperature and duration – see Figure 9.8), while it increases or remains approximately constant in most wood species.

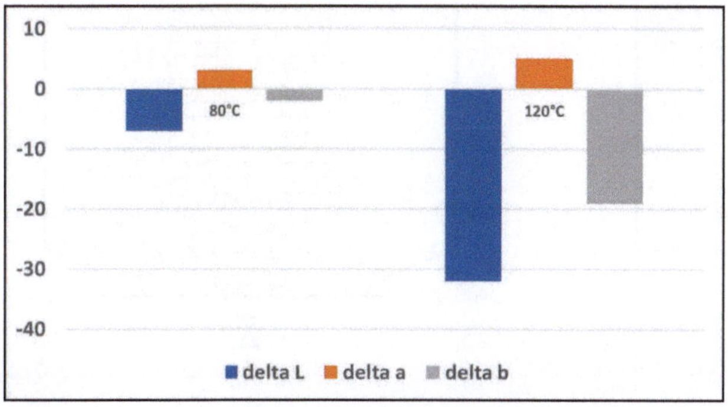

Figure 9.8 Changes in color coordinates during steaming of robinia wood (based on [8])

Under the direct influence of sunlight, steamed wood discolors over time (it becomes lighter), but only on the surface [12, 13, 14]. Robinia is the only wood species that shows better color stability after steaming [13, 15].

In addition to influencing wood color, steaming also has a positive effect on drying by equalizing the initial moisture content of the wood and opening the pores. This helps to shorten the drying time and improve drying quality. For this reason, with hard hardwood species, even when no color change is desired, a light steaming (for 2 – 5 hours) can be carried out prior to drying.

9.2.2 Processes and Equipment

The steaming of wood can be carried out by two different methods, namely:

- direct steaming (Figure 9.9 a) – in which saturated wet steam is introduced directly into the steaming chamber through spray pipes. In this case, the steam can come from a dedicated generator (fresh steam) or from power machines (degreased exhaust steam). Wood defects caused by local overheating are unavoidable and thermal efficiency is unfavorable, since no condensate recovery is possible;
- indirect steaming (Figure 9.9 b) – in which the necessary steam is generated within the steaming chamber by heating a water bath to evaporation temperature using a heating coil (Figure 9.10) located at the bottom of the steaming chamber. This method results in more uniform heating of the wood and is more economical, since the condensate can be recovered. However, the steaming time with the indirect method is longer.

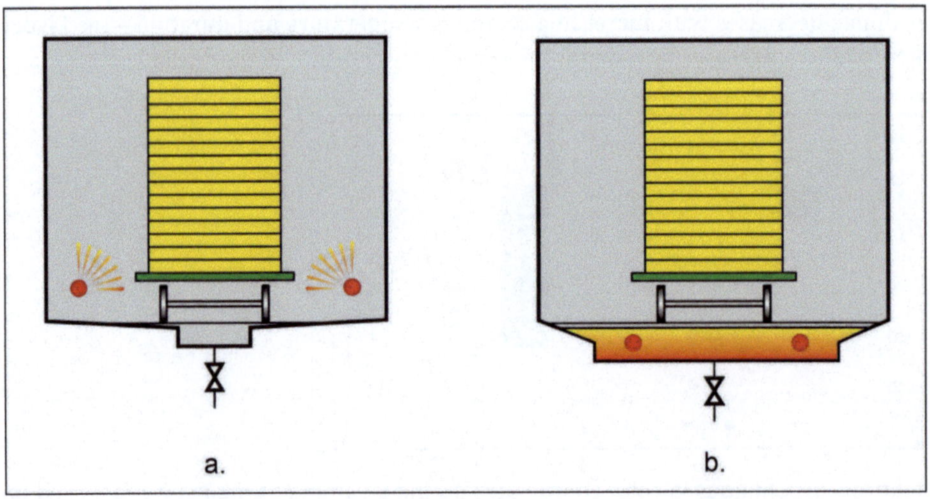

Figure 9.9 Steaming methods: a – direct steaming; b – indirect steaming

Figure 9.10 Heating coil in indirect steaming

The advantages of indirect steaming are:

- the complete recovery of the condensate heated to a temperature of 80 – 90 °C, which results in an immediate reduction of the required energy consumption;
- preheating of the water in the steam generator is not necessary (cost savings);
- the possibility of using any heating medium (steam, heated water under pressure, oil) in the heating coil without contaminating the wood. The heating medium does not come into contact with the wood. The process is slow, so that the wood is heated gradually and uniformly (better steaming quality)

Thanks to the simpler construction of the steaming chamber, the shorter duration, and the more precise control, direct steaming is, however, more frequently used.

The steaming of sawn timber can also be carried out under pressure (p = 1 – 2 bar). In this case, the steaming time can be significantly reduced [16]. Studies by Riehl et al. (2002) showed that the maximum color change of robinia wood during pressure steaming at a temperature of 130 °C can already be achieved after 10 hours, whereas steaming at atmospheric pressure and a temperature of 98 °C requires eight days [17].

In the case of oak wood as well, heat-pressure steaming is advantageous. The color change (darkening) is comparable to that obtained by fuming with ammonia, whereby heat-pressure steaming is a more environmentally friendly alternative for achieving these dark shades in oak wood [10].

Equipment. In the past, steaming chambers were built from masonry; today, as with drying, all-metal chambers (Figure 9.11) or autoclaves (Figure 9.12) made of stainless steel are mostly used. In the construction of all-metal chambers, prefabricated elements with a sandwich structure (mineral wool, polystyrene, or polyurethane as insulation between two aluminum sheets) are used, similar to drying chambers. Optionally, the inner panel can be made of stainless steel. This solution is recommended for direct steaming, because in this case the inner walls of the chamber are exposed to alkaline steam, against which aluminum is less resistant than stainless steel. It is also recommended to leave a certain air gap between the panels, which increases the thermal insulation capacity and at the same time allows the insulating material to compensate for changes in insulation thickness during temperature fluctuations.

In order to maintain a saturated steam atmosphere with the lowest possible energy consumption, the structure of the walls, floor, and ceiling must have a thermal transmittance coefficient $k < 1.163$ W/m^2. Particular attention must also be paid to the sealing at the transition area between the walls and the ceiling. In contrast to drying chambers, steaming chambers have no fans and no air ducts (Figure 9.11). Steaming installations should not be built too large, as otherwise the wood will not be steamed uniformly.

Figure 9.11 Metal chamber for steaming sawn timber

For pressure steaming of sawn timber, autoclaves are used (Figure 9.12, Figure 9.13). Stainless steel should be used to prevent corrosion.

Figure 9.12 Autoclave for pressure steaming of beech wood in Burgas (Bulgaria)

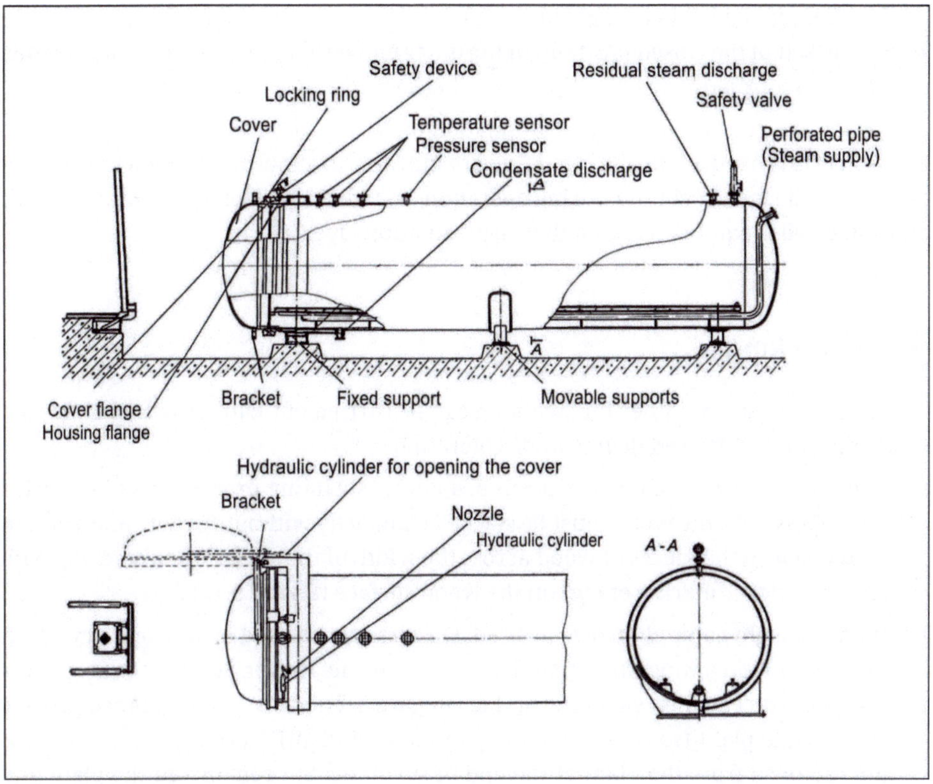

Figure 9.13 Overall view of an autoclave for pressure steaming of sawn timber (source: [16])

The autoclave is a horizontal cylindrical metal vessel with an inner diameter of approximately 2.4 m and a length of approximately 9.0 m. It is fitted with an elliptical door that is hinged to the cylindrical part (shell) of the autoclave by means of a bracket, allowing it to rotate about a vertical axis when closing and opening.

A special "bayonet lock" prevents the unintended opening of the door when steam pressure inside the autoclave is elevated. The bayonet lock consists of a bayonet ring and two flanges with interlocking teeth and working surfaces. The door flange is pressed against the housing flange, and the autoclave is sealed by means of a dovetail rubber ring gasket on the face of the housing flange. For this purpose, compressed air or steam at a pressure higher than the working pressure in the autoclave is fed through a nozzle into the chamber of this lock. By means of two diametrically arranged hydraulic cylinders, the bayonet ring can be rotated by a certain angle about its own axis into the respective end positions [18, 19]. In one of the end positions, the teeth of the door flange are located between the teeth of the housing flange and are tightened by the teeth of the bayonet ring. The face of the door flange is then pressed against the sealing rubber ring and the autoclave is pressurized. In this position, the door is locked in the closed position. In the other end position, the teeth of the bayonet ring are offset from the teeth of the door flange and the door can be opened.

The autoclave rests on a concrete foundation on three supports (Figure 9.13). One of the supports is fixed, while the other two are movable to allow unimpeded thermal expansion of the metal cylinder during operation. To reduce heat losses during steaming, the autoclave is equipped with a thermal insulation layer.

9.2.3 Stacking

The stacking of sawn timber for steaming can be carried out with or without stickers, depending on the desired degree of discoloration:

- If an intensive color change is desired, a longer steaming process must be carried out. In this case, the wood must be stacked compactly without stickers and without gaps between the pieces of wood across the width of the stack; when stacking with stickers, sticker marks remain on the wood surface (Figure 9.14).
- If only a slight color change is desired, stickers can be used as in wood drying. In this case, the steaming time should be 2 – 5 hours, no longer. Stacking with stickers is also practiced when steaming and drying are carried out in a combined process in the same plant (so-called steaming and drying plant). In this case, the wood is not removed from the plant at the end of steaming, but remains in the plant and the drying phase is started directly afterwards. This shortens the duration of the process by eliminating the cooling phase after steaming and the heating phase at the beginning of drying. However, sticker marks are unavoidable in this case.

Figure 9.14 Sticker marks in deep steaming of non-compactly stacked beech sawn timber

To avoid cracks, especially in thicker wood, it is advisable to nail stickers over the end grain surfaces or to drive in S-hooks. This absorbs the stresses from the end grain area and thus prevents crack formation.

9.2.4 Steaming Parameters and Process Phases

Initial moisture content of the wood. A basic prerequisite for the success of steaming is that the wood moisture content is as high as possible (green condition): 50 – 60 % in the heartwood zone and 60 – 100 % in the sapwood. To ensure this requirement, it is necessary that steaming in summer takes place no later than 24 hours after sawing the logs into sawn timber, and within 48 hours during the rest of the year. If this is not possible, the wood should be sprayed before steaming so that at least the wood moisture content does not fall below the fiber saturation point.

Temperature and time. In light steaming, the duration of the process is only 2 – 5 hours and the usual temperature is 65 – 80 °C. The wood is only lightly colored pink.

In intensive steaming, characterized by a pronounced color change, the temperature in the steaming chamber must be kept constant at 99 – 100 °C and the relative humidity must be 100 % throughout the entire process.

The three phases of intensive steaming are (Figure 9.15):

- heating of the chamber to the set temperature (99 – 100 °C) and partial warming of the wood; the duration of this phase is 4 – 6 hours and requires an average consumption of 30 kg of steam/m^3 h;
- the actual steaming phase with a duration of at least 24 hours, depending on the desired intensity of the color change, during which the temperature is maintained at 99 – 100 °C, with an average consumption of 5 – 7 kg of steam/m^3 h;
- the cooling period with a duration of four hours, during which the doors remain closed and the steam supply is stopped. The purpose of this phase is to avoid the sudden transition of the wood temperature from 99 – 100 °C to the temperature of the external environment, which would lead to cracks.

The total duration of intensive steaming of beech wood is 38 – 48 hours, depending on the season but also as a function of the thickness of the wood.

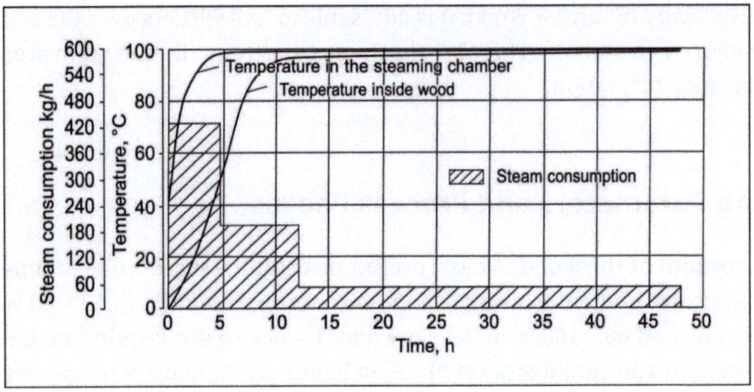

Figure 9.15 Steaming diagram for direct steaming (based on [20])

The thermal energy requirement during the heating period is 80 – 100 MJ/h m^3 of wood. In addition, 10 – 40 MJ/h m^3 of wood is required for the actual steaming [21].

Pressure steaming has the same three phases (Figure 9.16):

- in the heating phase the temperature is raised to 120 °C and the pressure to almost 2 bar. The duration of this phase is approximately 10 hours for oak wood;
- the actual steaming phase takes place at constant temperature and pressure and lasts approximately 50 hours for oak wood;
- in the cooling phase not only the temperature but also the pressure is reduced; the duration of this phase is also approximately 10 hours, like the heating phase.

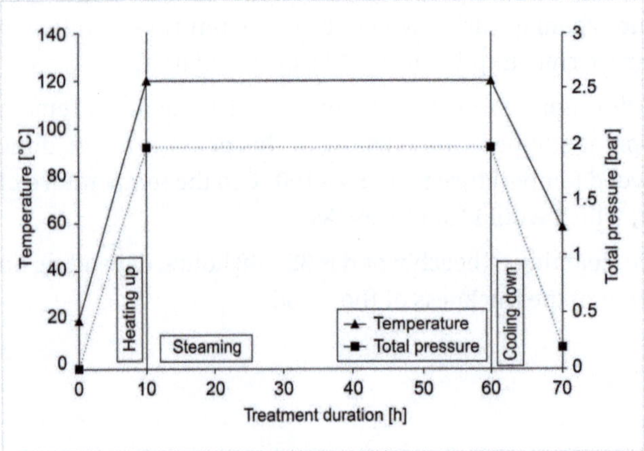

Figure 9.16 Process diagram for pressure steaming of oak wood (source: [10], with permission)

The total duration of heat treatment with saturated steam at elevated pressure compared to steaming at atmospheric pressure is given in Table 9.3.

Table 9.3 Duration of Steaming of Beech at Atmospheric and elevated Pressure to Obtain a Pinkish Color of the Wood [18]

Board thickness [mm]	Duration of the phase with steam supply [h]		Duration of the phase without steam supply [h]		Total duration [h]	
	t_{steam} = 80 °C	t_{steam} = 123 °C	atm. pressure	High pressure	atm. pressure	High pressure
20	90	6.5	10.0	7.0	100.0	13.5
30	93	7.0	10.5	7.2	103.5	14.2
40	96	7.5	11.0	7.4	107.0	14.9
50	99	8.0	11.5	7.7	110.5	15.7
60	103	8.5	12.0	8.0	115.0	16.5
70	107	9.1	12.5	8.4	119.5	17.5
80	111	9.7	13.0	8.9	124.0	18.6
90	115	10.3	13.5	9.4	128.5	19.7
100	119	11.0	14.0	10.0	133.0	21.0

After the cooling time in the system has elapsed, the steamed wood is removed from the system and must be stacked with stacking sticks within a maximum of 24 h (in April–October) or a maximum of 48 h (during the rest of the year).

9.2.5 Wood Protection Measures After Steaming

It should be noted that, due to the high temperatures, steaming does sterilize the wood (fungi and insects are killed), but does not ensure long-term wood protection. To prevent infection of the wood during storage, especially during prolonged outdoor storage before drying or during sea transport, it is recommended to carry out an antiseptic treatment immediately after steaming. For this purpose, the wood stacks are dipped in a biocidal solution (e.g., Wolsin® FL-35 – a concentrated wood preservative based on propiconazole, fenpropimorph, and boric acid, miscible with water, prepared as a solution with a concentration of 0.8–3%) or impregnated under pressure in an autoclave.

9.2.6 Quality Control and Steaming Defects

Quality control in steaming relates primarily to the color obtained:

- a color that is too light or uneven indicates insufficient steaming, either due to too low a temperature, too short a steaming time, or too low an initial moisture content of the wood;
- a color that is too dark indicates excessive steaming due to too long a steaming duration;
- uneven color can be caused by too long a storage of the wood before steaming or by the presence of decay;
- the presence of black spots on the material during direct steaming indicates either direct contact between steam and wood (there is a need to realign the nozzle pipes) or the oil content of the steam.

End-grain checks in the wood are also frequently occurring defects, especially in thicker sawn timber [22].

9.3 Steaming of Wood for Bending

9.3.1 Objective

Wood has relatively low plastic properties in the dry state. Therefore, heat treatment before bending (e.g., for chair seat elements or barrel staves, Figure 9.17) is necessary.

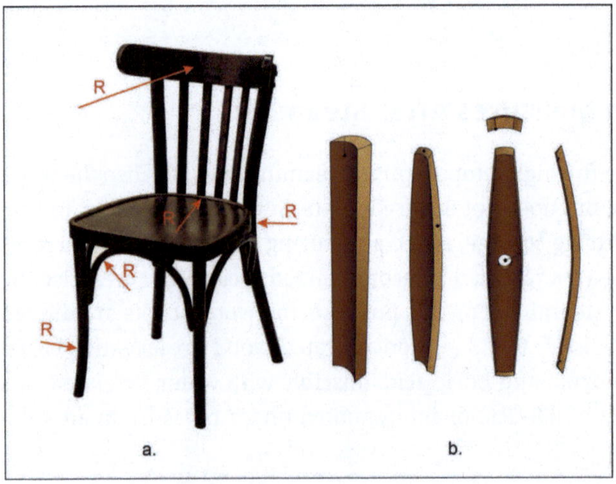

Figure 9.17 Wood products that must be steamed before bending: a – bent chair elements (bending radius *R*); b – barrel staves

The objective of steaming before bending is to increase the plasticity of the wood under the simultaneous action of heat and moisture (Figure 9.18). The wood is softened, thereby imparting greater bendability, so that bending can be performed with little force and without defects (cracks, splits, wrinkles).

When the bending radius is large, e.g., for chair legs, bending can be carried out without difficulty, but with small radii of curvature, e.g., for seat frames, steaming plays an essential role.

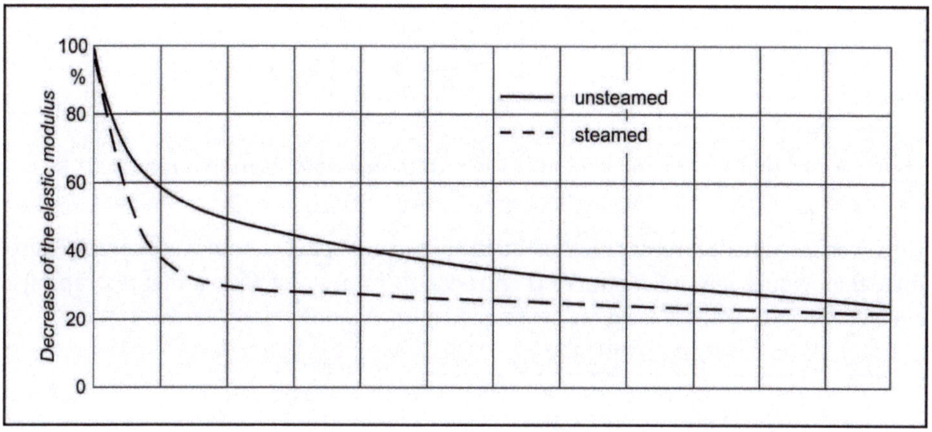

Figure 9.18 Change in stiffness of wood (beech) during steaming (source: [21], with permission)

9.3.2 Bendability of Wood and Influencing Factors

When bending wood, a complex stress distribution prevails. Compressive stresses occur in the concave region (inner side of the wood part, in contact with the template) and tensile stresses in the convex region (outer side) (Figure 9.19).

Due to the cell structure of wood, compressive loads can easily be absorbed by deformation [21]. This means that bending occurs mainly due to compressive deformations, which are significantly increased (by approximately 15 times) in heat-treated wood compared to untreated wood. The dimension of the bent piece decreases on the inside, which simultaneously leads to an increase in density in this zone.

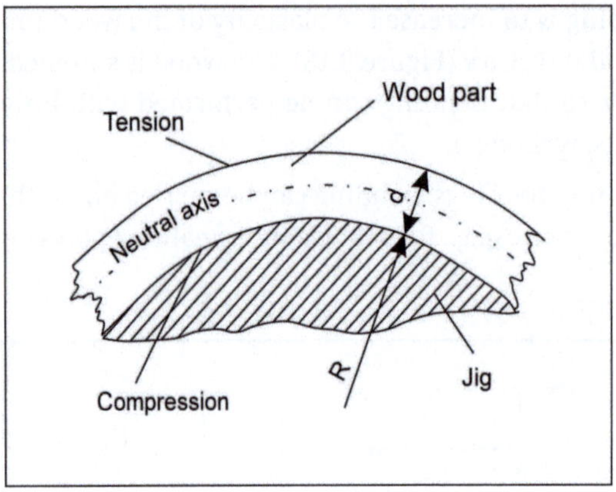

Figure 9.19 Bending of solid wood: *d* – wood thickness; *R* – bending radius

However, the tensile stresses on the outside of the bent part cannot be relieved. Therefore, they would lead to fractures (Figure 9.20). To prevent this, a metal strap (also called a tension strap) is used on the outside of the wooden part (Figure 9.21).

Figure 9.20 Fracture on the outside of the bent part in the absence of the metal strap (source: [21], with permission)

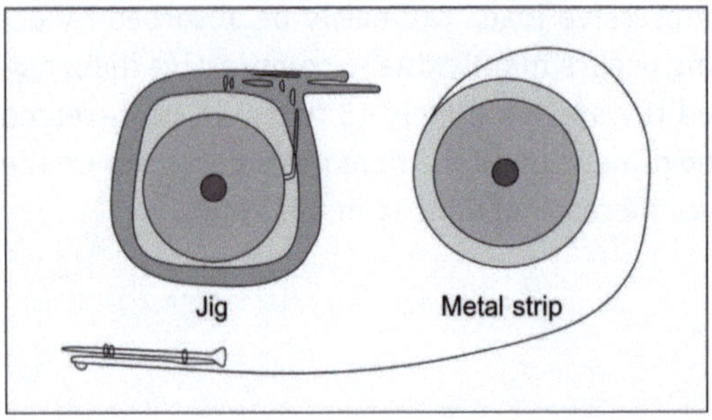

Figure 9.21 Use of a metal strap in bending (source: [20], with permission)

The use of the metal strap shifts the neutral axis outward, toward the tension-loaded zone (Figure 9.22). In doing so, the metal strap covers nearly the entire tensile load that occurs during the bending process (Figure 9.22 bottom). It is, however, very important that this strap fits the workpiece perfectly before the bending process begins. Tolerances between the strap and the workpiece lead to damage [21].

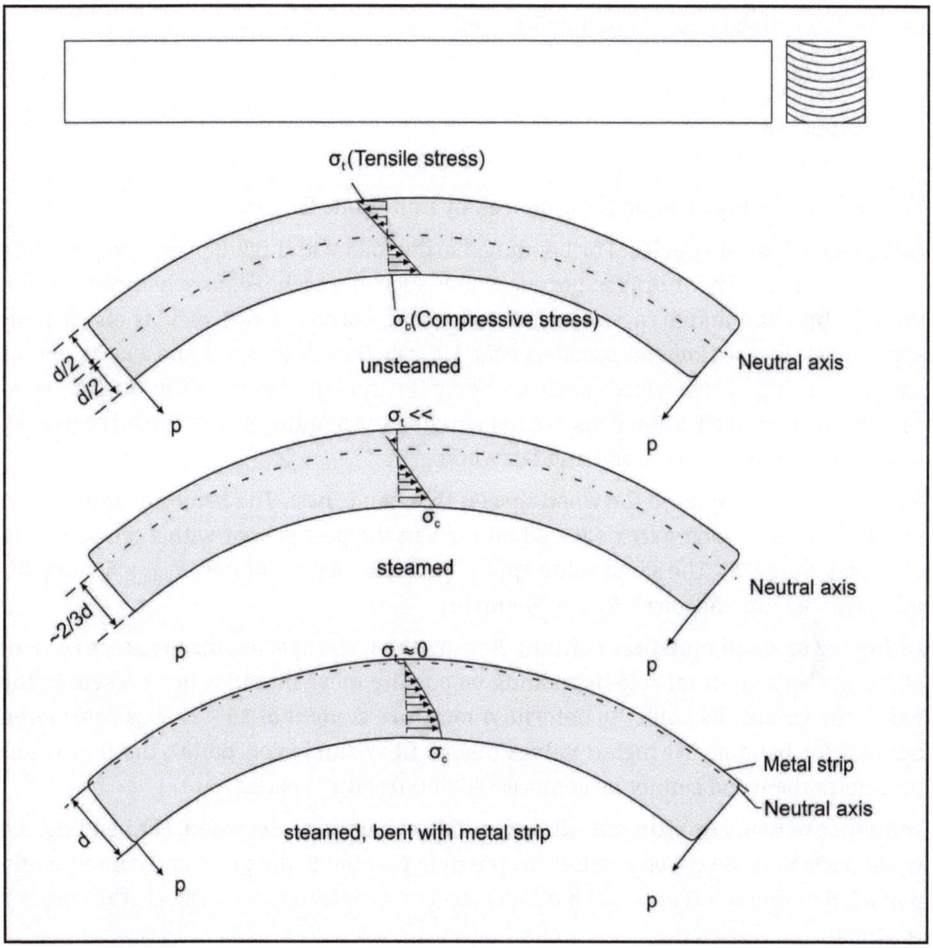

Figure 9.22 Shift of the neutral axis during steaming of wood (adapted from [1, 23]

In addition to the classic steaming process prior to bending, the so-called process for producing patent bent wood was developed [24]. In this case, the wood is first steamed at low pressure and then fixed in a device to be compressed in the fiber direction without being bent. The formability of pre-compressed solid wood is higher and can be attributed to the folding of the cell walls and the additional expansion paths thereby released [25]. Eggert (1995) distinguishes between pre-compression with and without springback. Pre-compression without elastic recovery, for example by drying and cooling in a fixed position that does not allow recovery in the longitudinal direction, leads to the highest degree of compression and the lowest bending stiffness or elasticity [21].

The bending ability of wood is characterized by the ratio *d/R*. Depending on the thickness of the wooden part, the wood species, and the treatment process prior to bending, the minimum bending radius at which the wooden part can be bent without fractures can be determined using Equation 9.1 [25]:

$$\frac{d}{R} = \frac{\varepsilon_t - \varepsilon_c}{1 + \varepsilon_c} \tag{9.1}$$

ε_t tensile strain
ε_c compressive strain (negative)

The bending ability of wood is influenced by numerous factors.

Influence of wood species. For bending, hardwoods with high density are generally preferred, in particular diffuse-porous woods such as beech (*Fagus sylvatica* L.), birch (Betula spp.) and maple (Acer spp.), but also ring-porous woods such as elm (Ulmus spp.), black locust (*Robinia pseudoacacia* L.), ash (Fraxinus spp.) and oak (Quercus spp.). Semi-ring-porous woods such as cherry (Prunus spp.) and walnut (Juglans spp.) can also be bent well. Softwoods are not suitable for bending due to the large density difference between earlywood and latewood.

Beech wood is considered the wood species that bends best. The bending radius for 25 mm thick beech wood parts is $R > 43$ mm, when the part is bent with a tension strap after steaming [25]. The same value applies to black locust. For oak, $R_{min} = 50$ mm, for ash $R_{min} = 64$ mm, for birch $R_{min} = 70$ mm [3].

Influence of wood moisture content. Besides the wood species, the moisture content also plays an important role in bending wood. The more bound water present in the wood, the greater its ability to deform. A moisture content of 15 – 25 % is considered optimal for bending. At higher values (above fiber saturation point), the free water present in the wood cannot be compressed and fractures occur.

Influence of wood quality, cut, shape, and dimensions of the wood. For bending, the wood parts must be cut as precisely as possible parallel to the grain and subsequently planed. Bending wood must not have any structural defects (knots, spiral grain, cracks, rot, etc.).

Regarding the shape of the cross-section, it has been found that wood parts with a circular cross-section require a shorter treatment time for plasticization than square cross-sections, due to the easier penetration of steam between the stacked parts and the constant distance between the surfaces. For example, the steaming time for backrest components with a square cross-section (20 × 20 mm) is 40 – 60 minutes, compared to only 30 – 40 minutes for a circular cross-section (Ø 20 mm) [20].

The dimensions of the wood part influence the treatment time in direct proportion: the greater the thickness (or diameter), the longer the steaming time.

Influence of temperature. Temperature also plays an important role in the heat treatment of wood prior to bending. As temperature increases, most mechanical properties decrease. The deformability of the wood increases, while the compressive strength decreases.

For efficient steaming, the temperature at the center of the wood part must reach approximately 70 °C. For this, the steaming temperature must be 105 – 110 °C. At this temperature, the hydrolysis of pentosans to pentoses takes place, followed by a conversion into thermoplastic furfural resins, which enable crack-free bending. Therefore, hard hardwood species that are rich in pentosans exhibit very good bending capacity. Temperatures above 160 °C should be avoided, because beyond this point the strength of the wood decreases significantly.

Influence of steam pressure. During steaming, there is a relationship between the temperature and the pressure of the steam. Its optimal value is 0.2 – 0.34 bar. At lower pressures, the wood is not properly plasticized, which results in an extension of the steaming time. The application of higher pressures leads to an increase in rejects (bending defects) of approximately 30 %. These defects are usually only detectable after drying.

Influence of steaming duration. The duration of heat treatment must be adjusted so that uniform heating of the parts and the achievement of a temperature of approximately 70 °C in the interior of the wood are possible. In general, a treatment time of 1 minute/mm of thickness is considered optimal for steaming wood. However, this value must be adjusted individually. For beech wood, the steaming time can be determined using the diagram in Figure 9.23.

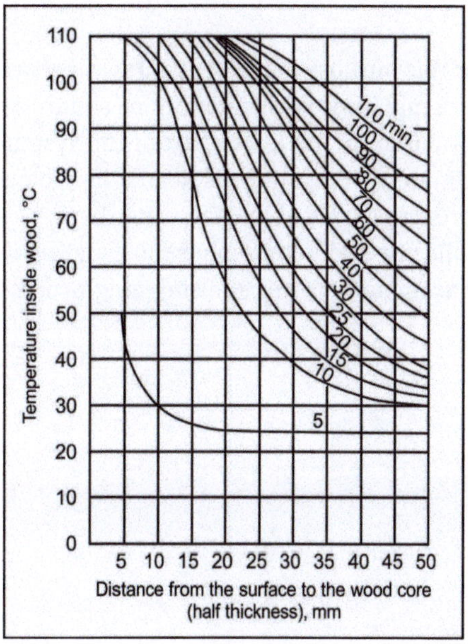

Figure 9.23 Determination of steaming time for bending wood parts made of beech (adapted from [20])

An insufficient treatment duration leads to inadequate plasticization and high material loss (fractures during bending) (Figure 9.24 a) [25]. An excessively long treatment duration leads to over-plasticization of the wood, which usually results in wrinkling in the compression zone (Figure 9.24 b), but also in local fractures and tensile cracks perpendicular to the grain [26].

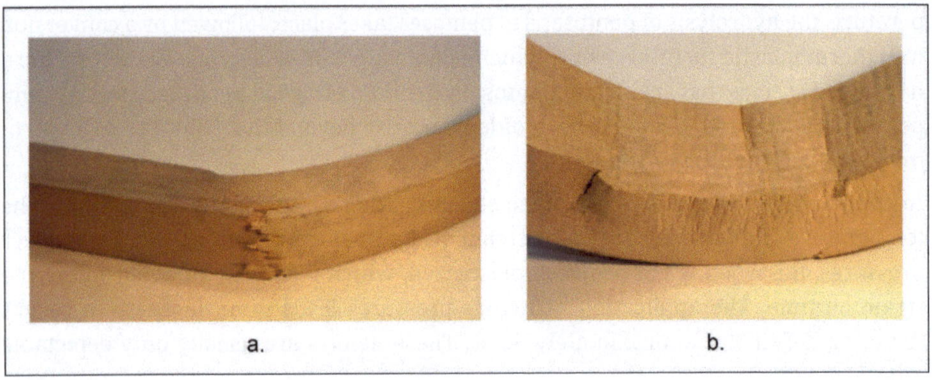

Figure 9.24 Defects during steaming of wood prior to bending: a – cracks; b – wrinkling (source: [25], with permission)

9.3.3 Equipment

Steaming is carried out in a moist, saturated steam environment, at a pressure of 0.2 – 0.39 bar, at a temperature of 105 – 110 °C.

Small autoclaves are most commonly used for this purpose made of stainless steel are used. They have a length of approximately 1 – 2 m. The cross-section can be square or round, with a side length or diameter of 0.5 – 0.6 m. In the lower part of the system there is a water bath with a depth of approximately 5 cm, which is heated by means of a perforated pipe with steam. The wood parts are placed above the water bath. The system is also equipped with a condensate collector and with temperature and pressure sensors to continuously monitor both parameters during the steaming process (Figure 9.25).

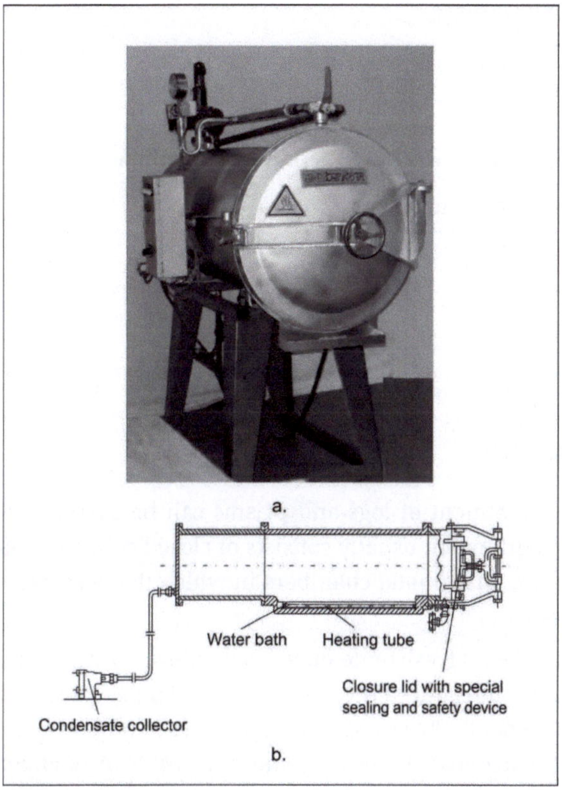

Figure 9.25 Steaming unit for bent wood parts: a – HODA compact steamer (source: © GHEBavaria, with permission); b – schematic diagram, structure (based on [20])

For a chair model, such as the one shown in Figure 9.15 a, approximately 60 to 120 pieces can be steamed simultaneously depending on the thickness of the parts (Table 9.4).

Table 9.4 Loading Capacity of a Steaming Unit for a Bent Wood Chair [20]

Furniture part	Length [mm]	Width [mm]	Thickness [mm]	Number of pieces
Seat frame	1,700	40	40	60
Backrest	508	105	24	30
Front leg	480	Diameter Ø 28 – 32		70
Rear leg	1,000	Diameter Ø 30 – 32		60
Leg arches	600	Diameter Ø 25		120

9.4 Steaming of Veneer Blocks

9.4.1 Objective

The heat treatment of prisms and logs for veneer production has the following tasks:

- to plasticize the wood for peeling or slicing,
- to reduce cutting forces,
- to reduce tool wear, and
- to obtain good veneer quality.

9.4.2 Processes and Equipment

As with sawn timber, the thermal treatment of logs and prisms can be carried out by direct or indirect steaming. The equipment usually consists of closed concrete pits (covered at the top) (Figure 9.26 a and b) or metal chambers in which the heat treatment is carried out at normal pressure.

In direct steaming, the wood is treated with fresh or de-oiled, used saturated steam at a pressure of 1.06 – 1.08 bar. The steam is introduced through perforated pipes located at the bottom of the pit (Figure 9.26 a) beneath the crossbeams. The holes in the steaming pipes must be oriented so that the steam is distributed as evenly as possible throughout the equipment without the hot steam coming into direct contact with the wood, as this would cause intense local heating, leading to black spots or even scorching. It is therefore advisable to orient the holes at an acute angle toward the bottom of the pit.

The main advantage of direct steaming is the relatively low steam consumption (on average 250 kg/m^3).

Indirect steaming takes longer but provides more uniform heating of the material compared to direct steaming. The equipment (Figure 9.26 b) is structurally similar to the direct steaming equipment, but it has a water bath at the bottom of the pit that is heated. This bath has a depth of 30 – 60 cm. Inside it there is a heating coil that heats and evaporates the water bath. Heating can be achieved using spent steam at higher pressure (1.27 – 1.37 bar), fresh steam, or pressurized hot water. The evaporation of the water causes the temperature in the steaming pit to rise slowly and uniformly, ensuring gradual and even heating and protecting the quality of the wood. Since the heating medium does not come into contact with the wood, it can be reheated and reused. This leads to savings in thermal energy.

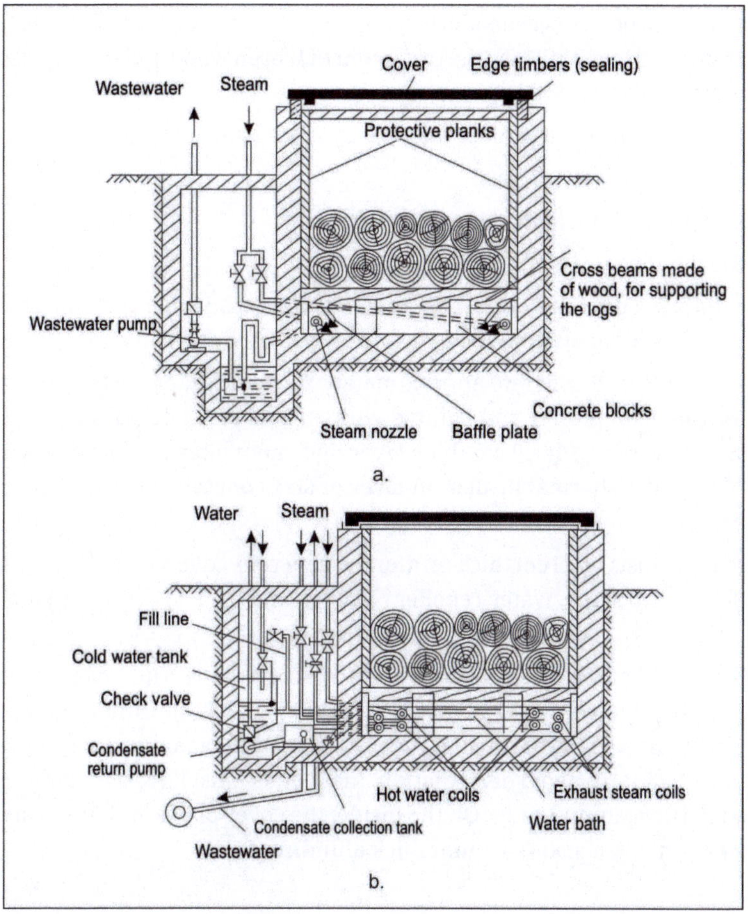

Figure 9.26 Steaming pits: a – with direct heating; b – with indirect heating (adapted from [26])

Steaming of veneer logs can also be carried out in autoclaves at overpressure (p = 2 – 3 bar). This reduces the required time to approximately half.

As an alternative to steaming, the heat treatment of veneer logs can also be carried out by immersion in hot water (boiling). The boiling pits are constructed analogously to the steaming pits. Usually one third cold water and two thirds hot water are introduced over the logs so that the initial water temperature is approximately 50 °C. The pit is supplied with water until it is completely filled. The water is then heated to the operating temperature (60 – 80 °C, depending on the wood species) by means of a heating coil installed at the bottom of the equipment.

Since the heating of the wood proceeds more slowly with hot water treatment than with steaming, this method is suitable for the treatment of frozen wood and temperature-sensitive wood species with a strong tendency to crack.

Compared to steaming, hot water treatment has the following advantages:

- lower investment and operating costs;
- less color change occurs (due to the lower temperature).

The disadvantage is the longer treatment duration.

Construction of steaming pits. Industrial steaming pits are mostly made of concrete and are often installed outdoors within the working area of an overhead crane.

The walls consist of reinforced concrete approximately 25 cm thick, protected on the inside with a water-repellent plaster and painted with an acid-resistant paint. On the outside, the walls are covered with a 2 cm thick layer of cement mortar, followed by a vapor barrier, a 20 cm thick thermal insulation layer of slag concrete, and two layers of asphalt bitumen.

The floor of the basin consists of 35 cm thick reinforced concrete, covered on the inside with concrete with a 20 cm slope, water-repellent plaster, and an acid-resistant paint layer.

Each pit is closed with four cover plates that are moved using a crane. The cover plates consist of a metal frame and softwood planks (Figure 9.27). Each has a mass of approximately 400 kg. Larch wood is particularly well suited for the manufacture of the covers, as it is acid-resistant. To avoid deformation, the planks must have a maximum width of 20 cm and vertical annual rings. On the inside, the cover plates are fitted with two layers of roofing felt and a sheet of 2 mm thick aluminum.

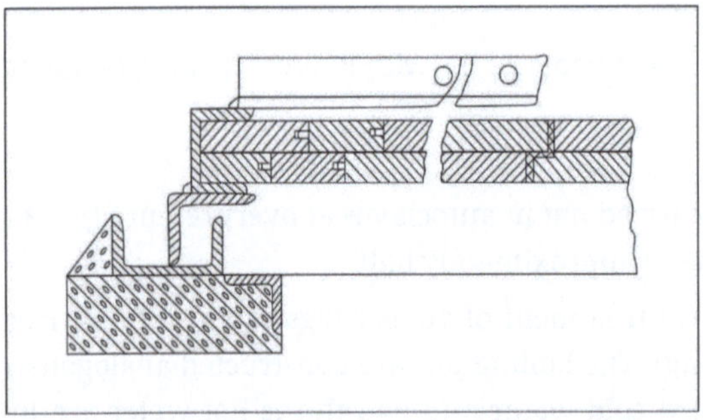

Figure 9.27 Construction of the cover plate for steaming pits (adapted from [20])

The typical dimensions of a steaming pit are: 11.8 m length; 3 m width and 2.8 m height (of which 1 m above the floor as accident protection). Such a steaming pit has a volume of approximately 44 m^3, with a utilization rate of 44 %.

At least three pits must be provided per slicing machine [27]. The workflow must be arranged so that the steamed logs reach the slicing machine as quickly as possible to prevent them from cooling down.

9.4.3 Steaming Parameters and Process Phases

The **moisture content** of the logs when placed in the pit should be as high as possible (but above the fiber saturation point). If this is not observed, inadequate heat treatment and cracking in the veneers will occur. In some cases, heat treatment by storage in hot water is more favorable.

The **temperature.** The optimal temperature for steaming logs depends mainly on the wood species (Table 9.5), but must also take into account the diameter and initial temperature of the logs, especially in winter.

Table 9.5 Steaming Temperature for the Heat Treatment of Veneer Logs [20]

Wood species	Temperature [°C]
Fir, poplar	25 - 50
Spruce	29 - 55
Douglas fir	37 - 63
Pine, alder	40 - 67
Larch	50 - 67
Birch, pear tree	59 - 84
Maple	60 - 85
Walnut	62 - 87
Oak, ash	63 - 88
Beech	66 - 90

If the temperature is too low, cracks occur during veneer production, as well as vibrations in the knife holder. This effect is more pronounced the greater the density difference between earlywood and latewood.

If the temperature is too high, the logs split severely at the ends and the surface of the veneers appears peeled and wavy.

Lower temperatures are used for steaming softwoods and soft hardwood species.

When producing thick veneers (> 1.5 mm), a temperature increase of 5 °C per each additional mm of veneer thickness is recommended, without exceeding the value of 90 °C.

The **duration** of the heat treatment of veneer logs correlates with the temperature, the treatment method, and the diameter group. It is given in Table 9.6 for some domestic wood species.

Table 9.6 Temperature and Duration of Steaming of Veneer Logs of Various Wood Species [20]

Wood species	Optimal temperature in the pit [°C]	Target temperature inside the wood [°C]	Duration [h] for diameter group (in cm):				
			20 – 30	31 – 40	41 – 50	51 – 60	61 – 70
Beech	60 – 70	60	–	–	24 – 36	36 – 60	–
	70 – 90	65	–	–	40 – 56	56 – 80	–
Oak	70 – 90	65	16 – 24	24 – 32	32 – 40	40 – 60	–
Ash	70 – 80	65	16 – 24	24 – 30	30 – 40	40 – 50	–
Maple	60 – 70	60	10 – 14	14 – 20	20 – 30	30 – 40	40 – 70
Birch	60 – 80	70	8 – 10	10 – 16	16 – 24	24 – 32	32 – 48
Cherry	40 – 50	40	8	8 – 16	16 – 20	20 – 30	30 – 36
	60 – 70	60	30 – 36	36 – 42	42 – 50	50 – 56	56 – 64
Linden, alder	40 – 50	40	8	8 – 10	16 – 20	20 – 30	30 – 36
Walnut	80 – 100	75	48	48 – 60	60 – 72	72 – 86	–

The steaming of veneer logs comprises three phases (Figure 9.28):

- the heating phase, in which the temperature is gradually raised to the intended value, giving the material time to adapt to the steam environment. Depending on the method applied, the duration of this phase is minimal with direct steaming (6 – 10 hours);
- the actual steaming, characterized by maintaining the temperature at a constant value. During this phase, the aim is to reach the recommended temperature value

(see Table 9.7) up to the center of the log. The duration of the actual steaming varies depending on the wood species and the diameter of the material (see Table 9.7);

- the cooling time, which is required to reduce the wood temperature in order to avoid thermal shock when emptying the pit. The duration of this phase is 2 – 6 hours.

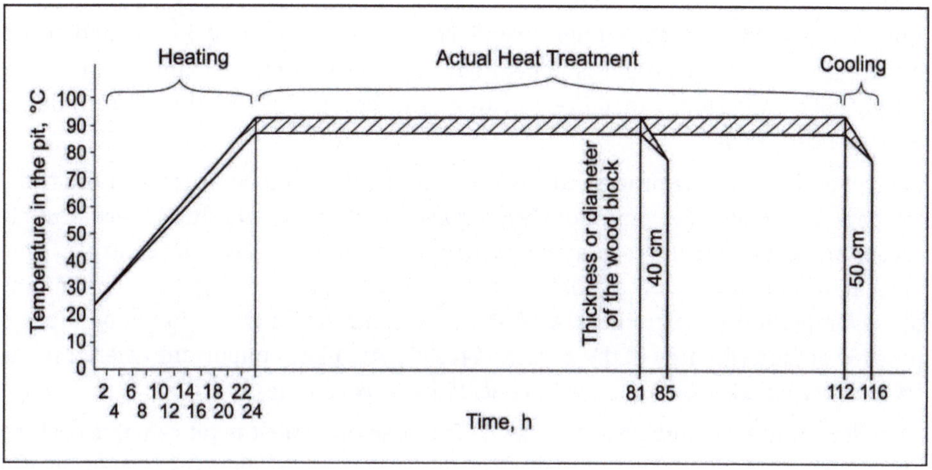

Figure 9.28 Steaming diagram for the heat treatment of veneer logs (based on [20])

The total duration of steaming of beech logs for veneers as a function of the method applied is given in Table 9.7.

Table 9.7 Total Duration of Heat Treatment of Veneer Logs of Domestic Wood Species [20]

Method	Warm-up time [h]	Duration of actual steaming [h]	Cooling [h]	Total duration [h]
Direct steaming	6 - 10	15 - 30	4 - 6	25 - 46
Indirect steaming	9 - 16	36 - 48	2 - 3	47 - 67

9.4.4 Technology of Heat Treatment of Veneer Logs

The preparation for heat treatment of veneer logs and veneer prisms consists of cleaning the pit, loading the wood material, and setting the process parameters (temperature and time).

After each batch, once the pit has been emptied, it is checked whether each individual component and the temperature sensor are in perfect condition. Any defects found are remedied.

When loading the pit with a new batch of material, it is ensured that the logs belong to the same wood species and the same diameter class. For prisms, overlapping of flat surfaces is avoided; if necessary, spacers are used.

After loading, the pit is sealed tightly with covers and, depending on the method applied, the treatment medium is generated: in direct steaming, steam is introduced through perforated pipes; in indirect steaming, the heat transfer medium is introduced into the heating coils.

Control of the steaming process consists of regulating the temperature over time. The temperature in the pit is continuously monitored. It is measured using a thermocouple sensor and automatically regulated by means of an electric valve. The control system must allow not only the programming of the desired value, but also the programming of the temperature rise gradient, so that a slow, controlled increase in temperature is possible at the beginning of the process. This is particularly important when treating frozen logs, but also with temperature-sensitive wood species.

After steaming, the condensate is drained from the pit while the pit remains covered and the wood stays in the pit for approximately 2 – 3 hours to allow the temperature in its interior to equalize. Only then are the cover plates removed and the wood unloaded, so that it can reach the slicing machines while still warm.

Possible treatment defects can only be detected after the veneers have been cut:

- veneers with cracks at the ends, with a peeled and wavy surface indicate the application of too high a temperature;
- fiber cracks indicate either too short a treatment time, too low a temperature, or too low an initial moisture content of the logs.

References

[1] Vorreiter, L.; Fromme, G. (1958): *Holztechnologisches Handbuch. Band II: System Holz-Wasser-Wärme, Holztrocknung, Dämpfen und Kochen, Spanlose Holzverformung*. Verlag Georg Fromme & Co., Wien, Munich, Germany

[2] Marinescu, I. (1982): *Proiectarea instalațiilor de tratare termică şi ameliorarea lemnului*. University Brasov, Romania

[3] Kollmann, F. F. P.; Cote, W. A. (1968): *Principles of Wood Science and Technology I: Solid Wood*. Springer Verlag, Berlin/Göttingen/Heidelberg

[4] Geffert, A.; Geffertová, J.; Vybohová, E.; Dudiak, M. (2020): *Impact of Steaming Mode on Chemical Characteristics and Colour of Birch Wood*. In: Forests, 11, 478. *https://doi.org/10.3390/f11040478*

[5] Geffert, A.; Vybohová, E.; Geffertová, J. (2019): *Changes in the Chemical Composition of Oak Wood Due to Steaming*. In: Acta Facultatis Xylologiae Zvolen, 61(1), p. 19 – 29. Technical University of Zvolen, Slovakia *https://doi.org/10.17423/afx.2019.61.1.02*

[6] Kollmann, F. F. P. (1939): *Vorgänge und Änderungen von Holzeigenschaften beim Dämpfen*. In: Holz als Roh- und Werkstoff, 2, p. 1 – 11

[7] Hacibektasoglu, M. (2020): *Researches Concerning the Manufacturing and Some Mechanical and Technological Properties of Solid Wood Panels Made from Heat-Treated Beech Wood*. Doctoral thesis, Transilvania University, Brasov, Romania

[8] Németh, R.; Hill, C.; Takáts, P.; Tolvaj, L. (2016): *Chemical Changes of Wood During Steaming Measured by IR Spectroscopy*. In: Wood Material Science & Engineering, 11(2), p. 95 – 101. *https://doi.org/10.1080/17480272.2014.961169*

[9] Dzurenda, L. (2022): *Range of Color Changes of Beech Wood in the Steaming Process*. In: BioResources, 17(1), p. 1690 – 1702. *https://doi.org/10.15376/biores.17. 1. 1690* – 1702

[10] Volkmer, T.; Moser, R.; Lorenz, T.; von Arx, U.; Hass, P.; Niemz, P. (2014): *Einfluss des Hitze-Druck-Dämpfens auf das optische Erscheinungsbild und das Emissionsverhalten von Eichenholz*. In: Bauphysik, 36(2), p. 80 – 86

[11] Dzurenda, L.; Dudiak, M. (2024): *Homogenization of the Color of Beech Sapwood and False Heartwood by the Steaming Process*. In: Forests, 15, 1009. *https://doi.org/10.3390/f15061009*

[12] Banadics, E. A.; Tolvaj, L.; Varga, D. (2019): *Colour Stability of Steamed Poplar Wood During Short-Term Photodegradation*. In: BioResources, 14(4), p. 8250 – 8256

[13] Varga, D.; Tolvaj, L.; Preklet, E. (2021): *Colour Stability of Steamed Black Locust, Beech and Spruce Timbers*. In: Wood Research, 66(4), p. 544 – 555

[14] Dzurenda, L.; Dudiak, M.; Banski, A. (2020): *Influence of UV Radiation on Color Stability of Natural and Thermally Treated Maple Wood with Saturated Water Steam*. In: 10th International Scientific and Technical Conference „Innovations in Forest Industry and Engineering Design" (INNO 2020), 1 – 3 Oktober 2020, Vitosha Park Hotel, Sofia, Bulgaria, p. 219 – 224

[15] Hofmann, T.; Tolvaj, L.; Visi-Rajczi, E.; Varga, D. (2022): *Chemical Changes of Steamed Timber During Short-Term Photodegradation Monitored by FTIR Spectroscopy*. In: European Journal of Wood and Wood Products, 80, p. 841 – 849. *https://doi.org/10.1007/s00107-022-01814-6*

[16] Deliiski, N.; Niemz, P.; Dzurenda, L.; Angelski, D.; Campean, M.; Vitchev, P. (2024): *A Methodology for Computing the Thermal Balance and Energy Consumption of Autoclaves During Steaming of Wood*. In: Wood Material Science and Engineering. *https://doi.org/10.1080/17480272.2024.2429772*

[17] Riehl, T.; Welling, J.; Frühwald, A. (2002): *Druckdämpfen von Schnittholz*. Arbeitsbericht 2002/01, Bundesforschungsanstalt für Forst- und Holzwirtschaft, Institut für Holzphysik, Hamburg, Germany

[18] Deliiski, N.; Dzurenda, L. (2010): *Modelling of the Thermal Processes in the Technologies for Wood Thermal Treatment*. Technical University, Zvolen, Slovakia

[19] Ouyang, X.; Ye, Z.; Lin, P.; Li, Y. (2023): *The Optimization of the Fillet Size of Autoclave Tooth Based on the Experiment and Numerical Simulation*. In: Advances in Materials Science and Engineering, 8051118. *https://doi.org/10.1155/2023/8051118*

[20] Marinescu, I. (1980): *Uscarea şi tratarea termică a lemnului. Vol. 2*. Editura Tehnică, Bucharest, Romania

[21] Eggert, O. T. (1995): *Untersuchung der Einflussgrößen beim Biegen von Vollholz*. Dissertation, Fakultät für Konstruktions- und Fertigungstechnik, Universität Stuttgart, Germany

[22] Trübswetter, T. (2009): *Holztrocknung. Verfahren zur Trocknung von Schnittholz – Planung von Trocknungsanlagen*. 2nd ed., Carl Hanser Verlag, Munich, Germany

[23] Hanemann, M. (1917): *Holzaufbereitungsverfahren*. Deutsches Reichspatentamt, Patentschrift Nr. 318197

[24] Sandberg, D.; Eggert, O.; Haider, A.; Kamke, F.; Wagenführ, A. (2023): *Forming, Densification and Molding*. Chapter 18 in *Springer Handbook of Wood Science and Technology*. Springer Nature Switzerland AG

[25] Eggert, O. (2014): *Bentwood Chair Production – A Stunning Production System*. COST Action FP0904 Final Conference, 19. – 20. Mai 2014, Skellefteå, Schweden. Online: *https://www.ghebavaria.de/fileadmin/user_upload/Benwood-ProductionSystem.pdf*, retrieved 05. 08. 2025

[26] Doffiné, E. (1950): *Dämpfgruben für die Furniererzeugung*. In: Norddeutsche Holzwirtschaft, 4(22), p. 6 – 7

[27] Lohmann, U. (1998): *Holz Handbuch*. DRW Verlag, Leinfelden-Echterdingen, Germany

10 Modeling of the Thermal Treatment of Veneer Logs and Veneer Prisms

For the optimization of the heat treatment of wood by steaming or boiling, it is necessary to know the distribution of the temperature field in the wood and the energy consumed for heating the wood at any point in time. The intensity of heating and energy consumption depend on the dimensions of the wood part, on the wood properties (moisture content, anisotropy, density), on the initial temperature of the wood, and on the temperature during the heating process.

A correct and effective control of the heating process is only possible if its physical principles, the sorption behavior, and the effect of the factors influencing the process are known. The identification of the most important influencing factors and their mathematical and physical modeling are an important basis for simulating the process mathematically and physically [1–3].

To avoid the disadvantage that the sign of the temperature in the frozen wood material changes when it exceeds 0 °C during heat treatment, it is useful to develop the models using the absolute temperature (T) [3–5].

10.1 Mechanism of Heat Distribution in Wood During Thermal Treatment

During the thermal treatment of frozen and non-frozen wood materials and the subsequent conditioning in air, in addition to purely thermal processes, mass exchange occurs through diffusion. Mass diffusion in wood is approximately one hundred times smaller than thermal diffusivity. This results in a relatively small change in the mass of the materials, which is slower than the heat distribution through heating or cooling. This makes it possible to treat the mass exchange between wood and steam or hot water as well as the temperature changes as the result of a pure heat transfer. It is assumed that heat in wood is transferred only by thermal conductivity. For this reason, the mechanism of heat distribution in the materials can be described by the heat conduction equation (Fourier–Kirchhoff equation). Its most compact form is represented by Equation 10.1:

$$c_w \cdot \rho_w \frac{\partial T}{\partial \tau} = -div(\lambda_w \cdot gradT) \pm q_{f-lat} \quad (10.1)$$

c_w specific heat capacity of wood [J/kgK]
ρ_w density of wood [kg/m^3]
λ_w thermal conductivity of wood [W/mK]
T temperature [K]
τ time [s]
$q_{f\text{-}lat}$ flux of the internal source of latent heat of water [W/m^3]

In the general case, c_w and λ_w depend on T and the wood moisture content u, and ρ_w depends on u. In Equation 10.1, the dependencies $c_w(T,u)$, $\rho_w\ (u)$ and $\lambda_w(T,u)$ must therefore be taken into account.

Ice in wood can form through the freezing of bound water contained in the wood (in the hygroscopic range) or of both bound and free water (above the hygroscopic range). The phase transition from water to ice and vice versa is expressed by the latent heat.

In the context of heat conduction in frozen materials, it is useful to include the latent heat in the so-called effective specific heat capacity $c_{w\text{-}eff}$ [5]. This takes into account the influence of the flow of the internal latent heat source in water, $q_{f\text{-}lat}$ via the change in the temperature field.

This means that the effective specific heat capacity of frozen wood, $c_{w\text{-}eff}$, equals the sum of the specific heat capacity of wood, c_w, and the specific heat capacity of the bound frozen wood water, $c_{ice\text{-}bw}$, or of both bound, $c_{ice\text{-}bw}$, and free water, $c_{ice\text{-}fw}$.

When modeling heat conduction processes in wood, it must also be taken into account that the thermal conductivity λ_w in addition to T and u additionally depends on the cutting direction of the wood (radial, tangential, and along the grain).

10.2 Modeling of the 3D, 2D, and 1D Temperature Distribution in Veneer Prisms

If the anatomical directions (radial, tangential, and parallel to the wood fibers) of the veneer prisms are assigned to the coordinate axes x, y and z, one can derive from Equation 10.1 [3]:

$$c_{w-eff}(T,u) \cdot \rho_w(u) \frac{\partial T(x,y,z,\tau)}{\partial \tau} = \frac{\partial}{\partial x}\left[\lambda_{w-r}(T,u)\frac{\partial T(x,y,z,\tau)}{\partial x}\right] + \frac{\partial}{\partial y}\left[\lambda_{w-t}(T,u)\frac{\partial T(x,y,z,\tau)}{\partial y}\right] + \frac{\partial}{\partial z}\left[\lambda_{w-l}(T,u)\frac{\partial T(x,y,z,\tau)}{\partial z}\right] \quad (10.2)$$

u wood moisture content [kg/kg]
$\lambda_{w\text{-}r}$ thermal conductivity of wood in the radial direction [W/mK]

$\lambda_{w\text{-}t}$ thermal conductivity of wood in the tangential direction [W/mK]
$\lambda_{w\text{-}l}$ thermal conductivity of wood along the grain [W/mK]

After differentiating the right-hand side of Equation 10.3 with respect to the spatial coordinates x, y and z, the following is obtained for transient heating and cooling (excluding the bracket arguments):

$$c_{w-\text{eff}} \cdot \rho_w \frac{\partial T}{\partial \tau} = \lambda_{w-r} \frac{\partial^2 T}{\partial x^2} + \frac{\partial \lambda_{w-r}}{\partial T} \left(\frac{\partial T}{\partial x} \right)^2$$

$$+\lambda_{w-t} \frac{\partial^2 T}{\partial y^2} + \frac{\partial \lambda_{w-t}}{\partial T} \left(\frac{\partial T}{\partial y} \right)^2 + \lambda_{w-l} \frac{\partial^2 T}{\partial z^2} + \frac{\partial \lambda_{w-l}}{\partial T} \left(\frac{\partial T}{\partial z} \right)^2 \tag{10.3}$$

with the initial condition

$$T(x,y,z,0) = T_{w0} \tag{10.4}$$

T_{w0} initial value of the wood temperature [K]

and the boundary conditions

- During the time τ of the heat treatment, the surface temperature equals the temperature of the medium (saturated steam or hot water), T_m:

$$T(0,y,z,\tau) = T(x,0,z,\tau) = T(x,y,0,\tau) = T_m(\tau) \tag{10.5}$$

- During conditioning with the air temperature T_{air}:

$$\frac{\partial T(0,y,z,\tau)}{\partial x} = -\frac{\alpha_{w-r}(0,y,z,\tau)}{\lambda_{w-r}(0,y,z,\tau)} [T(0,y,z,\tau) - T_{air}(\tau)] \tag{10.6}$$

$$\frac{\partial T(x,0,z,\tau)}{\partial y} = -\frac{\alpha_{w-t}(x,0,z,\tau)}{\lambda_{w-t}(x,0,z,\tau)} [T(x,0,z,\tau) - T_{air}(\tau)] \tag{10.7}$$

$$\frac{\partial T(x,y,0,\tau)}{\partial z} = -\frac{\alpha_{w-l}(x,y,0,\tau)}{\lambda_{w-l}(x,y,0,\tau)} [T(x,y,0,\tau) - T_{air}(\tau)] \tag{10.8}$$

$\alpha_{w\text{-}r}$, $\alpha_{w\text{-}t}$, $\alpha_{w\text{-}l}$ convective heat transfer coefficients of the material surfaces in radial, tangential and parallel fiber direction, respectively [W/m^2K]

The system of equations (Equation 10.3–Equation 10.8) represents a mathematical 3D model that describes the temperature change in the prismatic wood piece during heat treatment and the resulting conditioning under corresponding initial and boundary conditions.

If the length of the wood prisms l_w is at least 4 to 5 times greater than their thickness d_w and the width b_w does not exceed the thickness by more than 3 times, the following mathematical 2D model can be used:

$$c_{w-\text{eff}} \cdot \rho_w \frac{\partial T}{\partial \tau} = \lambda_{w-r} \frac{\partial^2 T}{\partial x^2} + \frac{\partial \lambda_{w-r}}{\partial T} \left(\frac{\partial T}{\partial x} \right)^2 + \lambda_{w-t} \frac{\partial^2 T}{\partial y^2} + \frac{\partial \lambda_{w-t}}{\partial T} \left(\frac{\partial T}{\partial y} \right)^2 \tag{10.9}$$

with the initial condition:

$$T(x,y,0) = T_{w0} \tag{10.10}$$

and the boundary conditions:

- During heat treatment

$$T(0,y,\tau) = T(x,0,\tau) = T_{\mathrm{m}}(\tau) \tag{10.11}$$

- During conditioning:

$$\frac{\partial T(0,y,\tau)}{\partial x} = -\frac{\alpha_{\mathrm{w-r}}(0,y,\tau)}{\lambda_{\mathrm{w-r}}(0,y,\tau)}[T(0,y,\tau) - T_{\mathrm{air}}(\tau)] \tag{10.12}$$

$$\frac{\partial T(x,0,\tau)}{\partial y} = -\frac{\alpha_{\mathrm{w-t}}(x,0,\tau)}{\lambda_{\mathrm{w-t}}(x,0,\tau)}[T(x,0,\tau) - T_{\mathrm{air}}(\tau)] \tag{10.13}$$

If the length of the wood is 4 to 5 times greater than its thickness and the thickness is less than one third of the width, the following 1D model can be applied:

$$c_{\mathrm{w-eff}} \cdot \rho \frac{\partial T}{\partial \tau} = \lambda_{\mathrm{w-r}} \frac{\partial^2 T}{\partial x^2} + \frac{\partial \lambda_{\mathrm{w-r}}}{\partial T} \left(\frac{\partial T}{\partial x} \right)^2 \tag{10.14}$$

with the initial condition:

$$T(x,0) = T_{\mathrm{w0}} \tag{10.15}$$

and the boundary conditions:

- During heat treatment:

$$T(0,\tau) = T_{\mathrm{m}}(\tau) \tag{10.16}$$

- During conditioning:

$$\frac{\partial T(0,\tau)}{\partial x} = -\frac{\alpha_{\mathrm{w-r}}(0,\tau)}{\lambda_{\mathrm{w-r}}(0,\tau)}[T(0,\tau) - T_{\mathrm{air}}(\tau)] \tag{10.17}$$

10.3 Modeling of the 2D and 1D Temperature Distribution in Veneer Logs

The mechanism of heat distribution in roundwood (logs) can be described by the following nonlinear partial differential equation, which is obtained from Equation 10.2 after the transition from rectangular to cylindrical coordinates:

$$c_{\mathrm{w-eff}} \cdot \rho_{\mathrm{w}} \frac{\partial T}{\partial \tau} = \lambda_{\mathrm{w-r}} \left(\frac{\partial^2 T}{\partial r^2} + \frac{1}{r}\frac{\partial T}{\partial r} \right) + \frac{\partial \lambda_{\mathrm{w-r}}}{\partial T} \left(\frac{\partial T}{\partial r} \right)^2 + \lambda_{\mathrm{w-l}} \frac{\partial^2 T}{\partial z^2} + \frac{\partial \lambda_{\mathrm{w-l}}}{\partial T} \left(\frac{\partial T}{\partial z} \right)^2 \tag{10.18}$$

with the initial condition:

$$T(r,z,0) = T_{\mathrm{w0}} \tag{10.19}$$

and the boundary conditions:

- During heat treatment:

$$T(0,z,\tau) = T(r,0,\tau) = T_{\mathrm{m}}(\tau) \tag{10.20}$$

- During conditioning:

$$\frac{\partial T(0,z,\tau)}{\partial r} = -\frac{\alpha_{\mathrm{w-r}}(0,z,\tau)}{\lambda_{\mathrm{w-r}}(0,z,\tau)}[T(0,z,\tau) - T_{\mathrm{air}}(\tau)] \tag{10.21}$$

$$\frac{\partial T(r,0,\tau)}{\partial z} = -\frac{\alpha_{\mathrm{w-l}}(r,0,\tau)}{\lambda_{\mathrm{w-l}}(r,0,\tau)}[T(r,0,\tau) - T_{\mathrm{air}}(\tau)] \tag{10.22}$$

If the length of the log exceeds its diameter by at least 4 to 5 times, heat transfer through its end faces can be neglected, as it has no influence on the temperature change of the cross-section that is equidistant from the end faces. In such cases, the following 1D model can be used to calculate the temperature change only along the coordinate r of this cross-section:

$$c_{\mathrm{w-eff}} \cdot \rho_{\mathrm{w}} \frac{\partial T}{\partial \tau} = \lambda_{\mathrm{w-r}} \left(\frac{\partial^2 T}{\partial r^2} + \frac{1}{r}\frac{\partial T}{\partial r} \right) + \frac{\partial \lambda_{\mathrm{w-r}}}{\partial T} \left(\frac{\partial T}{\partial r} \right)^2 \tag{10.23}$$

with the initial condition

$$T(r,0) = T_{\mathrm{w0}} \tag{10.24}$$

and with the boundary condition analogous to Equation 10.17, but with the derivative along the r-direction instead of the x-direction.

10.4 Mathematical Description of the Thermophysical Properties of Wood

The mathematical description of the thermophysical properties of wood is based on the use of knowledge of the influencing factors and the use of experimental data for modeling.

The models are based on the following physical properties:

- the specific heat capacity,
- the thermal conductivity, and
- the density.

Table 10.1 shows the bulk density ρ_{b}, standardized values of the fiber saturation point at 20 °C (i.e., at 293.15 K), $u_{\mathrm{FSP(293.15)}}$ and the volumetric shrinkage β_{v} of wood species that are frequently steamed in practice.

Table 10.1 Bulk Density ratio, Fiber Saturation Point at T = 293.15 K (i.e. at t = 20 °C), and Volumetric Shrinkage of Selected Wood Species (source: based on [6–8])

No.	Wood species	Bulk density ρ_b [kg/m³]	Fiber saturation point $u_{FSP(293.15)}$ [kg/kg]	Volumetric shrinkage β_v [%]
1.	Poplar	355	0.35	13.3
2.	Spruce	380	0.26 – heartwood; 0.32 – sapwood	11.4
3.	Pine	430	0.30 – heartwood	11.8
4.	Chestnut	525	0.22 – heartwood	10.4
5.	Maple	525	0.28	11.2
6.	Birch	515	0.30	13.2
7.	Walnut	555	0.24 – heartwood	12.9
8.	European beech	560	0.31	17.3
9.	Ash	565	0.25	12.8
10.	Black locust	660	0.20 – heartwood; 0.30 – sapwood	11.2
11.	Oak	670	0.22 – heartwood; 0.29 – sapwood	11.9
12.	Hornbeam	670	0.31	18.0

10.4.1 Mathematical Description of Specific Heat Capacity

As stated in Section 7.1.5, specific heat capacity is the amount of heat that one kilogram of wood must absorb in order to raise its temperature by one degree. In calculating various processes of thermal treatment of wood materials, it is common practice to use the specific heat capacity of wood at constant pressure [9–13].

An overview of the applicability of published equations for frozen wood at constant pressure is published in [14]. Considerable differences in the published characteristic values were found.

An investigation of the specific heat capacity of beech and pine wood with different wood moisture contents and temperatures between –140 °C and 50 °C was carried out by means of differential scanning in [15]. It was found that the change in specific heat capacity of both wood species in the investigated temperature range is linear and that the values above 0 °C are consistent with the existing literature.

In addition to the specific heat capacity of non-frozen $c_{\text{w-nfr}}$ and of frozen wood $c_{\text{H-fr}}$, when calculating the steaming duration and energy consumption during the thawing of frozen wood, the additional heat capacities must also be taken into account (heat required for the phase transition, i.e., for melting the ice). This is necessary for free and bound water in the frozen state. If these heat consumptions are related to 1 K, they can be expressed by their corresponding specific heat capacities.

Experimental investigations into the thawing process of logs of various wood species and various moisture contents above the hygroscopic range were carried out in [16]. It was found that the melting of frozen free water in wood occurs in the range between 272.15 K and 273.15 K (i.e., between –1 °C and 0 °C). The melting of frozen bound water ends at T = 272.15 K (i.e., at –1 °C).

The thawing process can be divided into three regions:

- in the first region, the frozen wood is heated and the bound frozen water melts gradually;
- in the second region, between –1 °C and 0 °C, further heating of the wood takes place until the state is reached at which melting of the free water occurs;
- in the third region, above 0 °C, further heating of the wood takes place, with the entire water content in the wood being liquid.

In Figure 10.1, the symbols of the thermophysical properties of frozen and non-frozen wood are shown, which are required for calculating the temperature distribution.

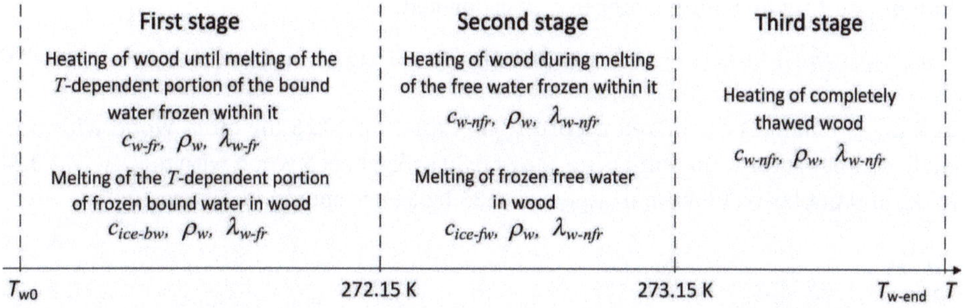

Figure 10.1 Use of different specific heat capacities, thermal conductivities, and densities for calculating the temperature distribution in frozen and non-frozen wood assortments above the hygroscopic range (source: based on [16])

$c_{w\text{-}fr}$ specific heat capacity of frozen wood [J/kgK]
$c_{w\text{-}nfr}$ specific heat capacity of non-frozen wood [J/kgK]
$c_{ice\text{-}bw}$ specific heat capacity of frozen bound water in wood [J/kgK]
$c_{ice\text{-}fw}$ specific heat capacity of frozen free water in wood [J/kgK]
ρ_w bulk density of wood [kg/m^3]
$\lambda_{w\text{-}fr}$ thermal conductivity of frozen wood [W/mK]
$\lambda_{w\text{-}nfr}$ thermal conductivity of non-frozen wood [W/mK]

The specific heat capacity of frozen wood during the three regions of its thawing process, $c_{w\text{-}eff} = c_{w\text{-}eff1,2,3}$, formulated in Equation 10.3, Equation 10.9, Equation 10.14, Equation 10.18 and Equation 10.23, can be represented as follows:

$$c_{w-eff1} = c_{w-fr} + c_{ice\text{-}bw} \tag{10.25}$$

at $u > u_{bw\text{-}nfr}$ as well as 213.15 K $\leq T \leq$ 272.15 K

$$c_{w-eff2} = c_{w-nfr} + c_{ice-fw} \tag{10.26}$$

at $u > u_{FSP(272.15)}$ as well as 272.15 K $< T <$ 273.15 K

$$c_{w-eff3} = c_{w-nfr} \tag{10.27}$$

at $u \leq u_{bw\text{-}nfr}$ as well as $T \geq$ 273.15 K or at $u > u_{bw-nfr}$ as well as $T > T_{bw\text{-}nfr}$
$u_{bw\text{-}nfr}$ non-frozen, bound water in wood (see Figure 10.2) [kg/kg]

In [9] it was determined that even at a temperature T < 272.15 K a certain portion of the water bound in wood, $u_{bw\text{-}nfr} = f(T)$, remains in an non-frozen state. For the mathematical description of $u_{bw\text{-}nfr}$, the data derived in [9] on the change of $u_{bw\text{-}nfr}$ were used.

Taking into account the thawing temperature of frozen bound water of 272.15 K and also the influence of the fiber saturation point of individual wood species, the proportion of non-frozen, bound water $u_{bw\text{-}nfr}$ is derived.

$$u_{bw\text{-}nfr} = 0.12 + (u_{FSP(T)} - 0.12) \cdot \exp[0.0567 \cdot (T - 272.15)] \tag{10.28}$$

The $u_{bw\text{-}nfr}$ changes calculated according to Equation 10.28 for birch wood with normalized fiber saturation point $u_{FSP(293.15)}$ = 0.30 kg/kg, beech wood with $u_{FSP(293.15)}$ = 0.31 kg/kg and poplar wood with $u_{FSP(293.15)}$ = 0.35 kg/kg are shown in Figure 10.2.

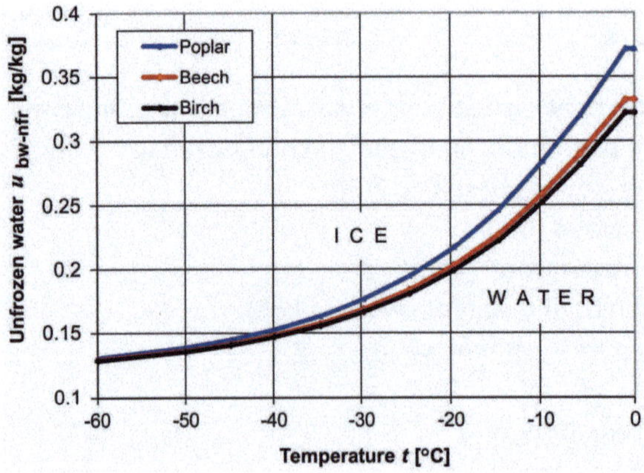

Figure 10.2 Unfrozen, bound water in wood as a function of temperature (source: [3, 5])

The values of $u_{FSP(T)}$ in Equation 10.28 are determined as a function of temperature using the following equation developed in [17]:

$$u_{FSP(T)} = u_{FSP(293.15)} - 0.001(T - 293.15) \tag{10.29}$$

The temperature required for this is derived:

$$T_{bw\text{-}nfr} = 272.15 + \frac{\ln \frac{u_{bw\text{-}nfr} - 0.12}{u_{FSP(T)} - 0.12}}{0.0567} \tag{10.30}$$

at $0.12 \text{ kg/kg} \le u = u_{bw\text{-}nfr} < u_{FSP(272.15)}$

$$T_{bw\text{-}nfr} = 272.15 \text{ K} \tag{10.31}$$

at $u \ge u_{FSP(272.15)}$

Here, $u_{FSP(272.15)}$ is the fiber saturation point of wood at T = 272.15 K (i.e., at t = –1 °C), which is calculated according to Equation 10.29.

The change in $t_{bw\text{-}nfr}$ for birch, beech, and poplar wood as a function of u, calculated according to Equation 10.30 and Equation 10.31, is shown in Figure 10.3.

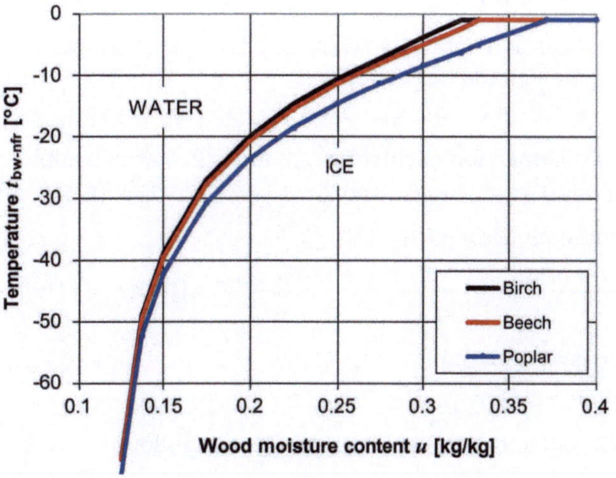

Figure 10.3 Melting temperature of frozen bound water in wood as a function of wood moisture content (source: [3, 5])

The specific heat capacity of frozen and non-frozen wood, $c_{w\text{-}fr}$ and $c_{w\text{-}nfr}$, respectively, (see Equation 10.25–Equation 10.27) is described mathematically with the results shown in Figure 10.4. In the dissertations [9] and [18], experimental data on the influence of temperature t and wood moisture content u were determined for birch wood with $u_{FSP(293.15)}$ = 0.3 kg/kg. The experimental results are shown in Figure 10.4. They are frequently cited or used in European [4, 6–8, 13, 18–20] and American [21, 22] literature to calculate the thawing and/or heating of various wood materials.

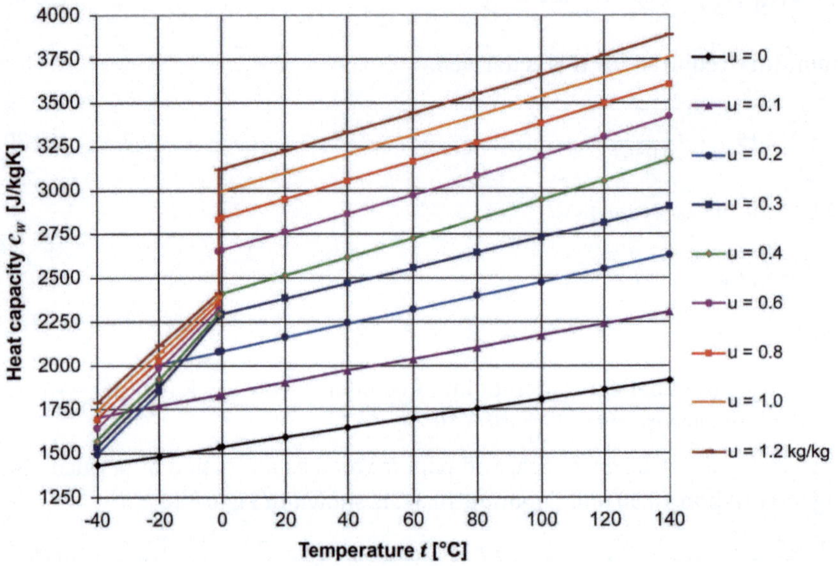

Figure 10.4 Heat capacity of frozen and non-frozen wood as a function of temperature and wood moisture content (source: based on [9, 18])

The following equations were obtained for calculating $c_{\text{w-fr}}$ and $c_{\text{w-nfr}}$, which apply to all wood species for the intervals $0 \leq u \leq 1.6$ kg/kg and $223.15 \leq T \leq 423.15$ K [3–5]:

- When $T \leq 272.15$ K and simultaneously $u > u_{\text{bw-nfr}}$:

$$c_{\text{w-fr}} = K_{\text{c-fr}} \frac{526 + 2.95 \cdot T + 0.0022 \cdot T^2 + 2261 \cdot u + 1276 \cdot u_{\text{bw-nfr}}}{1 + u} \text{ [J/kgK]} \tag{10.32}$$

$$K_{\text{c-fr}} = 1.06 + 0.04 \cdot u + \frac{0.00075 \cdot (T - 272.15)}{u_{\text{bw-nfr}}} \tag{10.33}$$

- For $T > 272.15$ K or $T \leq 272.15$ K and simultaneously $u \leq u_{\text{bw-nfr}}$ it follows:

$$c_{\text{w-nfr}} = \frac{1}{1 + u} \cdot (2097 \cdot u + 2.55 \cdot T + 9.92 \cdot u \cdot T + 0.0002 \cdot T^2 + 826) \tag{10.34}$$

at $u \leq u_{\text{FSP}(272.15)}$

$$c_{\text{w-nfr}} = \frac{1}{1 + u} \cdot (2862 \cdot u + 2.95 \cdot T + 5.49 \cdot u \cdot T + 0.0036 \cdot T^2 + 555) \tag{10.35}$$

at $u > u_{\text{FSP}(272.15)}$

$$u_{\text{FSP}(272.15)} = u_{\text{FSP}(293.15)} + 0.021 \tag{10.36}$$

Figure 10.4 shows that at t = –1 °C and all values of u > 0.3 kg/kg, a sudden increase in $c_w$values occurs during the thawing of wood. This is due to the phase transition of the frozen free water in the wood. As can be seen from Figure 10.4, the $c_w$jump for u = 0.2 kg/kg occurs at $t \approx$ –20 °C. Both at u = 0 kg/kg and at u = 0.1 kg/kg, a jump in the values of $c_w$is absent, since no ice is present in the wood even at –40 °C (see Figure 10.2).

For the mathematical description of the specific heat capacity of frozen free water in wood, $u_{\text{ice-fw}}$, the amount of free water contained in 1 kg of wood and the latent heat of the phase transition of free water from the solid to the liquid state must be determined (also called heat of fusion $L_{\text{f-fw}}$). The values of $u_{\text{ice-fw}}$ and $L_{\text{f-fw}}$ are obtained according to [9, 16]:

$$u_{\text{ice-fw}} = \frac{u - u_{\text{FSP}(272.15)}}{1 + u} \tag{10.37}$$

$$L_{\text{f-fw}} = 3.34 \cdot 10^5 \text{ [J/kg]} \tag{10.38}$$

The amount of heat required to melt the frozen free water in wood consists of the product of $u_{\text{ice-fw}}$ and $L_{\text{f-fw}}$. If this amount of heat is related to a temperature range of 1 K, in which the entire amount of frozen free water melts (between –1 °C and 0 °C), the specific heat capacity of the frozen free water in wood $c_{\text{ice-fw}}$ is obtained as:

$$c_{\text{ice-fw}} = 3.34 \cdot 10^5 \frac{u - u_{\text{FSP}(272.15)}}{1 + u} \tag{10.39}$$

at $u > u_{\text{FSP}(272.15)}$ as well as $272.15 \text{ K} < T < 273.15 \text{ K}$

In Figure 10.5, the change in $c_{\text{ice-fw}}$ calculated according to Equation 10.39 for oak, pine, beech, and poplar wood as a function of u in the range of $u = u_{\text{FSP}(272.15)}$ to u = 1.2 kg/kg is shown.

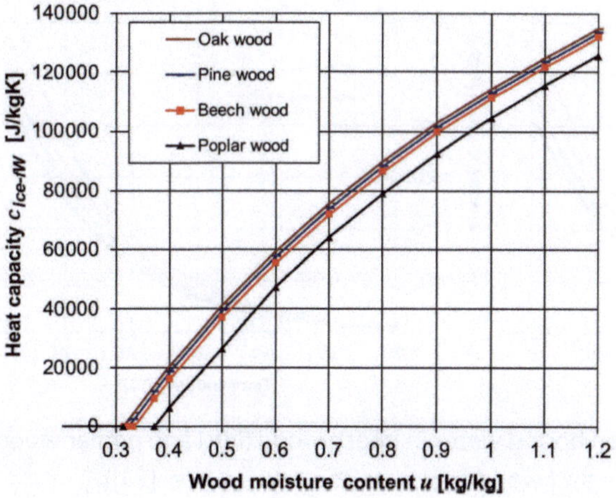

Figure 10.5 Heat capacity of frozen free water in wood as a function of wood moisture content (source: [3, 4])

The increase in $u_{FSP(293.15)}$ leads to a reduction in $c_{ice\text{-}fw}$, since the amount of free water decreases. Each increase in $u_{FSP(293.15)}$ by 0.01 kg/kg reduces $c_{ice\text{-}fw}$ by approximately 2000 J/kgK. The increase in u leads to a weakly nonlinear increase in $c_{ice\text{-}fw}$, which is caused by the increase in the amount of frozen free water in the wetter wood.

Taking into account the equations formulated in [23] for the latent heat of fusion of water bound in wood, $L_{f\text{-}bw}$, it follows:

$$L_{f-bw} = 1.223 \cdot 10^3 T + 2.102 \cdot 10^3 T \cdot \ln \frac{T}{273.15} \tag{10.40}$$

To use Equation 10.28 to determine the freezing of bound water in the range 213.15 K < T < 272.15 K, Equation 10.41 is used to calculate the specific heat capacity of frozen bound water in wood [24].

$$c_{ice\text{-}bw} = \left(69.344 \cdot T + 119.183 \cdot T \cdot \ln \frac{T}{273.15}\right) \cdot (u_{FSP(T)} - 0.12) \cdot \frac{\exp[0.0567 \cdot (T - 272.15)]}{1 + u} \tag{10.41}$$

at $u > u_{bw\text{-}nfr}$ as well as $213.15\text{ K} \le T \le 272.15\text{ K}$

It is evident that an increase in u leads to a decrease in $c_{ice\text{-}bw}$. When u increases from 0.4 to 1.0 kg/kg at a given t, the $c_{ice\text{-}bw}$ for both beech and poplar wood by approximately 28%. Each increase of $u_{FSP(293.15)}$ by 0.01 kg/kg for individual wood species leads to a linear increase of $c_{ice\text{-}bw}$ by approximately 4%.

If T and u are outside the ranges specified in Equation 10.39 and Equation 10.41, it was assumed in solving the mathematical models that the values of $c_{ice\text{-}fw}$ and $c_{ice\text{-}bw}$ are equal to zero.

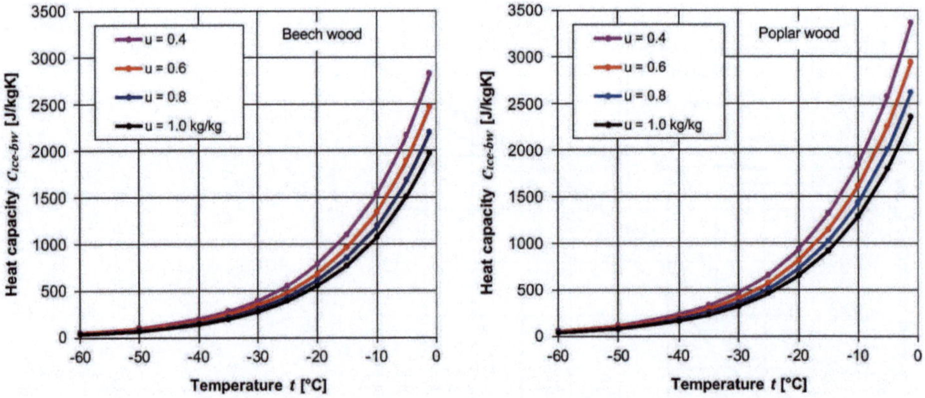

Figure 10.6 Heat capacity of frozen bound water in beech wood (left) and poplar wood (right) as a function of temperature and wood moisture content (source: [24])

10.4.2 Mathematical Description of Thermal Conductivity

The thermal conductivity of wood describes the intensity of heat distribution in wood. For this reason, knowledge of λ_W in the non-frozen and frozen state is also necessary for calculating the thawing and heating processes of wood.

From a thermodynamic perspective, moist wood is a capillary-porous body consisting of three components:

- wood substance,
- water,
- air.

The thermal conductivity of each of the three components and that of ice are different.

The thermal conductivity of water at t = 0 °C is 0.551 W/mK; that of ice at the same temperature is 1.047 W/mK, almost twice as much, and increases at t = –50 °C to 2.780 W/mK [9].

The wood moisture content u is one of the most important factors influencing thermal conductivity. The replacement of air in the wood pores by water leads to a significant increase in thermal conductivity λ_W. The relationship is very complex.

The thermal conductivity of wood increases when the water in the wood is in a frozen state, and decreases when the ice in the wood melts [6–13, 18–20, 25–27]. The thermal conductivity of frozen and non-frozen wood is different.

The influence of temperature t on the thermal conductivity of ice-containing wood differs from that of ice-free wood. As $t$decreases, the thermal conductivity of ice-containing wood increases. The reason for this is, on the one hand, the decreasing amount of non-frozen bound water in the wood as t decreases, whose λ has a significantly lower value compared to ice, and on the other hand, the increase of λ of the ice in the wood as $t$decreases. As t increases in ice-free wood, its λ_W increases as well [6–13, 26].

The bulk density ρ_W, which correlates inversely with the porosity of the wood, also has a considerable influence on the thermal conductivity λ_W. With increasing density (i.e., with decreasing porosity), λ_W increases [7–9, 25, 28]. Not only the porosity, but also the shape, dimensions, and arrangement of the pores influence the thermal conductivity λ_H. Anisotropy also leads to significant differences in heat conduction along the three principal cutting directions (longitudinal, radial, tangential).

In the mathematical description of the thermal conductivity of wood, in addition to the parameters t and u, the fiber saturation point moisture content u_{FSP} is also included. This depends on the wood species. The description is based on the experimentally determined data presented in Figure 10.7 for the change of λ_W as a function of t and u (see [9] and [18]).

The data in Figure 10.7 refer to the change in thermal conductivity perpendicular to the fibers $\lambda_{w\text{-}p}$ of frozen and non-frozen birch wood with the bulk density ρ_b = 515 kg/m^3 and $u_{FSP(293.15)}$ = 0.30 kg/kg (see Table 10.1).

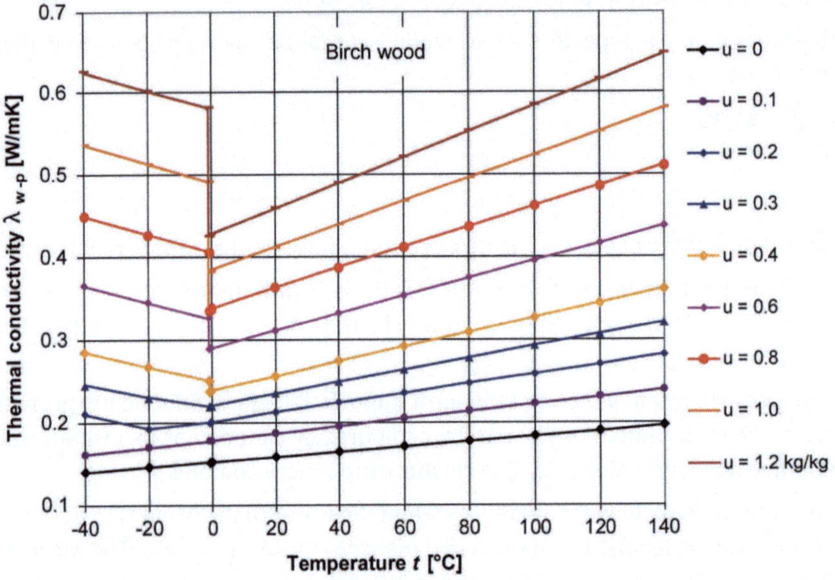

Figure 10.7 Thermal conductivity of frozen and non-frozen birch wood as a function of temperature and wood moisture content (source: based on [9, 18])

From Figure 10.7 it can be seen that above the hygroscopic range, i.e. at $u > u_{FSP(272.15)}$ the temperature t = –1 °C is critical for the thawing of wood, since at this temperature a jump in the values of $\lambda_{w\text{-}p}$ is present and must be taken into account. This jump increases with increasing moisture content u. The reason is the significant difference between the thermal conductivity of water in the liquid and frozen state. The data presented in Figure 10.7 for $\lambda_{w\text{-}p}(t, u)$ as well as the data for $c_{w(t,u)}$ in Figure 10.4 are frequently cited in European and US-American literature.

In the description of λ_w the fiber saturation point of wood $u_{FSP(272.15)}$ as an independent variable for various types of wood. The following equations for calculating λ_w for the ranges $0 \leq u \leq 1.2$ kg/kg and 213.15 K $\leq T \leq$ 413.15 K (i.e. –60 °C $\leq t \leq$ 140 °C) were derived [3–5]:

$$\lambda_w = \lambda_{w0} \cdot \xi \cdot [1 + \mu \cdot (T - 273.15)] \text{ [W/mK]} \tag{10.42}$$

$$\lambda_{w0} = K_{aw} \cdot \upsilon \cdot \left[0.165 + (1.39 + 3.8 \cdot u) \cdot \left(3.3 \cdot 10^{-7} \rho_b^2 + 1.015 \cdot 10^{-3} \rho_b\right)\right] \tag{10.43}$$

$$v = 0.15 - 0.07 \cdot u \tag{10.44}$$

at $u \leq u_{FSP(272.15)} + 0.1$

$$v = 0.1284 - 0.013 \cdot u \tag{10.45}$$

at $u > u_{FSP(272.15)} + 0.1$
λ_{w0} thermal conductivity of wood at T = 273.15, i.e. at t = 0 °C [W/mk]
$u_{FSP(272.15)}$ fiber saturation point at T = 272.15, i.e. at t = -1 °C [kg/kg]
ρ_b bulk density of wood (see Table 10.45 and Section 10.4.3) [kg/m³]

The coefficients ξ and μ in Equation 10.42 are calculated using the following equations:

- If $T \leq 272.15$ K and simultaneously $u > u_{FSP(272.15)}$ or if $T \leq T_{bw\text{-}nfr}$ and simultaneously 0.13 kg/kg $\leq u < u_{FSP(272.15)}$:

$$\xi = 1 + 0.34 \cdot [1.15 \cdot (u - u_{FSP(272.15)})] \tag{10.46}$$

$$\mu = 0.002 \cdot (u - u_{FSP(272.15)}) - 0.0038 \cdot \left(\frac{579}{\rho_{wb}} - 0.124\right) \tag{10.47}$$

- If $T > 272.15$ K at any u or if $T \leq 272.15$ K and simultaneously $0 \leq u < u_{bw\text{-}nfr}$:

$$\xi = 1 \tag{10.48}$$

$$\mu = (2.05 + 4u) \cdot \left(\frac{579}{\rho_{wb}} - 0.124\right) \cdot 10^{-3} \tag{10.49}$$

at $u \leq u_{FSP(272.15)} + 0.1$

$$\mu = 3.65 \cdot \left(\frac{579}{\rho_{wb}} - 0.124\right) \cdot 10^{-3} \tag{10.50}$$

at $u > u_{FSP(272.15)} + 0.1$

For the calculation of λ_w in the case that $T \leq T_{bw\text{-}nfr}$ (see Equation 10.30) and simultaneously 0.13 kg/kg $\leq u < u_{FSP(272.15)}$ is, the value of $\lambda_{w\text{-}nfr}$ is first determined by substituting $T = T_{bw\text{-}nfr}$ into Equation 10.42. The values of ξ are determined from Equation 10.48 and those of μ from Equation 10.49. Then, using the coefficient μ from Equation 10.47 and the already determined $\lambda_{w\text{-}nfr}$, the values of λ_w are calculated as follows:

$$\lambda_w = \lambda_{w-nfr} \cdot [1 + \mu \cdot (T - T_{bw\text{-}nfr})] \tag{10.51}$$

for $T < T_{bw\text{-}nfr}$ as well as 0.125 kg/kg $\leq u \leq u_{FSP(272.15)}$

The value $\lambda_w = \lambda_{w\text{-}nfr}$ forms the inflection point of the curve in the slope of the dependencies $\lambda_w = f(T)$ in Figure 10.7, if 0.13 kg/kg $\leq u < u_{FSP(272.15)}$. Equation 10.51 together with Equation 10.42 enables the calculation of λ_w of frozen wood for all temperatures greater than –60 °C (i.e., T = 213.15 K), since at t = –60 °C the value of $u_{bw\text{-}nfr}$ is approximately equal to 0.13 kg/kg (see Figure 10.2).

The equations developed in [9] and in [1, 3–5, 28] can be used to determine the values of the coefficient K_{aH} in Equation 10.43. This accounts for the influence of the heat flux along the three principal anatomical axes on λ_H at 0 °C, λ_{H0}, as well as on $\lambda_{H\text{-}r}$, $\lambda_{H\text{-}t}$ and $\lambda_{H\text{-}l}$. In [3–5], values for K_{aw} were determined experimentally for selected wood species. In Table 10.2 are the determined values of K_{aw} for four wood species that are frequently steamed or boiled.

Table 10.2 Coefficients K_{aw} for Calculating λ_{w0} According to Equation 10.43 along the Three Anatomical Principal Axes of Wood (source: [3–5])

Wood species	Radial K_r	Tangential K_t	Longitudinal to the grain direction K_l	Perpendicular to the grain direction K_p
Oak	1.22	1.04	1.99	1.13
Pine	1.25	1.37	2.96	1.31
Beech	1.35	1.21	2.40	1.28
Poplar	1.48	1.36	2.88	1.42

Figure 10.8 shows the change in thermal conductivity of frozen and non-frozen beech wood as a function of t and u perpendicular to the grain, $\lambda_{w\text{-}p}$, which is calculated using Equation 10.42–Equation 10.51 where in Equation 10.473 the coefficient for K_p = 1.28 from Table 10.2 is used.

The calculated change in thermal conductivity of non-frozen beech wood as a function of t and u in the radial direction, $\lambda_{w\text{-}r}$, at K_r = 1.35 is shown in Figure 10.9.

In Figure 10.8 the difference between free and bound water over the temperature range on the thermal conductivity is recognizable. An increase in temperature at a given u leads to an increase in $\lambda_{w\text{-}p}$ of non-frozen wood over the entire investigated temperature range from –60 °C to 140 °C. At the same time, with increasing u the slope of the dependencies $\lambda_{w\text{-}p} = f(t)$ and $\lambda_{w\text{-}r} = f(t)$ of non-frozen wood increases (Figure 10.8 and Figure 10.9).

From the diagrams in Figure 10.8 it is evident that an increase in t leads to a decrease in $\lambda_{w\text{-}p}$, both for wood containing only frozen bound water and for wood with frozen bound and free water.

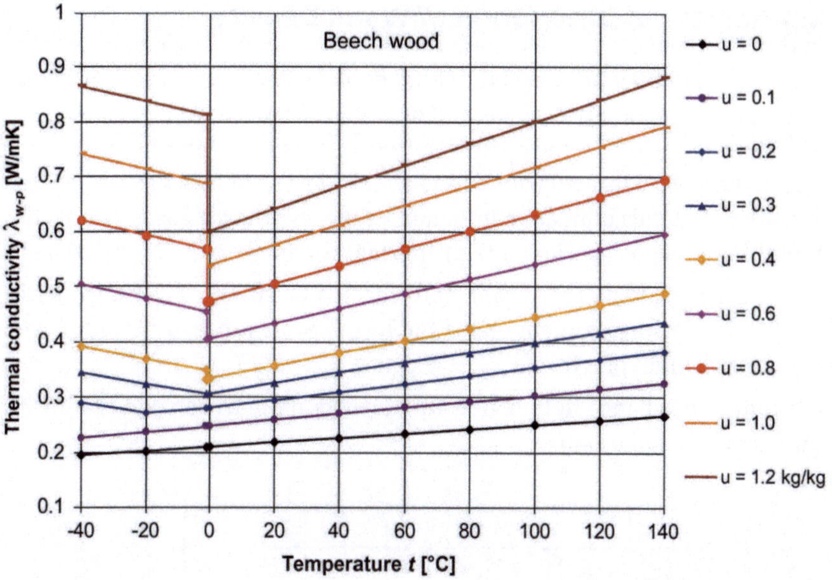

Figure 10.8 Thermal conductivity of frozen and non-frozen beech wood perpendicular to the grain as a function of temperature and wood moisture content (source: [4])

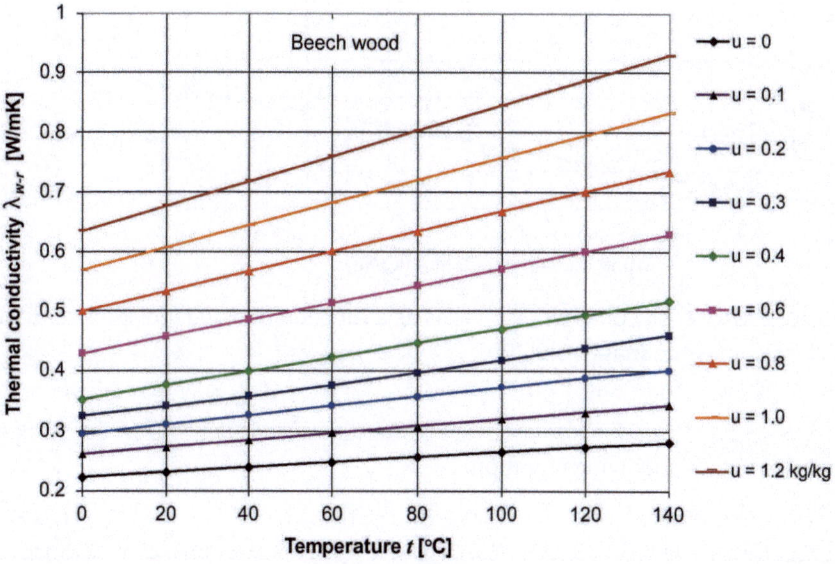

Figure 10.9 Thermal conductivity of non-frozen beech wood in the radial direction as a function of temperature in the range 0 °C $\leq t \leq$ 140 °C and wood moisture content

10.4.3 Mathematical Description of Wood Density

In the models for calculating the thermal treatment of veneer logs and veneer prisms, the bulk density ρ_b is frequently used instead of the density at a given wood moisture content (ρ_w). The bulk density is the ratio of the wood mass in the absolutely dry state to its volume in the state saturated with bound water. The use of bulk density makes it possible to account for the cumulative influence of the specific microstructural characteristics of different wood species on their thermal conductivity [6–13, 19, 20, 29].

To ensure good plasticization of the wood, non-frozen and frozen veneer logs and veneer prisms with a moisture content above the hygroscopic range are typically subjected to heat treatment in practice.

Figure 10.10 shows the change in bulk density of wood in the frozen and non-frozen state as a function of u and ρ_b [29].

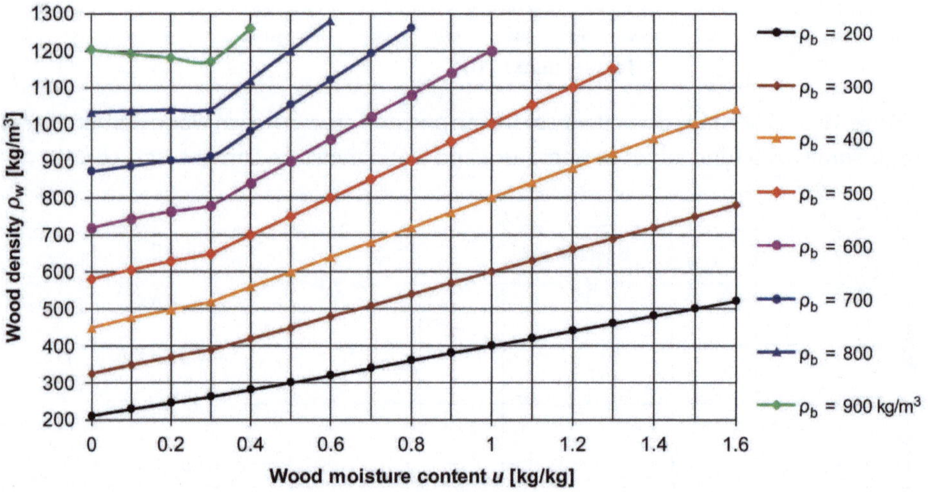

Figure 10.10 Bulk density of all wood species as a function of wood moisture content and bulk density of wood (based on [29])

The bulk density ρ_w of wood whose moisture content u exceeds the hygroscopic range is calculated using the following equation:

$$\rho_w = \rho_b \cdot (1 + u) \text{ [kg/m}^3\text{]} \qquad (10.52)$$

for $u > u_{FSP(272.15)}$

Figure 10.11 shows the calculated change in density of oak, beech, pine, and poplar in the range $u_{FSP(272.15)} \leq u \leq 1.0$ kg/kg according to Equation 10.52 and Table 10.1.

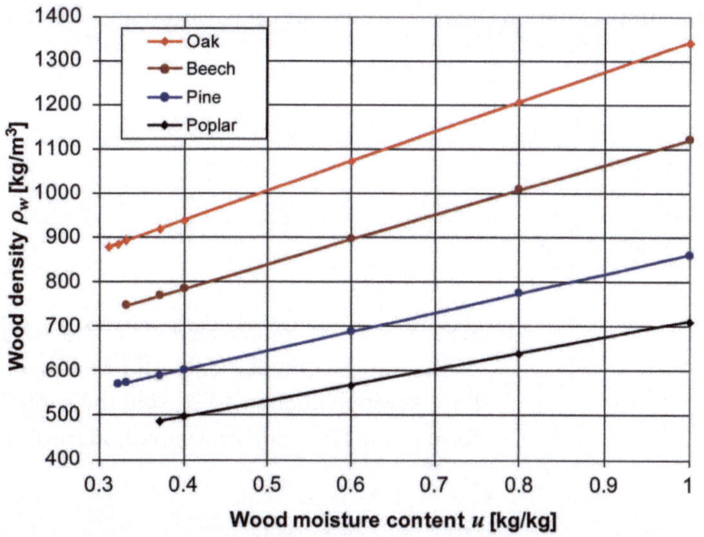

Figure 10.11 Bulk density of oak, beech, pine, and poplar wood as a function of wood moisture content above the hygroscopic range

10.5 Conversion of Mathematical Models into a Form Suitable for Programming Suitable Form

The analytical solution of mathematical models containing nonlinear partial differential equations in the form of Equation 10.3, Equation 10.9, Equation 10.14, Equation 10.18 and Equation 10.23 is impossible under practical boundary conditions without significant simplifications of these equations.

For the numerical solution of models with such equations, finite difference or finite element methods are used. If the wood has prismatic dimensions or a cylindrical cross-section, the finite difference method is preferred.

For the numerical solution of the models presented above, it is advisable to use the explicit form of the finite difference method, which allows the exclusion of any simplifications. With the computing capacities of modern computers, this is easily possible.

According to the basic idea of the finite difference method, the temperature, which is a function of space and time, is represented using a grid vector, and the derivatives $\frac{\partial T}{\partial x}$, $\frac{\partial T}{\partial y}$, $\frac{\partial T}{\partial z}$, and $\frac{\partial T}{\partial \tau}$ are replaced by their finite difference (discrete) analogues along the spatial and temporal coordinates using the computational grid created [30].

For this purpose, the wood prism or one eighth of it, in the presence of mirror symmetry with respect to the other seven eighths, is "pierced" by a system of mutually perpendicular lines running parallel to the three coordinate axes. The distance between the lines, also called the differentiation step, is constant for each coordinate direction (Figure 10.12).

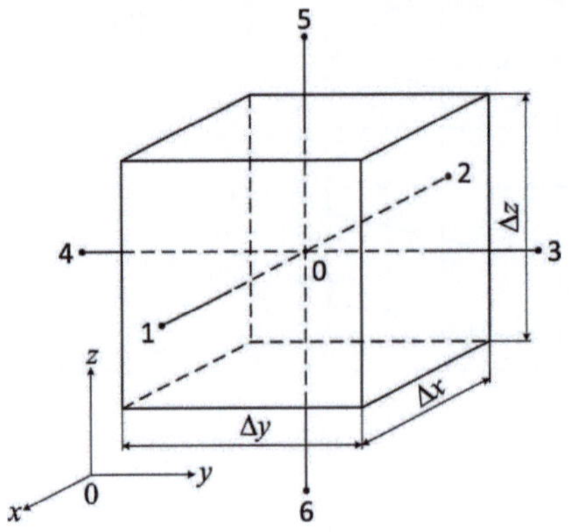

Figure 10.12 A volume element of the wood prism with a computational grid built upon it for solving Equation 10.3 using the explicit form of the finite difference method (source: [1, 3])

The nodes from 0 to 6 with the corresponding temperatures from T_0 to T_6 are centers for the illustrated and adjacent volume elements. The lengths of the sides of the volume element (Δx, Δy and Δz) are differentiation steps whose size determines the distance between the individual nodes of the mesh. The total time of the thermal treatment of the body is also divided into a certain number n of intervals (steps) of equal duration. A volume element of a heated body used for the solution of Equation 10.3 is shown in Figure 10.12 together with its associated part of the rectangular calculation mesh.

Through the described methods, the process of solving the nonlinear partial differential equation (Equation 10.3) is converted into the solution of the equivalent system of linear finite difference equations. This requires carrying out numerous steps, but no complicated algebraic operations. In the embedded 3D rectangular mesh of the prism, heat transfer is considered only along its gradient lines. For each intersection point of three mesh lines, a so-called node, a finite difference equation is derived, by means of which the temperature of this node is calculated.

Finite difference equations are obtained by replacing derivatives in the differential equation (Equation 10.3) with their approximate expressions, which are differences between the values of the function at selected nodes of the calculation mesh. The temperature at node 0 with coordinates (x_i, y_J, z_k, τ_n) in Figure 10.12 is denoted as $T_0 = T^n_{i,j,k}$, and the temperature at adjacent nodes correspondingly as $T_1 = T^n_{i+1,j,k}$ (node 1), $T_2 = T^n_{i-1,j,k}$ (node 2), $T_3 = T^n_{i,j+1,k}$ (node 3), $T_4 = T^n_{i,j-1,k}$ (node 4), $T_5 = T^n_{i,j,k+1}$ (node 5) and $T_6 = T^n_{i,j,k-1}$ (node 6). To calculate the temperature at each subsequent point in time, the temperature values up to the previous point in time must be known.

10.5.1 3D Model for the Temperature Distribution in Prismatic Timber Assortments

The following system of equations as a discrete equivalent of Equation 10.3 was established using the finite difference method and taking into account Equation 10.42 for the thermal conductivity, λ_w, in the various anatomical directions:

$$T_{i,j,k}^{n+1} = T_{i,j,k}^{n} + \frac{\xi \cdot \Delta\tau}{c_{\text{w-eff1,2,3}} \cdot \rho_{\text{w}}} \left\{ \begin{array}{l} \frac{\lambda_{\text{w0-r}}}{\Delta x^2} \left[\left[1 + \mu \cdot \left(T_{i,j,k}^{n} - 273.15\right)\right] \cdot \left(T_{i+1,j,k}^{n} + T_{i-1,j,k}^{n} - 2T_{i,j,k}^{n}\right) + \mu \cdot \left(T_{i,j,k}^{n} - T_{i-1,j,k}^{n}\right)^2 \right] + \\ \frac{\lambda_{\text{w0-t}}}{\Delta y^2} \left[\left[1 + \mu \cdot \left(T_{i,j,k}^{n} - 273.15\right)\right] \cdot \left(T_{i,j+1,k}^{n} + T_{i,j-1,k}^{n} - 2T_{i,j,k}^{n}\right) + \mu \cdot \left(T_{i,j,k}^{n} - T_{i,j-1,k}^{n}\right)^2 \right] + \\ \frac{\lambda_{\text{w0-l}}}{\Delta z^2} \left[\left[1 + \mu \cdot \left(T_{i,j,k}^{n} - 273.15\right)\right] \cdot \left(T_{i,j,k+1}^{n} + T_{i,j,k-1}^{n} - 2T_{i,j,k}^{n}\right) + \mu \cdot \left(T_{i,j,k}^{n} - T_{i,j,k-1}^{n}\right)^2 \right] \end{array} \right\} \quad (10.53)$$

Since in practice no clearly defined radial or tangential annual ring orientation exists, the mean value between the radial and tangential thermal conductivity at 0 °C is calculated, which determines the value perpendicular to the grain direction, $\lambda_{\text{w0-p}}$, as follows:

$$\lambda_{\text{w0-p}} = \frac{\lambda_{\text{w0-r}} + \lambda_{\text{w0-t}}}{2} \quad (10.54)$$

The thermal conductivity at 0 °C parallel to the grain direction can be expressed as follows:

$$\lambda_{\text{w0-l}} = K_{\text{l/p}} \cdot \lambda_{\text{w0-p}} \quad (10.55)$$

where the coefficient $K_{\text{l/p}} = \frac{\lambda_{\text{w0-l}}}{\lambda_{\text{H0-p}}}$ depends on the wood species (see Table 10.2).

To standardize the calculations, it is convenient to use such a step of the calculation grid along the spatial coordinates $\Delta x = \Delta y = \Delta z$ (see Figure 10.13). Taking this condition into account, as well as Equation 10.54 and Equation 10.55, the system of equations (Equation 10.53) is as follows:

$$\begin{aligned} T_{i,j,k}^{n+1} = T_{i,j,k}^{n} + \frac{\lambda_{\text{w0-s}} \cdot \xi \cdot \Delta\tau}{c_{\text{w-eff1,2,3}} \cdot \rho_{\text{H}} \cdot \Delta x^2} \Big\{ & \left[1 + \mu \cdot \left(T_{i,j,k}^{n} - 273.15\right)\right] \\ & \left[T_{i+1,j,k}^{n} + T_{i-1,j,k}^{n} + T_{i,j+1,k}^{n} + T_{i,j-1,k}^{n} + K_{\text{l/p}} \cdot \left(T_{i,j,k+1}^{n} + T_{i,j,k-1}^{n}\right) - (4 + 2K_{\text{l/p}}) \cdot T_{i,j,k}^{n}\right] \\ & + \mu \cdot \left[\left(T_{i,j,k}^{n} - T_{i-1,j,k}^{n}\right)^2 + \left(T_{i,j,k}^{n} - T_{i,j-1,k}^{n}\right)^2 + K_{\text{l/p}} \cdot \left(T_{i,j,k}^{n} - T_{i,j,k-1}^{n}\right)^2\right] \Big\} \end{aligned} \quad (10.56)$$

The initial condition (Equation 10.4) in the 3D model is represented using the following finite difference equation:

$$T^{0}_{i,j,k} = T_{w0} \tag{10.57}$$

The boundary conditions (Equation 10.5) take the following easy-to-program form:

$$T^{n+1}_{1,j,k} = T^{n+1}_{i,1,k} = T^{n+1}_{i,j,1} = T^{n+1}_{m} \tag{10.58}$$

The representation of the nonlinear partial differential equation (Equation 10.3) from the mathematical model by its discrete analogue (Equation 10.56) corresponds to the coordinate system setup and the positioning of the nodes in the mesh shown in Figure 10.13, in which the 3D non-stationary temperature distribution in prismatic wood materials during their steaming is calculated. The calculation mesh for solving the model by the finite difference method is constructed on one eighth of the prism volume due to its mirror symmetry, with the remaining seven-eighths parts of the prism volume.

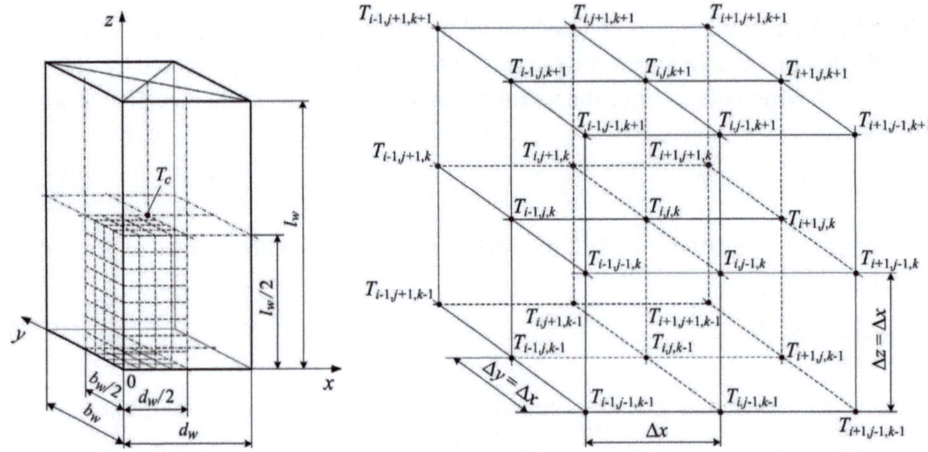

Figure 10.13 Positioning of the nodes of the 3D calculation mesh in a discretized wood prism

The coordinate system setup shown in Figure 10.13 makes it possible, using only one system of equations (Equation 10.56 – Equation 10.58), to calculate the temperature change at each nodal point of the volume of the steamed frozen wood prism at the moment $(n + 1) \cdot \Delta\tau$ using the $n \cdot \Delta\tau$ values already calculated at the previous moment. In [3], extensive experimental investigations were carried out to determine the spatial distribution of temperature in the volume of frozen and non-frozen prismatic and cylindrical wood assortments of oak, beech, poplar, and pine during their steaming. The values of all coefficients in Table 10.2 and also the coefficient $K_{l/p} = \frac{\lambda_{w0-l}}{\lambda_{w0-p}}$ in Equation 10.56 were determined by solving the model with the same initial and boundary conditions in order to achieve maximum agreement between the calculated and experimental results.

Using Equation 10.56 – Equation 10.58, the integration of the differential equation (Equation 10.3) is converted into a consistent determination of the values of T at the nodes of the calculation mesh. For the calculations, the distribution of T at the previous moment, which is separated from the current moment by $\Delta\tau$, is used.

The value of the step along the time coordinate, $\Delta\tau$, is determined by the stability requirement for the solution of the corresponding system of equations. It must not be greater than the smaller of the two values that can be calculated [3, 30]:

$$\Delta\tau = \frac{c_{w-eff1,2,3}(T_{min},u) \cdot \rho_w(u) \cdot \Delta x^2}{K_{aw} \cdot \lambda_w(T_{min},u)} \tag{10.59}$$

$$\Delta\tau = \frac{c_{w-eff1,2,3}(T_{max},u) \cdot \rho_w(u) \cdot \Delta x^2}{K_{aw} \cdot \lambda_w(T_{max},u)} \tag{10.60}$$

Here, T_{min} and T_{max} are the smallest and largest of all temperature values that appear in the initial and boundary conditions of heat transfer during thermal treatment. The coefficient K_{aw}, which accounts for the dimensioning of the heat flux in the calculation of the time step $\Delta\tau$, is as follows: for 3D calculation: $K_{aw} = 6K_{l/p}$; for 2D calculation: $K_{aw} = 4$, and for 1D calculation: $K_{aw} = 2$.

10.5.2 2D Model for Temperature Distribution in Cylindrical Wood Assortments

To obtain a discrete analogue of Equation 10.18–Equation 10.20, the explicit form of the finite difference method is used in a manner comparable to the case of 2D heat treatment of prismatic bodies described above.

The transformation of the nonlinear partial differential equation (Equation 10.18) into its discrete analogue is carried out using the coordinate system shown in Figure 10.14 for the positioning of the nodes of the computational grid, in which the temperature distribution in the longitudinal cross-section of the wood to be steamed is calculated.

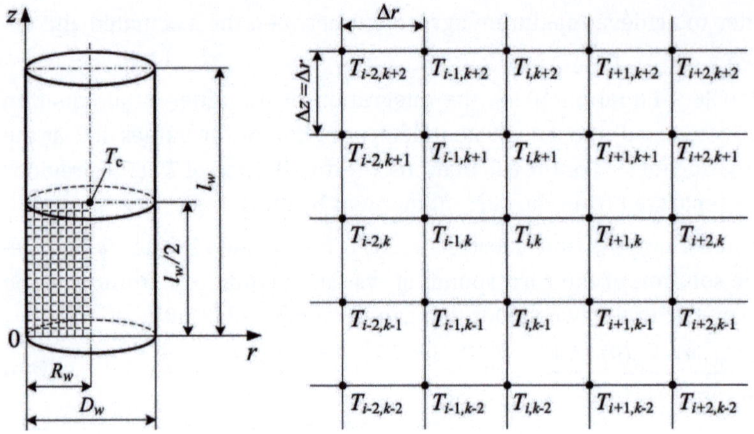

Figure 10.14 Positioning of the nodes of the 2D computational grid in a discretized cylindrical wood assortment

The computational grid for solving the 2D model using the finite difference method is constructed on 1/4 of the longitudinal cross-section of the log, since this is mirror-symmetric to the remaining three quarters of the same cross-section.

Taking into account Equation 10.41 and using the coefficient $K_{\mathrm{l/r}} = \frac{\lambda_{\mathrm{w0-l}}}{\lambda_{\mathrm{w0-r}}}$ are Equation 10.18–Equation 10.20 converted according to the finite difference method:

$$T_{i,k}^{n+1} = T_{i,k}^{n} + \frac{\lambda_{\mathrm{w0-r}} \cdot \xi \cdot \Delta\tau}{c_{\mathrm{w-eff1,2,3}} \cdot \rho_{\mathrm{w}} \cdot \Delta r^2} \left\{ \begin{array}{l} \left[1 + \mu \cdot \left(T_{i,k}^{n} - 273.15\right)\right] \\ \left[\begin{array}{l} T_{i-1,k}^{n} + T_{i+1,k}^{n} + K_{\mathrm{l/r}}\left(T_{i,k-1}^{n} + T_{i,k+1}^{n}\right) - \\ (2 + 2K_{\mathrm{l/r}})T_{i,k}^{n} + \frac{1}{i-1}\left(T_{i-1,k}^{n} - T_{i,k}^{n}\right) \end{array} \right] + \\ \mu \cdot \left[\left(T_{i-1,k}^{n} - T_{i,k}^{n}\right)^2 + K_{\mathrm{l/r}}\left(T_{i,k-1}^{n} - T_{i,k}^{n}\right)^2\right] \end{array} \right\} \quad (10.61)$$

with the initial condition:

$$T_{i,k}^{0} = T_{\mathrm{w0}} \quad (10.62)$$

and the boundary condition:

$$T_{1,k}^{n+1} = T_{i,1}^{n+1} = T_{\mathrm{m}}^{n+1} \quad (10.63)$$

The value of $\Delta\tau$ is determined according to Equation 10.59a and Equation 10.59b. When solving the 2D and 1D models, $K_{\mathrm{aw}} = 4K_{\mathrm{l/r}}$ and $K_{\mathrm{aw}} = 1$ are used.

10.6 Calculation and Visualization of the Temperature Distribution in Heat-Treated Veneer Logs and Veneer Prisms

For the numerical solution of the above-presented models for transient heat conduction in prismatic and cylindrical wood materials during their thermal treatment, a software package was developed on the computing platform Visual FORTRAN Professional [3, 4, 30] developed by Microsoft.

The software package enables the calculation and color visualization (either as an animation of the entire process or as 3D, 2D, or 1D graphics of any desired process moment) of the non-stationary temperature distribution in ice-containing and non-ice-containing materials during steaming or boiling. The calculation of the temperature field change in the volume of wood assortments that contain ice at the beginning of the thermal treatment is correlated for the periods of ice melting and thereafter, describing the flexible spatial boundary of the melting ice.

The calculation of the temperature fields is performed continuously and for the processes of steam treatment and the subsequent cooling of the materials, i.e., the calculation of the non-stationary temperature change in the volume of the materials during the time of their cooling begins with the temperature distribution already reached at the end of the steam treatment during the calculation time.

Based on the calculations, it is possible to determine when the optimal temperatures required for veneer production [31] or the bending of wood parts are reached throughout the entire volume of the steamed wood.

10.6.1 Calculation of the Non-Stationary Temperature Distribution in the Heat-Treated Wood Assortments

Figure 10.15 shows, as an example, the change in the surface temperature of the prisms, calculated using the 3D models (Equation 10.3–Equation 10.5), which is equal to t_m, as well as the temperature at six representative points in the volume of oak and beech prisms with t_{w0} = –40 °C and u = 0.6 kg/kg during their 20-hour heat treatment at t_m = 80 °C. The prisms contain ice: approximately 0.15 kg/kg from the bound water at –40 °C (see Figure 10.3) and the total amount of free water. The coordinates of the representative points are given in the legend of the diagrams.

In Figure 10.16 to Figure 10.18, similar diagrams of temperature change under other heat treatment conditions for various wood species and timber assortments are illustrated.

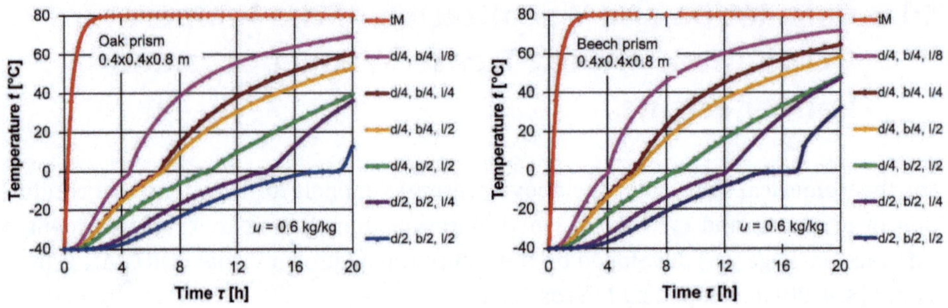

Figure 10.15 3D temperature distribution in oak prisms (left) and beech prisms (right) with dimensions 0.4 × 0.4 × 0.8 m and t_{w0} = –40 °C during 20 hours of heat treatment at t_m = 80 °C (source: [3, 4])

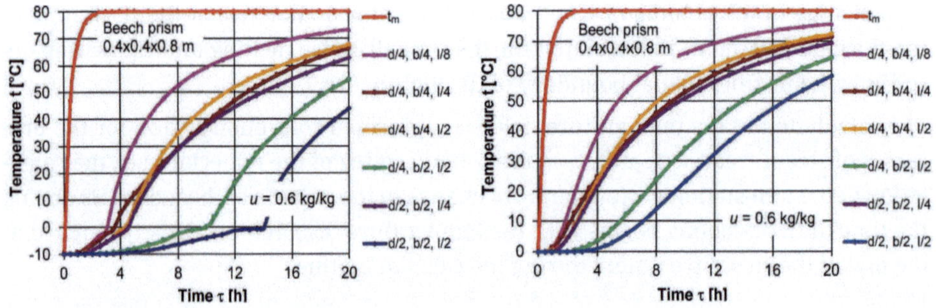

Figure 10.16 3D temperature distribution in beech prisms with dimensions 0.4 × 0.4 × 0.8 m, t_{w0} = –10 °C (left) and t_{w0} = 0 °C (right) during 20 hours of heat treatment at t_m = 80 °C (source: [3, 4])

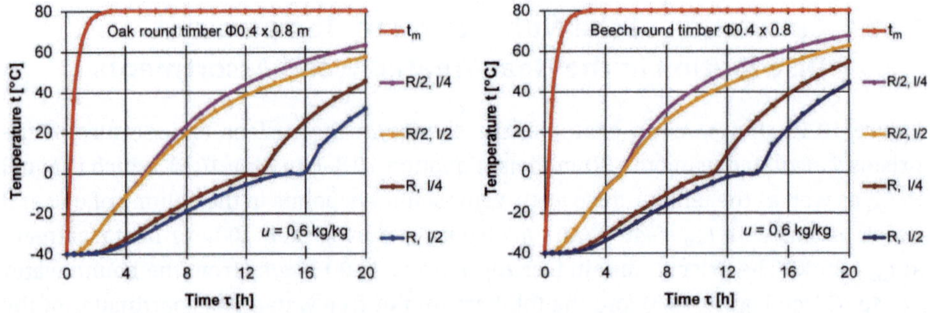

Figure 10.17 2D temperature distribution in oak round timber (left) and beech round timber (right) with dimensions Ø 0.4 × 0.8 m and t_{w0} = –40 °C during 20 hours of heat treatment at t_m = 80 °C (source: [3, 4])

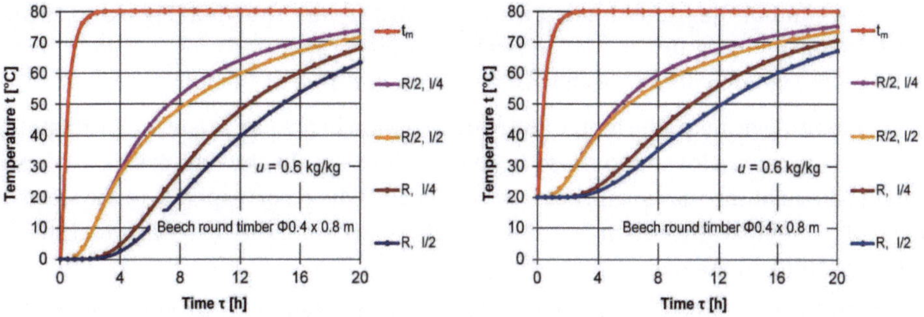

Figure 10.18 2D temperature distribution in beech roundwood with dimensions Ø 0.4 × 0.8 m, t_{w0} = 0 °C (left) and t_{w0} = 20 °C (right) during 20 hours of heat treatment at t_m = 80 °C (source: [3, 4])

In Figure 10.15, Figure 10.16 left and Figure 10.17, the specific, almost horizontal sections of the temperature over a long period in the range of –1 °C to 0 °C can be seen in the inner layers of the materials, while a complete melting of the frozen free water in the wood takes place at these points. The reason for the long-term temperature retention in this range is the very low thermal diffusivity of wood, which is caused by the extremely high specific heat capacity of the frozen free water in the wood during melting (see Figure 10.5).

Complete melting of the frozen water in the center of the investigated timber assortments with t_{w0} = –40 °C occurs after approximately 17 hours for beech prisms and after 19 hours for oak prisms, as well as after 14 hours for beech logs and after 16 hours for oak logs.

The slower thawing and heating of oak wood compared to beech wood is attributable to its lower thermal diffusivity. The main reason for this is, all other conditions being equal, the higher bulk density of oak wood (ρ_b = 670 kg/m^3), compared to beech wood (ρ_b = 560 kg/m^3) (see Table 10.1).

10.6.2 Visualization of the Temperature Distribution in the Steamed Timber Assortments

The results obtained with the 3D model (Equation 10.3–Equation 10.5) calculated transient temperature distribution in the volume of the wood prisms to be thawed, part of which is shown in Figure 10.15 on the right, was visualized using Microsoft Excel 2010 software as follows. The solution of the 3D model was carried out with a step of spatial coordinates $\Delta x = \Delta y = \Delta z$ = 0.01 m (see Figure 10.13), i.e., with a computational grid of 16,000 nodes in one eighth of the volume of each prism.

The color contour diagrams created with this software are shown in Figure 10.19 and illustrate the temperature change in twelve mutually separated cross-sections in one eighth of the volume of the beech prism after 5 h and 10 h of steaming at t_m = 80 °C.

The contour diagrams of the temperature distribution in the oak prism during thawing exhibit features analogous to those in the beech prism. In the oak prism, a certain slowing down of the thawing process is noticeable, which corresponds to the slowing down of the temperature change in denser oak wood compared to beech wood shown in Figure 10.15 on the left.

Figure 10.19 shows that the boundary line between the temperature zones from –8 °C to 0 °C and from 0 °C to 8 °C is distorted when the prism contains both frozen bound and free water [32]. One reason for this is the prolonged residence of t in the range from –1 °C to 0 °C over a long period of time at points located in the inner layers of the material, as shown in Figure 10.15 on the right. This temperature range between –1 °C and 0 °C persists as long as the ice formed by the freezing of free water in the wood is completely thawed.

While the points with not yet completely thawed ice are still within the color range from –8 °C to 0 °C, some of their neighboring points in the computational grid transition to the range from 0 °C to 8 °C after the ice has completely thawed. This explains the distortion of the smoothness of the boundaries between these regions in the contour diagrams at τ = 5 h and τ = 10 h.

In the contour diagrams, a significantly faster temperature rise along the prism length compared to the temperature rise transverse to the wood grain is clearly visible. One reason for this is the considerably higher thermal conductivity along the grain than in the cross-grain direction (see Table 10.2). For example, after 10 hours of thawing the beech prism, the temperature zone from 0 °C to 8 °C moves within the prism as follows:

- At a distance of $x = y$ = 120 mm from the surfaces forming the central cross-section at z = 400 mm, in which heat transfer in the cross-grain direction predominates;
- At a distance of $x = y$ = 180 mm from the surfaces forming the cross-section with z = 160 mm, in which heat transfer in the longitudinal direction to the fibers dominates.

The temperature distribution in the longitudinal section of the logs to be thawed, calculated using the 2D model (Equation 10.18 – Equation 10.20), part of which is shown in Figure 10.17 on the left, was visualized using MS Excel 2010 as follows. The solution of the 2D model was performed with a step along the planar coordinates $\Delta r = \Delta z$ = 0.01 m (see Figure 10.14), i.e., with a computational grid of 800 nodes in one quarter of the longitudinal section of each log.

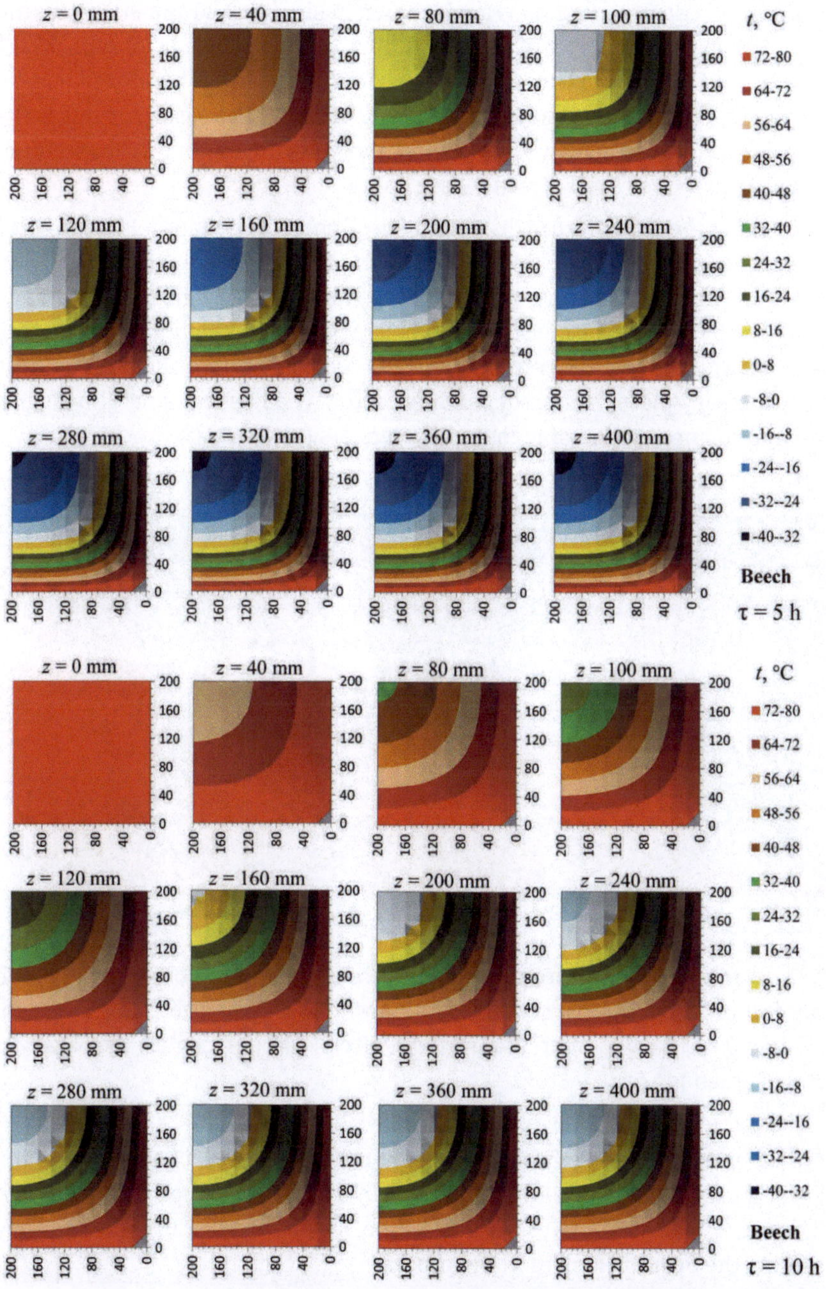

Figure 10.19 Contour diagrams for the temperature distribution in one eighth of the volume of the beech prism subjected to thawing with dimensions 0.4 × 0.4 × 0.8 m, t_{w0} = –40 °C and u = 0.6 kg/kg after 5 hours (top) and after 10 hours (bottom) of steaming at t_m = 80 °C (source: [32])

A part of the created color contour diagram is shown in Figure 10.20 and illustrates the temperature change in one quarter of the longitudinal section of an oak log with $u = 0.6$ kg/kg after 5, 10, and 15 hours of steaming at $t_m = 80$ °C. A faster temperature increase along the log length than along the radius is clearly visible. The reason for this is the approximately 1.63 times higher thermal conductivity coefficient in the longitudinal direction compared to the radial direction (see Table 10.2).

The visualization of the temperature distribution in wood-based materials using color contour diagrams enables the tracking and analysis of the movement of the thawing boundary within the wood volume during heat treatment. The temperature change across and along the fibers is also visible. Using the visualizing contour diagrams, the point in time at which the optimal temperature is reached can be determined. This ensures the necessary plasticization depending on the wood species and dimensions of the wood and the production of high-quality veneers [28, 29, 31].

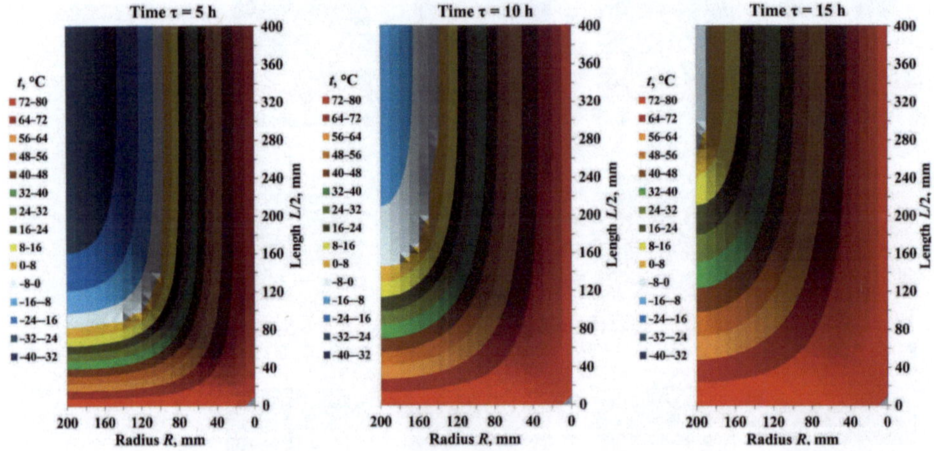

Figure 10.20 Contour diagrams for the temperature distribution in one quarter of the longitudinal section of the oak log subjected to thawing with dimensions Ø 0.4 × 0.8 m, $t_{w0} = -40$ °C and $u = 0.6$ kg/kg after 5 hours (left), after 10 hours (center), and after 15 hours (right) of steaming at $t_m = 80$ °C

The contour diagrams can not only be displayed individually for each time step of the steaming process for detailed examination, but can also be presented together as an animation to observe the overall trend. This can be very helpful for users in industry to easily predict the overall changes in the process.

10.6.3 Calculation of Technological Modes for Steaming Veneer Prisms in an Autoclave

The length of the wood prisms that are steamed for plasticization in veneer production is usually at least four to five times greater than their cross-section. As a result, heat transfer through their end faces can be neglected, since it does not affect the temperature change in the innermost cross-section of the prisms.

For the calculation of steaming modes in an autoclave for veneer prisms, the 2D model represented by Equation 10.9 – Equation 10.13 is suitable. Since the conditions of heat exchange between all surfaces of the prisms and the medium are equal, it is sufficient, when solving the model, to calculate the non-stationary temperature distribution in only one quarter of the prism cross-section. The temperature distribution in the remaining three quarters is mirror-symmetric.

For veneer production, two types of autoclave steaming of prisms are of interest: those with maximum autoclave productivity and those with a duration specified by the control system that is longer than the duration of the first type of mode. In the event of organizational or technical problems in the production lines for cutting and drying veneer, it is necessary to extend the current modes of steam treatment of the wood materials until the problem is resolved, while at the same time ensuring the required plasticity of the materials.

The changes in the temperature of the medium T_m during autoclave steaming of wood prisms with both modes (one basic mode without and three modes with dispatcher intervention) as well as during the subsequent conditioning, which were used in the numerical simulations with the model (Equation 10.9 – Equation 10.13), are shown in Figure 10.21.

The typical temperature–time profile of T_m in these modes comprises the following five sections:

I. Intensive increase of T_m during the time 0–τ_1 by fully opening the valve that directs the water steam into the autoclave;

II. Constant T_m of the steam medium during τ_1–τ_2 of the basic mode (without dispatcher intervention), caused by metered steam supply into the autoclave. In the mode with dispatcher intervention, the maximum temperature T_{m1} of the mode decreases after a certain time interval $\Delta\tau_a$, $\Delta\tau_b$ or $\Delta\tau_c$ from its beginning by $\Delta T_{m1\text{-}a}$, $\Delta T_{m1\text{-}b}$ or $\Delta T_{m1\text{-}c}$ respectively and remains unchanged until the point in time τ_2 = τ_{steam} is reached. Upon reaching $\tau_2 = \tau_{steam}$, $\tau_{2\text{-}a} = \tau_{steam\text{-}a}$, $\tau_{2\text{-}b} = \tau_{steam\text{-}b}$ or $\tau_{2\text{-}c} = \tau_{steam\text{-}c}$ the steam supply to the autoclave is stopped;

III. Decrease of T_m of the steam medium during τ_2–τ_3, caused by the utilization of the heat already accumulated in the autoclave for further heating of the prisms;

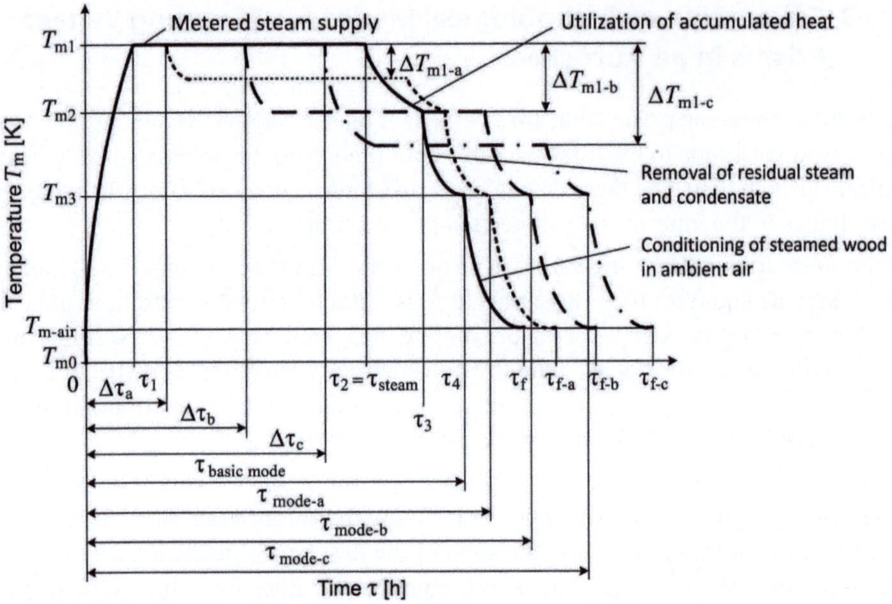

Figure 10.21 $T_m$change during steaming of prisms in an autoclave without and in the presence of dispatcher intervention and during their conditioning in an air environment (source: [33, 34])

IV. Decrease of T_m of the steam medium during τ_3–τ_4, caused by the opening of the valves that direct the residual steam and condensate out of the autoclave;

V. Decrease of T_m in the vicinity of the steamed prisms from the autoclave during τ_4–τ_f.

For the parameters of the modes, whose symbols are given in Figure 10.21, the following values were specified: T_{m0} = 253.15 K (i.e. t_{m0} = –20 °C); T_{m1} = 403.15 K (i.e. t_{m1} = 130 °C); $\Delta T_{m1\text{-}a}$ = 10 K = 10 °C, $\Delta T_{m1\text{-}b}$ = 20 K = 20 °C and $\Delta T_{m1\text{-}c}$ = 30 K = 30 °C; T_{m2} = 383.15 K (i.e. t_{m2} = 110 °C); T_{m3} = 353.15 K (i.e. t_{m3} = 80 °C); $T_{m\text{-}air}$ = 293.15 K (i.e. $t_{m\text{-}air}$ = 20 °C); $\Delta\tau_a$ = 3 h, $\Delta\tau_b$ = 5 h, $\Delta\tau_c$ = 7 h and $\Delta\tau_a$ = 3 h, $\Delta\tau_b$ = 7 h, $\Delta\tau_c$ = 11 h during the damping of non-frozen and frozen prisms, respectively.

With these values, basic modes (Mode 0 and Mode 10) without dispatcher intervention and nine modes with dispatcher intervention were calculated, which are given in Table 10.3 for prisms with t_{w0} = 0 °C and in Table 10.4 for prisms with t_{w0} = –20 °C as Mode 1 to 9 and Mode 11 to 19, respectively.

In Table 10.3 and Table 10.4, the temperatures and durations of the two phases (before and after the dispatcher intervention) of Modes 1 to 9 and 11 to 19, as well as the total duration of steam injection into the autoclave, τ_{steam}, and the duration, τ_{mode}, are given for all investigated modes.

Table 10.3 Change in Temperature t_{m1} and the Durations τ_{steam} and τ_{mode}, during Steaming of Beech Prisms with $d_w \times b_w$ = 0.4 × 0.4 m, t_{w0} = 0 °C and u = 0.6 kg/kg in an Autoclave (source: [33])

Steam mode	First section of t_{m1} [°C]	Second section of t_{m1} [°C]	Duration of the first section of t_m, τ_I [h]	Duration of the second section of t_m, τ_{II} [h]	Duration of steaming $\tau_2 = \tau_{steam} = \tau_I + \tau_{II}$ [h]	Duration of steaming mode $\tau_4 = \tau_{mode}$ [h]
Mode 0	130	–	9.15	–	9.15	12.65
Mode 1	130	120	3.0	6.90	9.90	13.20
Mode 2	130	120	5.0	4.65	9.65	12.95
Mode 3	130	120	7.0	2.40	9.40	12.70
Mode 4	130	110	3.0	9.70	12.70	14.20
Mode 5	130	110	5.0	7.10	12.10	13.60
Mode 6	130	110	7.0	4.50	11.50	13.00
Mode 7	130	100	3.0	11.15	14.15	15.65
Mode 8	130	100	5.0	8.15	13.15	14.65
Mode 9	130	100	7.0	5.15	12.15	13.65

In Figure 10.22 and Figure 10.23 the calculated change of the surface temperature, which equals t_m during heat treatment, and also of the average mass temperature $t_{w\text{-}avg}$ and t of two representative points t_1 (with coordinates $d_w/8$, $b_w/8$) and t_2 (with coordinates $d_w/2$, $b_w/2$) of the non-frozen prisms during mode 0, mode 5 and mode 8 and of the frozen prisms during mode 10, mode 15 and mode 18 at a loading degree of the autoclave with prisms of f = 50% and a limited heat output of the steam source of q = 500 kW are shown.

As can be seen from Figure 10.22 and Figure 10.23, the temperature change at the representative points of the prisms during the steaming and conditioning process follows very complex curves.

Table 10.4 Change in Temperature t_{m1} and the Durations τ_{steam} and τ_{mode}, when Steaming Frozen Beech Prisms with $d_w \times b_w$ = 0.4 × 0.4 m, t_{w0} = –20 °C and u = 0.6 kg/kg (Source: [34])

Steam mode	First section of t_{m1} [°C]	Second section of t_{m1} [°C]	Duration of the first section of t_m, τ_I [h]	Duration of the second section of t_m, τ_{II} [h]	Duration of steaming $\tau_2 = \tau_{steam} = \tau_I + \tau_{II}$ [h]	Duration of steam mode $\tau_4 = \tau_{mode}$ [h]
Mode 10	130	–	13.9	–	13.9	17.4
Mode 11	130	120	3.0	13.2	16.2	18.7
Mode 12	130	120	7.0	8.7	15.7	18.2
Mode 13	130	120	11.0	4.2	15.2	17.7
Mode 14	130	110	3.0	15.7	18.7	20.2
Mode 15	130	110	7.0	10.7	17.7	19.2
Mode 16	130	110	11.0	5.7	16.7	18.2
Mode 17	130	100	3.0	17.7	20.7	22.2
Mode 18	130	100	7.0	12.1	19.1	20.6
Mode 19	130	100	11.0	6.5	17.5	19.0

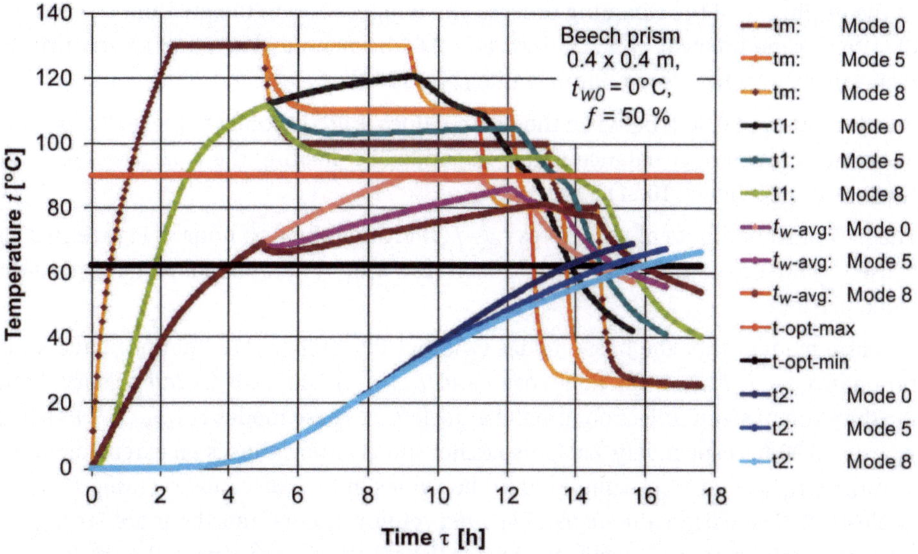

Figure 10.22 Changes in t_m, $t_{w\text{-avg}}$, t_1 and t_2 of non-frozen prisms with t_{w0} = 0 °C and u = 0.6 kg/kg during their steaming in an autoclave with dispatching intervention in basic mode in the 5th hour, depending on t_{m1} and τ (source: [33])

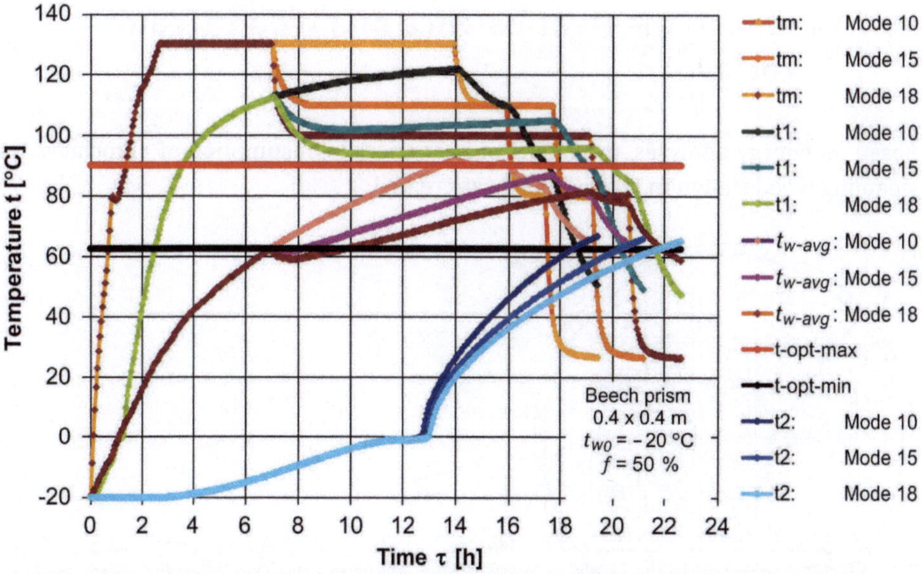

Figure 10.23 Changes in t_m, $t_{w\text{-avg}}$, t_1 and t_2 of frozen prisms with t_{w0} = –20 °C and u = 0.6 kg/kg during their steaming in the autoclave with dispatching intervention in basic mode in the 7th hour, depending on t_{w1} and τ (source: [35])

At the beginning of the steaming process, the temperature of the medium t_m rises until, after a time interval τ_1 = 2.35 h or τ_1 = 2.65 h for non-frozen and frozen prisms, respectively, reaches its maximum value t_{m1} = 130 °C.

The difference of 0.30 h between the two $\tau_1$values is attributable to the larger portion of the autoclave's heat balance that is expended on heating the frozen wood in the range 0–τ_1 (see Figure 10.21).

The lack of smoothness of the curves $t_m = f(\tau)$ around the first hour in Figure 10.23 is caused by the uneven melting of the frozen free water at individual points within the wood prisms.

It is also evident that the basic modes (without dispatcher intervention) have a duration of τ_{mode} = 12.65 h for non-frozen and τ_{mode} = 17.40 h for frozen prisms, while the duration of steam injection into the autoclave in these modes is τ_{steam} = 9.15 h and τ_{steam} = 13.90 h, respectively. Each dispatcher intervention causes an extension of the durations τ_{steam} and τ_{mode} compared to the values in the basic modes (Table 10.3 and Table 10.4). The earlier the dispatcher intervention occurs and the more strongly it leads to a reduction of t_{m1} leads, the longer the specified durations of the modes.

The longest duration τ_{mode} – 15.65 h each for non-frozen prisms and 22.20 h for frozen prisms – were obtained in mode 7 and mode 17.

10.7 Modeling the Heat Balance of a Wood Steam Autoclaveen

Based on energy analyses, the model of heat energy consumption of autoclaves for steaming wood shown in Figure 10.24 was created.

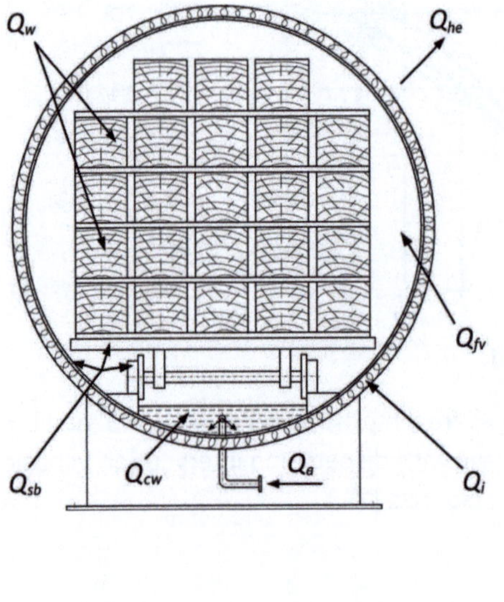

Figure 10.24 Structural model of heat distribution in an autoclave for steaming wood

The non-stationary heat balance of the steaming autoclave can be mathematically described by the model shown in Equation 10.64. This was experimentally validated in [3].

$$Q_a^n = Q_w^n + Q_{sb}^n + Q_i^n + Q_{he}^n + Q_{fV}^n + Q_{cw}^n \quad (10.64)$$

Q_a specific heat energy (heat energy required to heat 1 m^3 of wood) supplied to the autoclave [kWh/m^3]
Q_w energy used for heating the wood materials subjected to steaming [kWh/m^3]
Q_{sb} energy used for heating the steel body of the autoclave and the metal trolleys in the autoclave [kWh/m^3]
Q_i energy used for heating the insulation of the autoclave [kWh/m^3]
Q_{he} energy used for heat dissipation [kWh/m^3]
Q_{fv} energy used for filling the free volume of the autoclave with steam [kWh/m^3]
Q_{cw} Energy that accumulates in the condensate water that collects at the bottom of the autoclave during steaming [kWh/m^3]
n Time coordinate used to perform the numerical solution of the mathematical models [s]

The mathematical description of all components of the heat balance (Equation 10.64), depending on the influencing factors, were given in [3, 4] as follows:

$Q_w = f[c_w(x, y, \tau), \rho_w, T_{w0}, T_w(x, y, \tau), T_{w\text{-}avg}, S_w]$;

$Q_{sb} = f[c_{sb}, \rho_{sb}, V_{sb}, T_{m0}, T_m(\tau), V_a, f]$;

$Q_i = f[c_i(z, \tau), \rho_i, T_{I0}, T_i(z, \tau), d_i, V_a, f]$;

$Q_{he} = f[F_l, F_{ni}, T(d_l, \tau), T_{air}(\tau)\ \alpha_i(\tau), \alpha_{ni}(\tau), V_a, f]$;

$Q_{fv} = f[\rho_{steam}(\tau), V_a, f]$,

$Q_{cw} = f[m_{cw}(\tau), h_{cw}(T_m, \tau), V_a, f]$.

The meaning of the variables in these dependencies is as follows:

S_w cross-sectional area of the prism [m^2]

c_w, c_{sb} and c_{ila} heat capacities of the wood prisms, the steel body of the autoclave, and the insulating layer of the autoclave, respectively [J/kgK]

ρ_w, ρ_{sb} and ρ_{steam} density of the wood prisms, the steel body of the autoclave, and the water vapor in the autoclave, respectively [kg/m^3]

$T_{w\text{-}avg}$ average temperature of the wood [K]

T_{m0} and T_{l0} initial temperature of the medium in the autoclave and the insulating layer, respectively [K]

T_m current value of the temperature of the medium in the autoclave [K]

T_{air} temperature of the air near the autoclave [K]

V_a internal volume of the autoclave [m^3]

V_{Sk} volume of the steel body of the autoclave [m^3]

f loading degree (filling) of the autoclave with prisms relative to its internal volume [m^3/m^3]

x, y coordinates along the thickness and width of the wood prisms in the 2D model of their non-stationary heating (see Equation 10.9 – Equation 10.13) [m]

z coordinate along the thickness of the insulating layer in the 1D model of its non-stationary heating [m]

d_i thickness of the insulating layer of the autoclave [m]

F_i and F_{ni} area of the insulating layer and the non-insulated surfaces of the autoclave, respectively [m^2]

α_i and α_{ni} convective heat transfer coefficient between the ambient air and the insulated layer or the non-insulated surfaces of the autoclave [W/m^2 K]

m_{cw} mass of the condensate water collected in the autoclave [kg]

h_{cw} enthalpy of the condensate water collected in the autoclave [J/kg]

τ time [s].

The following equation can be used to calculate Q_w using the mathematical 2D model for steaming of frozen prismatic wood materials:

$$Q_w^n = \frac{\rho_w}{3.6 \cdot 10^6 F_f} \cdot \left\{ \iint_{F_w} \frac{c_{w-eff1,2,3} \text{ bei } T_{i,k}^n + c_{w-eff1,2,3} \text{ bei } T_{w0}}{2} \cdot \left(T_{i,j}^n - T_{w0}\right) dF_w \right\} \quad (10.65)$$

$$F_w = \frac{d_w \cdot b_w}{4} \quad (10.66)$$

ρ_w bulk density of the wood [kg/m^3]
F_w area of one quarter of the prism cross-section [m^2]
d_w thickness of the prisms [m]
b_w width of the prisms [m]
$c_{w\text{-eff1,2,3}}$ effective specific heat capacities of the prisms during their thawing [J/kgK]
i and j current values of the nodal points of the calculation grid along the thickness and width of the prisms, respectively [–]
n current value of the time coordinate when solving the model

With the results obtained by Equation 10.64 and Equation 10.65, it is possible to calculate the current value of the thermal efficiency of the various modes for steaming wood-based materials in the autoclave, η, according to the following equation:

$$\eta = 100 \frac{Q_{w-max}^n}{Q_{a-max}^n} [\%] \quad (10.67)$$

where $Q_{w\text{-max}}$ and $Q_{a\text{-max}}$ are the maximum values of Q_w and Q_a calculated with the models in a specific autoclave steaming mode.

10.8 Calculation of Energy Consumption and Efficiency of the Autoclave when Steaming Veneer Prisms

In Figure 10.25 and Figure 10.26, the calculated values of the surface temperature of the wood prisms, which equals t_m, and also of the energy consumption of the autoclave, Q_a, and the energy consumption for heating the wood, Q_w, during steaming modes 0, 4, 5, 6 and 10, 14, 15, 16, respectively (see Table 10.3 and Table 10.4), are presented.

In Table 10.5 and Table 10.6 is the variation of $\tau_2 = \tau_{steam}$ (see Figure 10.21), the average mass temperature of the wood, $t_{w\text{-}avg}$, the maximum values of Q_a and Q_w, $Q_{a\text{-}max}$ and $Q_{w\text{-}max}$, respectively, as well as the efficiency of the autoclave, η, during the investigated steaming modes. The data refer to the basic steaming modes 0 and 10 as well as to modes with dispatcher intervention, namely 4, 5 and 6 for non-frozen prisms with t_{w0} = 0 °C and 14, 15, 16 for frozen prisms with t_{w0} = –20 °C.

Figure 10.25 and Figure 10.26 show that the extension of the steaming time causes a gradual and uniform increase of Q_w and Q_a in the basic modes for steaming. Due to the influence of the last five terms on the right-hand side of Eq. 10.64, the rate of increase of Q_a is greater than that of Q_w. This means that the differences between the corresponding Q_a and Q_w equal the sum of the energies that supply the energy consumption described by the last five terms in Equation 10.64.

The uniform increase of Q_a and Q_w during the steaming process is disrupted at the moments when the dispatcher intervenes in the steaming modes of the prisms. The earlier such an intervention occurs, or the greater the temperature change Δt_{m1} in the mode, the more significant the difference between the values of Q_a and Q_w in the basic mode and in the corresponding mode with dispatcher intervention [33, 34].

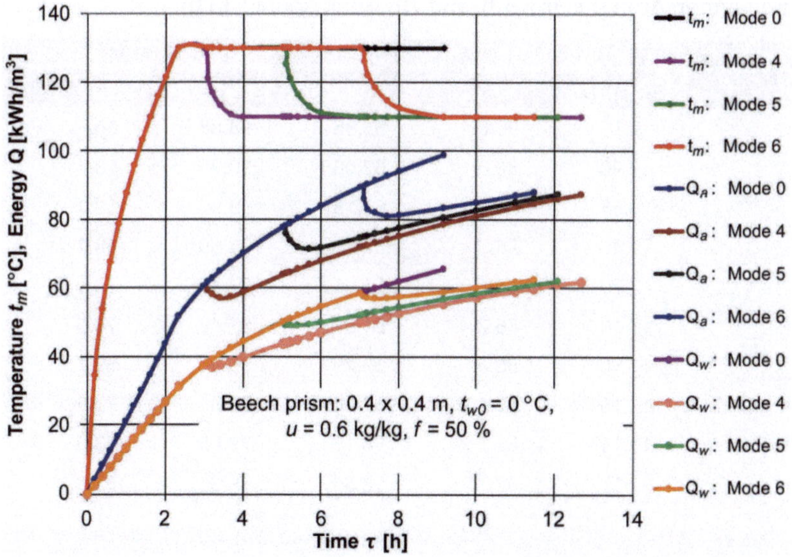

Figure 10.25 Changes of t_m, Q_a and Q_w during steaming modes 0, 4, 5 and 6, depending on τ (Source: [33])

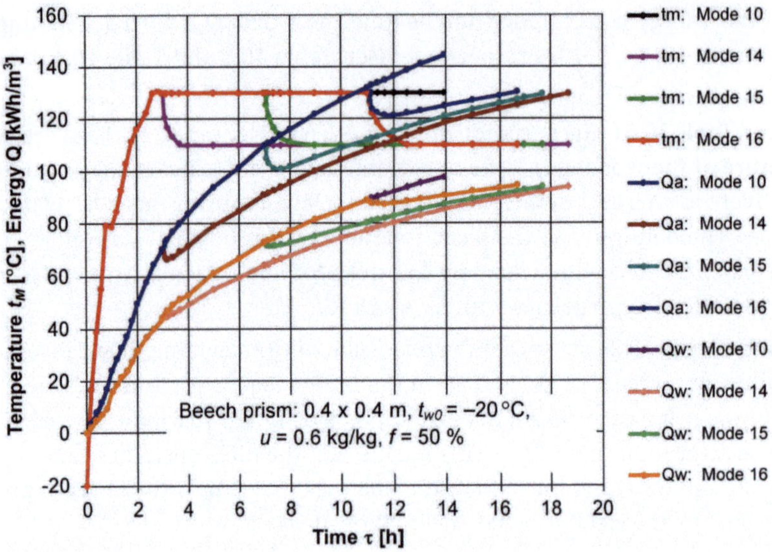

Figure 10.26 Changes of t_m, Q_a and Q_w during steaming modes 10, 14, 15 and 16, depending on τ (Source: [35])

Table 10.5 Change in Energy Consumption and Efficiency of the Damping Modes of Unfrozen Beech Prisms with t_{w0} = 0 °C and u = 0.6 kg/kg in an Autoclave, depending on Dispatcher Intervention $\Delta\tau$ in the 3rd, 5th, and 7th Hour (source: [33])

Steam mode	$\Delta\tau$ [h]	$\tau_2 = \tau_{steam}$ [h]	$t_{w\text{-}avg}$ at τ_{steam} [°C]	$Q_{w\text{-}max}$ [kWh/m³]	$Q_{a\text{-}max}$ [kWh/m³]	η [%]
Mode 0	0	9.15	90.0	65.35	98.88	66.1
Mode 1	3	9.90	86.0	62.06	90.76	68.4
Mode 2	5	9.65	86.2	62.18	90.91	68.4
Mode 3	7	9.40	86.7	62.61	91.50	68.4
Mode 4	3	12.70	86.1	61.91	87.36	70.9
Mode 5	5	12.10	86.2	61.97	87.43	70.9
Mode 6	7	11.50	86.6	62.29	89.35	69.7
Mode 7	3	14.15	81.1	57.96	79.44	73.0
Mode 8	5	13.15	81.3	58.15	79.65	73.0
Mode 9	7	12.15	82.2	58.84	89.35	65.9

Table 10.6 Change in Energy Consumption and Efficiency of the Drying Modes of Frozen Beech Prisms with t_{w0} = –20 °C and u = 0.6 kg/kg in an Autoclave, depending on the Dispatcher Intervention $\Delta\tau$ in the 3rd, 7th, and 11th hour (source: [35])

Steam mode	$\Delta\tau$ [h]	$\tau_2 = \tau_{steam}$ [h]	$t_{w\text{-}avg}$ at τ_{steam} [°C]	$Q_{w\text{-}max}$ [kWh/m³]	$Q_{a\text{-}max}$ [kWh/m³]	η [%]
Mode 10	0	13.9	91.7	97.95	144.08	68.0
Mode 11	3	16.2	90.2	96.44	137.19	70.3
Mode 12	7	15.7	90.0	96.34	137.06	70.3
Mode 13	11	15.2	90.5	96.69	137.49	70.3
Mode 14	3	18.7	87.3	94.01	129.65	72.5
Mode 15	7	17.7	87.1	93.86	129.44	72.5
Mode 16	11	16.7	87.9	94.50	131.21	72.0
Mode 17	3	20.7	81.8	89.60	120.11	74.6
Mode 18	7	19.1	81.7	89.53	119.98	74.6
Mode 19	11	17.5	83.0	90.55	131.75	68.7

In Figure 10.27, the change in thermal efficiency η calculated using Equation 10.67 is shown for the investigated modes. The efficiency varies between 66.1 and 73.0 % at t_{w0} = 0 °C and between 68.0 % and 74.6 % at t_{w0} = –20 °C.

If the dispatcher intervenes in the 3rd or 5th hour of the modes with t_{w0} = 0 °C and in the 3rd or 7th hour of the modes with t_{w0} = –20 °C, a greater reduction in t_{m1} leads to a larger increase in η compared to η of the basic mode.

When a dispatcher intervention is applied at the end of the modes, this dependency is interrupted in the modes with the smallest investigated reduction in t_{m1} between 130 and 100 °C. The efficiency then amounts to η 66.4 % at t_{h0} = 0 °C and 68.7% at t_{h0} = –20 °C, i.e., it is only slightly larger than η of the basic mode.

The reason for this are the much higher $Q_{a\text{-}max}$values in these cases compared to the $Q_{a\text{-}max}$values of the modes with earlier dispatcher interventions, when t_{m1} decreases between 130 and 100 °C (see the values of $Q_{a\text{-}max}$ for modes 9 and 19 respectively in Table 10.5 and Table 10.6).

The thermal efficiency for steaming frozen beech prisms is approximately 2% higher than that for non-frozen prisms.

The methods presented in Section 10.7 and Section 10.8 can be used in the development of software for energy-efficient automatic control of steaming.

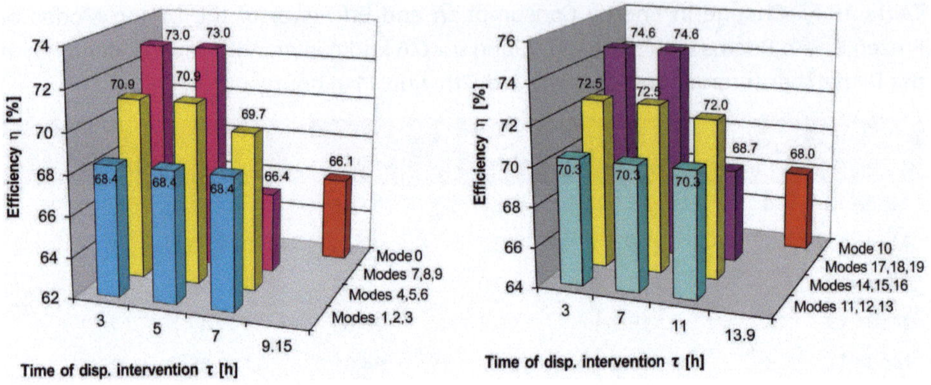

Figure 10.27 Change in efficiency of the investigated steaming modes for beech prisms with t_{w0} = 0 °C (left) and t_{w0} = –20 °C (right), depending on τ

10.9 Modeling the Heat Balance of a Concrete Pits for Steaming or Cooking of Veneer Logs

The heat balance of a concrete pit during steaming or cooking of veneer logs is mathematically represented by the following steady-state models [36–39]:

- when steaming logs

$$Q_{\text{pit-steaming}} = Q_{\text{wood}} + Q_{\text{construction}} + Q_{\text{condensation}} + Q_{\text{heating}} + Q_{\text{heat losses}} \tag{10.68}$$

- when cooking logs

$$Q_{\text{pit-steaming}} = Q_{\text{wood}} + Q_{\text{construction}} + Q_{\text{hot water}} + Q_{\text{heating}} + Q_{\text{heat losses}} \tag{10.69}$$

Figure 10.28 shows the structure of the concrete pit, whose heat balance is mathematically described in the modeling.

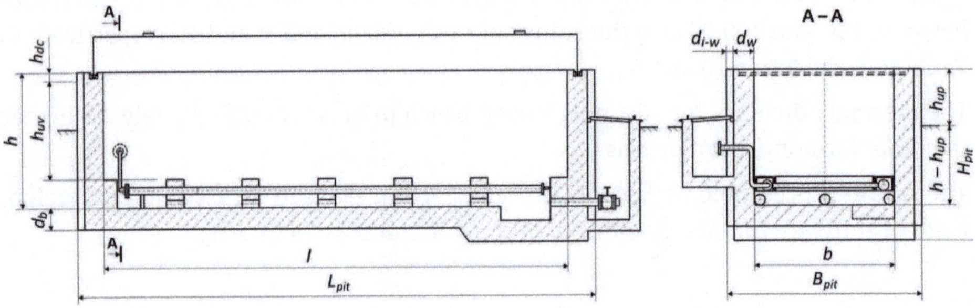

Figure 10.28 Longitudinal and cross-section of a concrete pit for steaming or cooking of wood materials (source: [37–39])

The required operating temperature t_m in the pit during steaming of the logs is achieved by direct saturated steam, which is introduced via perforated heating elements of a pipe heater in the lower part of the pit.

The heating of the water in the pit to the required operating temperature t_m during cooking of the logs is carried out indirectly via non-perforated heating elements.

The symbols, units and values of the main parameters of the concrete pit, which are also contained in the equations of the models of the individual components of the heat balance of the pit and were used for the computer simulations, are given in Table 10.7.

Table 10.7 Main Parameters of the Concrete Pit Used to Solve the Mathematical Models (source: [37]–[39])

Parameter	Symbol	Unit	Value
1. Length of the pit (usable volume)	l	m	6.6
2. Width of the pit (usable volume)	b	m	2.0
3. Depth of the pit (usable volume)	h_{uv}	m	1.52
4. Working volume of the pit, equal to $l \cdot b \cdot h_{uv}$	V_{pit}	m^3	20
5. Depth of the pit walls	h	m	2.0
6. Depth of the upper (above-ground) part of the pit	h_{up}	m	0.8
7. Distance of the drainage channel from the top edge of the pit	h_{dc}	m	0.13
8. Thickness of the walls and the floor of the pit	d_w, d_b	m	0.3
9. Thickness of the insulating layers of the walls and the steel cover	$d_{i\text{-}w}$, $d_{i\text{-}l}$	m	0.1
10. Thickness of the steel sheets of the pit cover	d_{Fe}	m	0.004
11. Density of the concrete walls and the floor of the pit	ρ_{wb}	kg/m^3	2,300
12. Density of the insulation of the walls and the cover, respectively	$\rho_{i\text{-}w}$, $\rho_{i\text{-}lid}$	kg/m^3	350
13. Density of steel parts	ρ_{Fe}	kg/m^3	7,850
14. Density of boiling water in the pit	ρ_{H2O}	kg/m^3	998
15. Initial temperature of the water or steam in the pit	t_{m0}	°C	10
16. Initial temperature of the concrete walls and the floor	t_{w0}, $t_{b0,\ soil0}$	°C	10
17. Initial temperature of the insulation of the walls and the floor	t_{iw0}, t_{ib0}	°C	10

Parameter	Symbol	Unit	Value
18. Maximum temperature of the water or steam in the pit	t_{m1}	°C	70, 80, 90
19. Specific heat capacity of water	c_{H2O}	J/kgK	4,180
20. Specific heat capacity of concrete (mean value)	c_{wb}	J/kgK	1,134
21. Specific heat capacity of the insulation and the lid	$c_{i\text{-}w}$, $c_{i\text{-}lid}$	J/kgK	850
22. Specific heat capacity of the steel lid	c_{Fe}	J/kgK	444
23. Thermal conductivity of concrete	$\lambda_{concr.}$	W/mK	1.28
24. Thermal conductivity of insulation	λ_{ilp}	W/mK	0.042
25. Thermal conductivity of the ground (sand, loam)	λ_{soil}	W/mK	1.4
26. Specific mass of the radiator (surface density)	$m_{radiator}$	kg/m^2	100
27. Temperature of the heat transfer medium in the radiator of the pit	$t_{radiator}$	°C	130
28. Temperature of the ambient air of the pit	t_{air}	°C	10
29. Loading degree of the pit	f	%	45, 60, 75
30. Volume of wood during steaming and cooking, equal to $f \cdot V_{pit}$	V_w	m^3	9, 12, 15

10.9.1 Mathematical Model of Q_{wood}

10.9.1.1 Modeling Q_{wood} for Frozen wood

Using the specific heat capacity of frozen wood listed in Section 10.4.1, the thermal energy Q_{wood} used for thawing and heating can be formulated as follows:

$$Q_{wood} = Q_{w-fr} + Q_{ice\text{-}bw} + Q_{ice\text{-}fw} + Q_{w-nfr}\ [\text{kWh/m}^3] \quad (10.70)$$

$Q_{w\text{-}fr}$ Energy required to heat the frozen wood so that the frozen bound water contained in it melts [kWh/m^3]
$Q_{ice\text{-}bw}$ Energy required to melt the frozen bound water in the wood [kWh/m^3]
$Q_{ice\text{-}fw}$ Energy required to melt the total amount of frozen free water in the wood [kWh/m^3]
$Q_{w\text{-}nfr}$ Energy required to heat the wood in the absence of ice until the set average temperature, $T_{avg\text{-}end}$ is reached, which corresponds to the desired state of the wood at the end of steaming or cooking, [kWh/m^3]

The individual components of the energy Q_{wood} required for the complete thawing and steaming or cooking of frozen wood can be determined using the following equations at the initial temperature of the wood $T_{w0} \leq 272.15$ K and $T_{avg\text{-}end} > 273.15$ K (e.g., $t_{w0} \leq -1$ °C and $t_{avg\text{-}end} > 0$ °C) can be calculated [3]:

$$Q_{\text{w–fr}} = \frac{\rho_{\text{w}} \cdot c_{\text{w–fr-avg}}}{3.6 \cdot 10^{6}} \cdot (272.15 - T_{\text{w0}}) \tag{10.71}$$

$$Q_{\text{ice-wb}} = \frac{\rho_{\text{ice}} \cdot c_{\text{ice-bw-avg}}}{3.6 \cdot 10^{6}} \cdot (272.15 - T_{\text{w0}}) \tag{10.72}$$

$$Q_{\text{ice-fw}} = \frac{\rho_{\text{ice}} \cdot c_{\text{ice-fw-avg}}}{3.6 \cdot 10^{6}} \cdot (273.15 - 272.15) \tag{10.73}$$

$$Q_{\text{w–nfr}} = \frac{\rho_{\text{w}} \cdot c_{\text{w–nfr-avg}}}{3.6 \cdot 10^{6}} \cdot (T_{\text{w–avg-end}} - 272.15) \tag{10.74}$$

ρ_{w} density of wet wood, calculated according to Equation 10.52 [kg/m^3]
ρ_{ice} density of frozen, bound and free water in wood: $\rho_{\text{ice}} = 917$ kg/m^3 [9, 23]
$c_{\text{w-fr-avg}}$ arithmetic mean of the specific heat capacity of frozen wood [J/kgK]
$c_{\text{w-nfr-avg}}$ arithmetic mean of the specific heat capacity of non-frozen wood [J/kgK]
$c_{\text{ice-bw}}$ arithmetic mean of the specific heat capacity of frozen bound water in wood [J/kgK]
$c_{\text{ice-fw-avg}}$ arithmetic mean of the specific heat capacity of frozen free water in wood [J/kgK]
T_{w0} initial temperature of frozen wood [K]
$T_{\text{w-avg-end}}$ average temperature of the wood at the end of thawing and subsequent heating, calculated using the following equation when the set temperature in the wood center is reached:

$$T^{n}_{\text{w–avg}} = \frac{1}{R_{\text{w}}} \int_{R} T(r, n \cdot \Delta\tau) \text{d}R_{\text{w}} \tag{10.75}$$

$T_{\text{w-avg}}$ current value of the average temperature of the logs during steaming or boiling [K]
R_{w} radius of the logs, equal to $D_{\text{H/2}}$ (see Figure 10.14) [m]
r coordinate along the radius of the logs in the 1D model of their non-stationary thawing (see Equation 10.23) [m]
$\Delta\tau$ step along the time coordinate in the numerical solution of Equation 10.23 [s]
n running number of the step along the time coordinate, with which $\Delta\tau$ is solved using Equation 10.23: $n = 0, 1, 2, 3, \ldots, \tau_{\text{steam}}/\Delta\tau$ or $\tau_{\text{boil}}/\Delta\tau$
τ_{steam}, τ_{boil} duration of the mode for steaming or boiling of frozen wood [s]

The arithmetic mean values of the individual specific heat capacities in Equation 10.71 – Equation 10.74 can be calculated using Equation 10.32, Equation 10.41, Equation 10.39 and Equation 10.35 respectively, in the temperature ranges in which they are valid, as follows:

$$c_{\text{w–fr-avg}} = \frac{c_{\text{w–fr}} \text{ at } T_{\text{w0}} + c_{\text{w–fr}} \text{ at } 272.15\text{K}}{2} \tag{10.76}$$

$$c_{\text{ice-bw-avg}} = \frac{c_{\text{ice-bw}} \text{ at } T_{\text{w0}} + c_{\text{ice-bw}} \text{ at } 272.15 \text{ K}}{2} \tag{10.77}$$

$$c_{\text{ice-fw-avg}} = c_{\text{ice-fw}} = 3.34 \cdot 10^{5} \cdot \frac{u - u_{\text{FSP}(272.15)}}{1 + u} \tag{10.78}$$

$$c_{\text{w–nfr-avg}} = \frac{c_{\text{w–nfr}} \text{ at } 272.15 \text{ K} + c_{\text{w–nfr}} \text{ at } T_{\text{w–avg-end}}}{2} \tag{10.79}$$

10.9.1.2 Modeling Q_{Wood} for Non-frozen Wood

If the logs are non-frozen and have an initial temperature $T_{w0} \geq 273.15$ K (i.e., $t_{w0} \geq 0$ °C), the thermal energy required to heat them to the temperature $T_{w\text{-avg-end}}$ can be calculated according to the following model:

$$Q_{wood} = Q_{w-nfr} = \frac{\rho_s \cdot c_{w-nfr\text{-avg}}}{3.6 \cdot 10^6} \cdot (T_{w-avg\text{-end}} - T_{w0}) \quad (10.80)$$

where the arithmetic mean of the specific heat capacity $c_{w\text{-nfr-avg}}$ is calculated as follows:

$$c_{w-nfr\text{-avg}} = \frac{c_{w-nfr} \text{ at } T_{w0} + c_{w-nfr} \text{ at } T_{w-avg\text{-end}}}{2} \quad (10.81)$$

The values of $c_{w\text{-nfr}}$ at T_{w0} and $T_{w\text{-avg-end}}$ are calculated using Equation 10.35.

10.9.1.3 Mathematical Model of $Q_{construction}$

The specific thermal energy required for heating the construction materials of the pit, $Q_{construction}$, can be determined by the following model:

$$Q_{construction} = Q_{constr.1} + Q_{constr.2} + Q_{constr.3} + Q_{constr.4} \quad (10.82)$$

where $Q_{constr.1}$ and $Q_{constr.2}$ are the energies required for heating the walls of the above-ground part and the below-ground part of the pit construction, respectively, and $Q_{constr.3}$, $Q_{constr.4}$ are the energies required for heating the pit floor and the pit cover, respectively.

These energies can be calculated using the following equations:

$$Q_{constr.1} = \frac{1}{3.6 \cdot 10^6 \cdot V_W} \cdot \left\{ [2(l+b+2d_W) \cdot d_W \cdot h_{up}] \cdot \rho_{wb} \cdot c_{wb} \cdot \left(\frac{t_{W-max} + t_{W1}}{2} - t_{W0} \right) + [2(l+b+2d_W+2d_{i-W}) \cdot d_{I-W} \cdot h_{up}] \cdot \rho_{i-W} \cdot c_{i-W} \cdot \left(\frac{t_{W1} + t_{i1}}{2} - t_{iW0} \right) \right\} \quad (10.83)$$

$$Q_{constr.2} = \frac{1}{3.6 \cdot 10^6 \cdot V_W} \cdot \left\{ [2(l+b+2d_W) \cdot d_W \cdot (h - h_{up})] \cdot \rho_{wb} \cdot c_{wb} \left(\frac{t_{W-max} + t_{W2}}{2} - t_{b0} \right) + [2(l+b+2d_W+2d_{i-W}) \cdot d_{i-W} \cdot (h + d_B - h_{up})] \cdot \rho_{i-W} \cdot c_{i-W} \cdot \left(\frac{t_{W2} + t_{i2}}{2} - t_{ib0} \right) \right\} \quad (10.84)$$

$$Q_{constr.3} = \frac{1}{3.6 \cdot 10^6 \cdot V_W} \cdot \left[(l + 2d_b) \cdot (b + 2d_b) \cdot d_b \cdot \rho_{wb} \cdot c_b \cdot \left(\frac{t_{b-max} + t_{be}}{2} - t_{b0} \right) \right] \quad (10.85)$$

$$Q_{constr.4} = \frac{1}{3.6 \cdot 10^6 \cdot V_W} \cdot [l \cdot b \cdot (2d_{Fe} \cdot \rho_{Fe} \cdot c_{Fe} + d_{i-1} \cdot \rho_{i-1} \cdot c_{i-1})] \cdot \left(\frac{t_{l1} + t_{l2}}{2} - t_{air} \right) \quad (10.86)$$

The meaning and values of the variables V_w, l, b, h_{uv}, h_{up}, d_w, $d_{i\text{-}w}$, $d_{i\text{-}l}$, d_{Fe}, ρ_{wb}, $\rho_{i\text{-}w}$, $\rho_{i\text{-}lid}$, ρ_{Fe}, t_{w0}, t_{I0}, t_{b0}, c_b, $c_{i\text{-}w}$, $c_{i\text{-}l}$, c_{Fe} and t_{air} are given in Table 10.7.

The remaining variables in these equations are:

$t_{w\text{-}max}$, $t_{B\text{-}max}$ maximum values of the inner surface temperatures of the concrete walls and the floor of the pit, which are equal to the operating temperature t_{m1} [°C]

t_{w1} surface temperature of the concrete walls of the upper (above-ground) part of the pit that are in contact with the insulating layer of the pit [°C]

t_{i1} surface temperature of the insulating layer of the walls of the upper part of the pit that are in contact with the atmospheric air [°C]

t_{w2} surface temperature of the concrete walls of the pit located in the soil zone that are in contact with the insulating layer of the pit [°C]

t_{i2} surface temperature of the insulating layer of the walls located in the soil zone [°C]

t_{be} surface temperature of the concrete base slab of the pit that is in contact with the ground [°C]

t_{d1} surface temperature of the pit cover located above the boiling water or saturated steam [°C]

t_{d2} surface temperature of the pit cover that is in contact with the atmospheric air [°C].

The surface temperatures t_{w1}, t_{i1}, t_{w2}, t_{i2}, t_{be}, t_{d1} and t_{d2} can be calculated according to the equations given in [36] as a function of the thermophysical properties of the relevant materials of the pit, the water, and the surrounding soil and air.

The volume of the wood V_w, in Equation 10.83 – Equation 10.86 is calculated as:

$$V_w = f \cdot V_{pit} \tag{10.87}$$

f loading degree (fill) of the pit [m³/m³]
V_{pit} working volume of the pit, depending on the design parameters (see also Figure 10.28 and Table 10.7):

$$V_{pit} = l \cdot b \cdot h_{uv} \tag{10.88}$$

10.9.1.4 Mathematical Model of $Q_{hot\ water}$

For modeling the energy $Q_{hot\ water}$ required to heat the water in the pit, knowledge of the water temperature t_m for cooking the wood is necessary.

Figure 10.29 shows the change in water or steam temperature t_m for cooking or steaming non-frozen and frozen wood. The process consists of two phases, in which t_m changes as follows:

- during phase I (0–τ_1), an increase of t_m from t_{m0} to t_{m1};
- during phase II, steam is introduced into the heating element in a metered manner. Upon reaching τ_2, the cooking or steaming process ends.

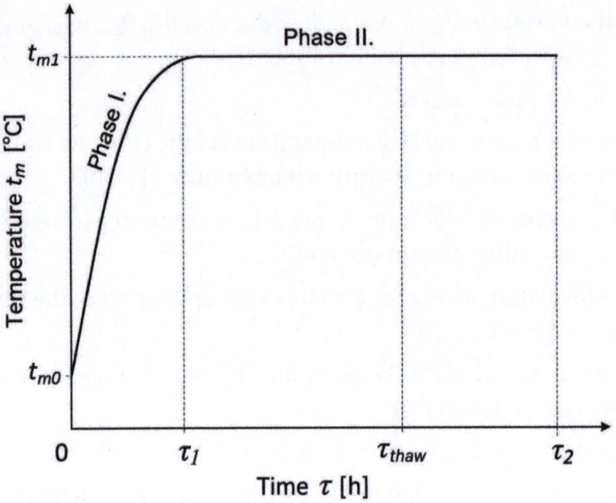

Figure 10.29 Typical change in water or steam temperature t_m during cooking or steaming of wood materials in pits (source: [37–39])

If the logs contain ice at the beginning of their steaming or cooking process, the time corresponding to the moment of thawing of the ice in the wood lies between τ_1 and $\tau_2 \tau_{thaw}$.

The specific thermal energy required to heat the water in the pit, $Q_{hot\,water}$, can be calculated as follows:

$$Q_{hot\,water} = \frac{1}{3.6 \cdot 10^6 \cdot V_w}[(l \cdot b \cdot (h_{uv} - h_{dc}) - V_w)] \cdot \rho_{H2O} \cdot c_{H2O} \cdot (t_{M1} - t_{M0}) \tag{10.89}$$

The values of all variables in Equation 10.89 are given in Table 10.7.

10.9.1.5 Mathematical Model of $Q_{condensation}$

The accumulated thermal energy in the condensate that collects during steaming of the logs in the pit can be calculated using the following model:

$$Q_{condensation} = m_{cw} \cdot \frac{h_{cw}^{\tau_2}}{3.6 \cdot 10^6 \cdot V_w} \tag{10.90}$$

$hcw\tau 2$ Enthalpy of the condensate in the pit at the end τ_2 (see Figure 10.29) of the steaming process [40] [J/kg]
V_w volume of the wood in the steaming pit (see Equation 10.87) [m^3]
m_{cw} mass of the condensate that accumulates in the pit during steaming [kg]

$$m_{cw} = 3.6 \cdot 10^6 \cdot V_w \cdot \frac{Q_{wood} + Q_{construction} + Q_{heat\,losses}}{r_{steam}^{\tau_2}} \tag{10.91}$$

Q_{wood} energy required for heating the wood [kWh/m^3]
$Q_{construction}$ energy required for heating the pit structure [kWh/m^3]

$Q_{\text{heat losses}}$ energy required to compensate for the heat losses of the pit during the entire steaming process (see Equation 10.94–Equation 10.97) [kWh/m^3]

The multiplier $3.6{\cdot}10^6$ in Equation 10.91 enables a conversion of the unit of the energies Q_{wood}, $Q_{\text{construction}}$ and $Q_{\text{heat losses}}$ from kWh/m^3 to the unit required for the calculation of m_{cw}: J/m^3.

10.9.1.6 Mathematical Model of Q_{radiator}

The specific thermal energy required for heating the metal radiator, Q_{radiator}, can be expressed by the following model:

$$Q_{\text{radiator}} = \frac{1}{3.6 \cdot 10^6 \cdot V_{\text{w}}}[l \cdot b \cdot m_{\text{radiator}} \cdot c_{\text{Fe}} \cdot (t_{\text{radiator}} - t_{\text{M0}})] \tag{10.92}$$

The values of all variables in Equation 10.91 and Equation 10.92 are given in Table 10.7.

10.9.1.7 Mathematical Model of $Q_{\text{heat losses}}$

The specific thermal energy required to cover the heat losses, $Q_{\text{heat losses}}$, can be determined using Equation 10.93:

$$Q_{\text{heat losses}} = Q_{\text{losses1}} + Q_{\text{losses2}} + Q_{\text{losses3}} + Q_{\text{losses4}} \tag{10.93}$$

Q_{losses1} and Q_{losses2} energies required to cover the heat losses caused by heat dissipation through the walls of the above-ground part and the walls in the below-ground part of the pit structure [kWh/m3]
Q_{losses3} and Q_{losses4} energies required to cover the heat losses caused by the bottom and the cover of the pit, respectively [kWh/m3]

These energies can be determined:

$$Q_{\text{losses1}} = \frac{1}{V_{\text{w}} \cdot 10^3} \cdot \left\{ \frac{1}{\frac{1}{\alpha_1} + \frac{d_{\text{w}}}{\lambda_{\text{B}}} + \frac{d_{\text{i-w}}}{\lambda_{\text{I}}} + \frac{1}{\alpha_2}} \cdot [2 \cdot (l + b) \cdot h_{\text{up}}] \cdot (t_{\text{M1}} - t_{\text{air}}) \cdot (\tau_2 - \tau_1) \right\} \tag{10.94}$$

$$Q_{\text{losses2}} = \frac{1}{V_{\text{w}} \cdot 10^3} \cdot \left\{ \frac{1}{\frac{1}{\alpha_1} + \frac{d_{\text{w}}}{\lambda_{\text{B}}} + \frac{d_{\text{i-w}}}{\lambda_{\text{Ip}}} + \frac{1}{\lambda_{\text{soil}}}} \cdot [2 \cdot (l + b) \cdot (h + d_{\text{b}} - h_{\text{up}})] \cdot (t_{\text{M1}} - t_{\text{soil0}}) \cdot (\tau_2 - \tau_1) \right\} \tag{10.95}$$

$$Q_{\text{losses3}} = \frac{1}{V_{\text{w}} \cdot 10^3} \cdot \left[\frac{1}{\frac{1}{\alpha_1} + \frac{d_{\text{b}}}{\lambda_{\text{b}}} + \frac{1}{\lambda_{\text{soil}}}} \cdot (l + 2d_{\text{b}}) \cdot (b + 2d_{\text{b}}) \cdot (t_{\text{M1}} - t_{\text{soil0}}) \cdot (\tau_2 - \tau_1) \right] \tag{10.96}$$

$$Q_{\text{losses4}} = \frac{1}{V_{\text{w}} \cdot 10^3} \cdot [\alpha_3 \cdot l \cdot b \cdot (t_{\text{D2}} - t_{\text{air}}) \cdot (\tau_2 - \tau_1)] \tag{10.97}$$

τ_1 Duration of the first phase of the steaming or cooking process of the wood materials [h]
τ_2 Duration of the entire mode for steaming or cooking the wood (see Figure 10.29) [h]

The durations τ_1 and τ_2 can be determined by solving the models specified in Section 10.3 for veneer logs.

The names and values of the variables l, b, h, h_{up}, d_w, d_b, $d_{i\text{-}w}$, t_{M1}, t_{soil0}, t_{air}, λ_B, λ_{ilp} and λ_{soil} in Equation 10.94 – Equation 10.97 are given in Table 10.7. The remaining variables can be calculated according to the equations given in [36], namely:

α_1 heat transfer coefficient from steam or boiling water to the walls of the pit [W/m^{2}K]

α_2 heat transfer coefficient from the surface of the upper part of the pit walls to the ambient air [W/m^{2}K]

α_3 heat transfer coefficient from the surface of the pit cover to the ambient air [W/m^{2}K]

t_{D2} surface temperature of the pit cover in contact with the atmospheric air [°C]

10.9.1.8 Determination of the Efficiency of Modes for Steaming or Boiling Wood Materials in a Concrete Pit

Using Equation 10.68 – Equation 10.70 and Equation 10.80, it is possible to determine the thermal efficiency (in %) of various modes for steaming or boiling wood as follows:

$$\eta_{\text{steaming}} = 100\frac{Q_{\text{wood}}}{Q_{\text{pit-steaming}}} \tag{10.98}$$

$$\eta_{\text{boiling}} = 100\frac{Q_{\text{wood}}}{Q_{\text{pit-boiling}}} \tag{10.99}$$

10.10 Calculation of the Heat Balance and the Efficiency of Concrete Pits During Steaming or Boiling of Beech Veneer Logs

A calculation of the heat balance and efficiency for frozen and non-frozen beech logs (*Fagus sylvatica* L.) (a wood species frequently used in veneer production) was carried out.

The symbols, units, and values of the parameters are given in Table 10.1 listed. The beech logs contain considerable amounts of free and bound water. Its heating or the melting of the ice cause significant differences in the duration between the individual steaming or cooking modes and also in the corresponding energy consumption of the pit.

Table 10.8 Main Parameters of Unfrozen and Frozen Beech Logs Used for Solving the Mathematical Models (source: [37–39])

Parameter	Symbol	Unit	Value
1. Diameter of the logs	D	m	0.4
2. Length of the logs	L	m	≥ 2.0
3. Volume of the logs in the pit	V_w	m	9; 12; 15
4. Moisture content of the wood	u	kg/kg	0.4; 0.6; 0.8
5. Fiber saturation point at 293.15 K (20 °C)	$u_{FSP(293.15)}$	kg/kg	0.31
6. Bulk density of the wood (beech)	ρ_b	kg/m³	560
7. Density of ice in the wood	ρ_{ice}	kg/m³	917
8. Initial temperature of the non-frozen wood	t_{w0}	°C	10
9. Initial temperature of the frozen wood	t_{w0}	°C	−10; −20; −30
10. Initial temperature in the pit	t_{w0}	°C	10
11. Maximum temperature of the steam or water	t_{m1}	°C	70; 80; 90
12. Time of temperature increase from t_{m0} to t_{m1}	τ_1	h	4
13. Minimum temperature in the center of the wood at the end	$t_{w\text{-}c}$	°C	62
14. Heat of condensation of the steam in the pit at 70 °C, 80 °C and 90 °C	r_{steam}	kJ/kg	2,333; 2,308; 2,282
15. Enthalpy of the condensate in the pit at 70 °C, 80 °C and 90 °C	h_{cw}	kJ/kg	293; 335; 377

The mathematical solution is carried out according to Equation 10.23. From the change in temperature as well as the average temperature of the wood calculated with Equation 10.75, the duration of the steaming or cooking process was determined.

For the joint solution of the equations in the models (Equation 10.69 – Equation 10.99), an Excel program was used.

The data used are given in Table 10.7 and Table 10.8 listed.

10.10.1 Influence of Steam or Hot Water Temperature on the Heat Balance of the Pit

Figure 10.30 and Figure 10.31 show the change in temperature at the center of the logs, $t_{\text{w-z}}$, and the average temperature $t_{\text{w-avg}}$, at operating temperature in the pit t_{m}, calculated using Equation 10.23 and Equation 10.75. The temperature t_{m} changes from an initial value of 10 °C to its maximum values of 70 °C, 80 °C and 90 °C. The steaming and cooking process ends for non-frozen/frozen wood after 27.0 h/34.0 h at t_{m1} = 70 °C, after 20.0 h/26.0 h at t_{m1} = 80 °C and after 17.5 h/22.0 h at t_{m1} = 90 °C.

For these values of the duration of the steaming or cooking process τ_2 (see Figure 10.29), the temperature at the center of the logs, $t_{\text{w-z}}$, reaches 62 °C, which corresponds to the minimum required temperature necessary for veneer production [3, 31, 37–39].

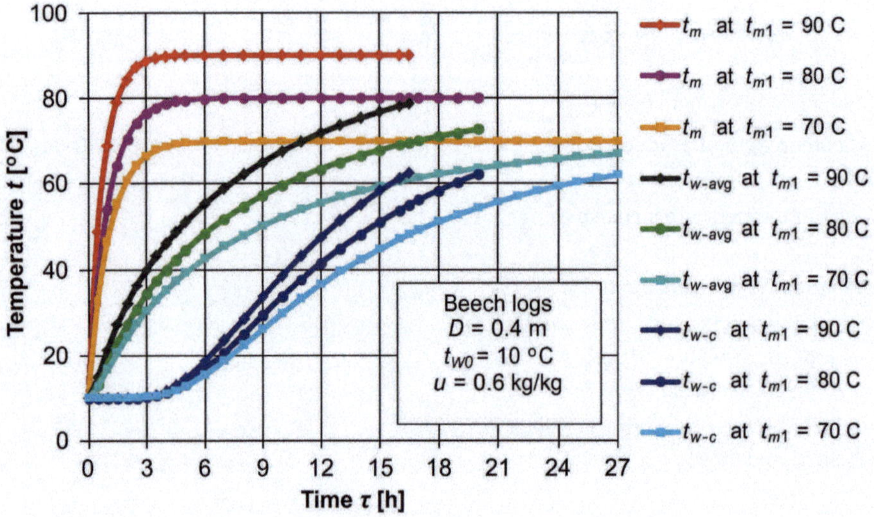

Figure 10.30 Change of t_{m}, $t_{\text{w-z}}$ and $t_{\text{w-avg}}$ in non-frozen beech logs with t_{w0} = 10 °C and u = 0.6 kg/kg during their steaming or boiling, depending on t_{m1} (source: [37])

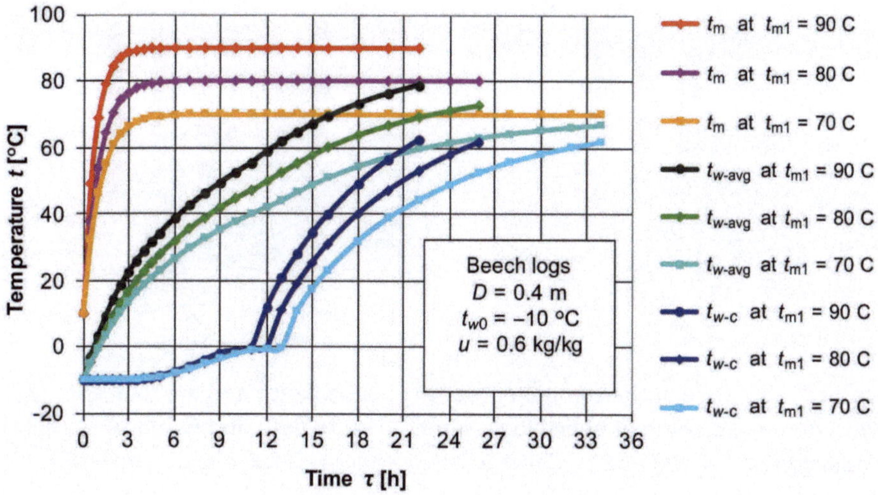

Figure 10.31 Change of t_m, $t_{w\text{-}z}$ and $t_{w\text{-}avg}$ in frozen beech logs with t_{h0} = –10 °C and u = 0.6 kg/kg during their steaming or boiling, depending on t_{m1} (source: [38])

The average temperature of the steamed or boiled, both non-frozen and frozen logs, $t_{w\text{-}avg\text{-}end}$, reaches the following values: 66.9 °C, 72.8 °C and 79.0 °C, respectively.

These values of $t_{w\text{-}avg\text{-}end}$ (in degrees Kelvin) were used in Equation 10.74 and Equation 10.80 for calculating the energy $Q_{w\text{-}nfr}$.

In Figure 10.31, a section of the maintenance of the temperature $t_{w\text{-}z}$ over a long period of time in the range of –1 °C to 0 °C at the log center is recognizable. This is due to the extremely low thermal diffusivity during the prolonged melting of the frozen free water in the wood [3, 4, 9, 38].

Figure 10.32 and Figure 10.33 show the change of all components of the heat balance as well as the total energy $Q_{\Sigma\text{-steaming}}$ and $Q_{\Sigma\text{-cooking}}$ (in kWh/m³), required to carry out the entire steaming and cooking process of the logs with u = 0.6 kg/kg at the maximum possible loading factor of the pit with logs (f = 75 %) as a function of the operating temperature t_{m1}.

Figure 10.34 and Figure 10.35 show the change of the individual components of the heat balance Q_i of the pit at u = 0.6 kg/kg and f = 75 %, as a percentage of the total energy consumption, $Q_{\Sigma\text{-steaming}}$ and $Q_{\Sigma\text{-cooking}}$, as a function of t_{m1}.

In Figure 10.32 and Figure 10.33 it can be seen that when expressing the heat balance of the pit in kWh/m³ an increase in t_{m1} from 70 °C to 90 °C causes an increase in the total energies $Q_{\Sigma\text{-steaming}}$ and $Q_{\Sigma\text{-cooking}}$, and their components Q_{wood}, $Q_{construction}$, $Q_{condensate}$ and $Q_{hot\ water}$, while at the same time $Q_{heat\ losses}$ decreases and $Q_{radiator}$ does not change.

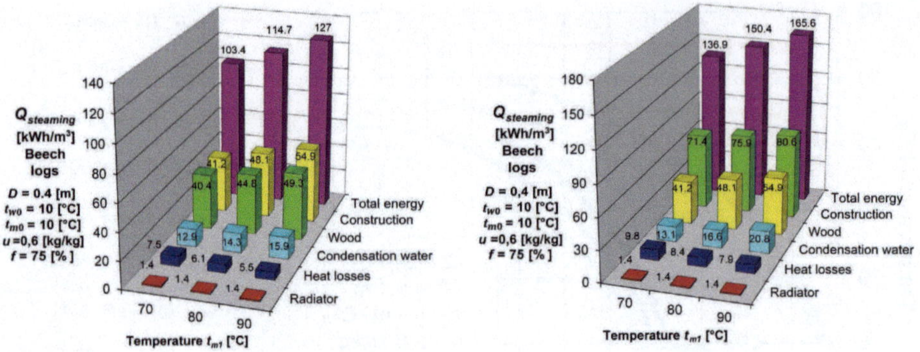

Figure 10.32 Change of the components of the heat balance and the total energy (in kWh/m^3) during steaming of beech logs with t_{h0} = 10 °C (left) and t_{h0} = –10 °C (right), depending on t_{m1}

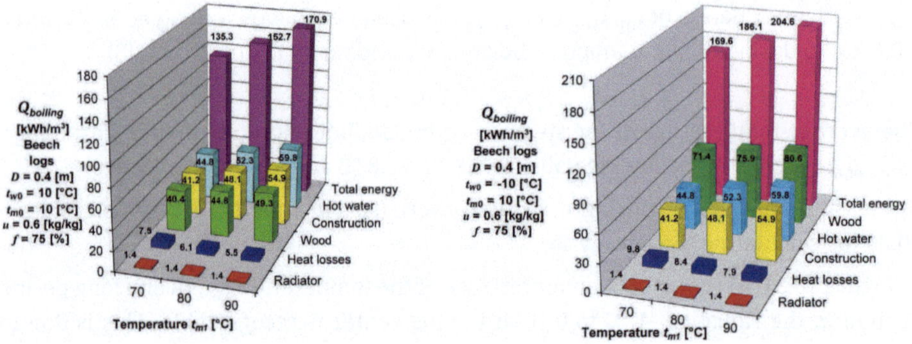

Figure 10.33 Change of the components of the heat balance and the total energy (in kWh/m^3) during cooking of beech logs with t_{h0} = 10 °C (left) and t_{h0} = –10 °C (right), depending on t_{m1} (source: [37, 38])

The total energy consumption of the pit, $Q_{\Sigma\text{-steaming}}$ and $Q_{\Sigma\text{-cooking}}$, increases in the following way:

- for steaming: from 103.4 kWh/m^3 to 127.0 kWh/m^3 for non-frozen and from 136.9 kWh/m^3 to 165.6 kWh/m^3 for frozen wood;
- for cooking: from 135.3 kWh/m^3 to 170.9 kWh/m^3 for non-frozen and from 168.6 kWh/m^3 to 204.6 kWh/m^3 for frozen wood.

The distribution is as follows:

- Q_{wood} increases from 40.4 kWh/m^3 to 49.3 kWh/m^3 and from 71.4 kWh/m^3 to 80.6 kWh/m^3 for both steaming and cooking of beech logs;

- $Q_{construction}$ increases from 41.2 kWh/m^3 to 54.9 kWh/m^3 for steaming and cooking due to the increase in the inner surface temperature of the concrete walls of the pit, which equals t_{m1} is;
- $Q_{hot\ water}$ increases during boiling due to the increase of t_{m1} from 44.8 kWh/m^3 to 59.8 kWh/m^3;
- $Q_{condensate}$ increases from 12.9 kWh/m^3 to 15.9 kWh/m^3 when steaming logs with an initial temperature of t_{h0} = 10 °C and from 13.1 kWh/m^3 to 20.8 kWh/m^3 at t_{h0} = –10 °C;
- $Q_{radiator}$ remains unchanged at a value of 1.4 kWh/m^3, since in the simulations the same heat carrier temperature $t_{radiator}$ = 130 °C is used for all values of t_{m1}:
- $Q_{heat\ losses}$ decreases from 7.5 kWh/m^3 to 5.5 kWh/m^3 and from 9.8 kWh/m^3 to 7.9 kWh/m^3 decreases for both steaming and cooking.
- Q_{wood} decreases from 39.1 % to 38.9 % and from 52.2 % to 48.7 % during steaming, and from 29.9 % to 28.8 % and from 42.4 % to 39.4 % during cooking;
- $Q_{construction}$ increases from 39.9 % to 43.2 % and from 30.1 % to 33.2 % during steaming, and from 30.4 % to 32.1 % and from 24.4 % to 26.8 % during cooking;
- $Q_{hot\ water}$ increases from 33.2 % to 35.1 % and from 26.6 % to 29.2 %;
- $Q_{condensate}$ remains unchanged at 12.5% when steaming non-frozen logs and increases from 9.6% to 12.5% when steaming frozen logs;
- $Q_{heat\ losses}$ decrease from 7.2% to 4.3% and from 7.1% to 4.8% when steaming, and from 5.5% to 3.2% and from 5.8% to 3.9% when cooking;
- $Q_{radiator}$ decrease from 1.3% to 1.1% and from 1.0% to 0.8% when steaming, and from 1.0% to 0.8% and from 0.8% to 0.7% when cooking.

If the individual components of the heat balance of the pit Q_i are expressed as percentages of the total energies $Q_{\Sigma\text{-steaming}}$ and $Q_{\Sigma\text{-cooking}}$respectively, an increase of t_{m1} from 70 °C to 90 °C causes the following change in the share of the individual components of this balance for non-frozen (the first two numbers) and frozen wood (second numbers), see also Figure 10.34 and Figure 10.35:

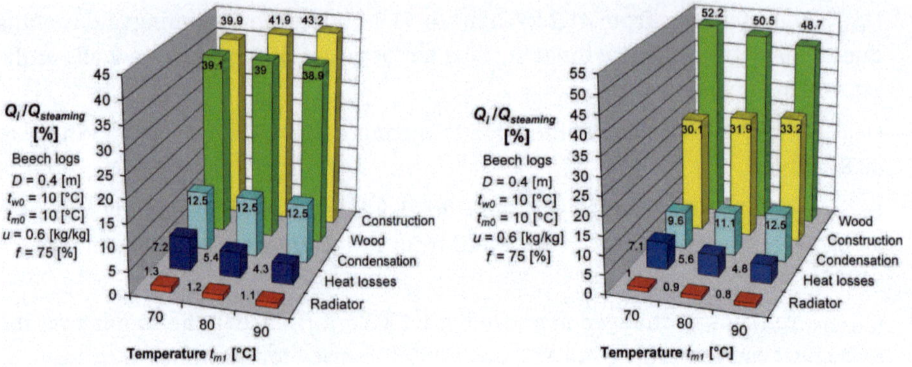

Figure 10.34 Change of the components of the heat balance of the pit in % of the total energy during steaming of beech logs with t_{h0} = 10 °C (left) and t_{h0} = –10 °C (right), depending on t_{m1}

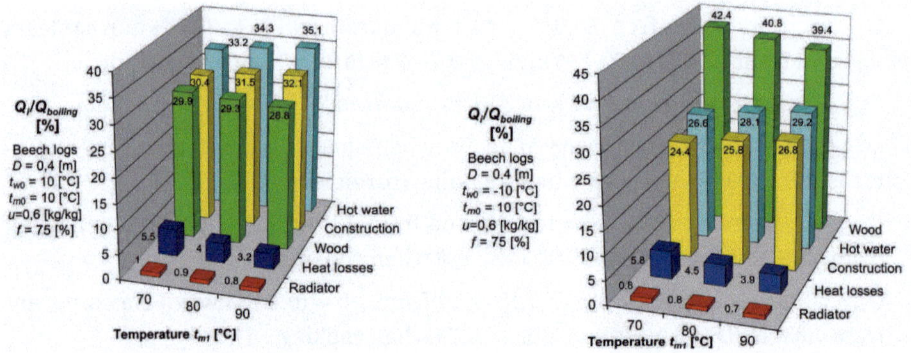

Figure 10.35 Change of the components of the heat balance of the pit in % of the total energy during cooking of beech logs with t_{h0} = 10 °C (left) and t_{h0} = –10 °C (right), depending on t_{m1} (Source: [37, 38])

10.10.2 Influence of the Initial Temperature of the Wood on the Heat Balance of the Pit

Figure 10.36 and Figure 10.37 show the calculated change of $t_{w\text{-}z}$ and $t_{w\text{-}avg}$ of non-frozen and frozen beech logs with u = 0.6 kg/kg during steaming/cooking at t_{m1} = 80 °C, depending on t_{h0}.

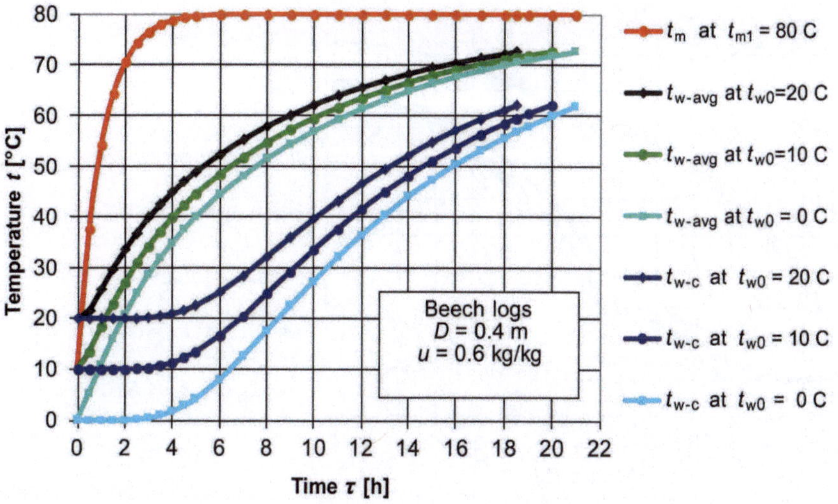

Figure 10.36 Change of t_m, $t_{w\text{-}z}$ and $t_{w\text{-}avg}$ in non-frozen beech logs with u = 0.6 kg/kg during their steaming or cooking, depending on t_{h0} (source: [39])

It can be seen that the duration of the treatment process τ_2 upon reaching $t_{w\text{-}z}$ = 62 °C, takes the following values:

- 21.0 h at t_{w0} = 0 °C, 20.0 h at t_{w0} = 10 °C and 18.5 h at t_{w0} = 20 °C;
- 26.0 h at t_{w0} = –10 °C, 27.0 h at t_{w0} = –20 °C and 27.5 h at t_{w0} = –30 °C.

A decrease in t_{w0} from 20 °C to 0 °C and from –10 °C to –30 °C leads to the following increase in the total energy consumption of the pit $Q_{\Sigma\text{-steaming}}$ and $Q_{\Sigma\text{-cooking}}$ [39]:

- for steaming: from 106.3 kWh/m^3 to 122.8 kWh/m^3 for non-frozen and from 150.4 kWh/m^3 to 166.0 kWh/m^3 for frozen wood;
- for cooking: from 145.3 kWh/m^3 to 159.7 kWh/m^3 for non-frozen and from 186.1 kWh/m^3 to 199.8 kWh/m^3 for frozen wood;

The individual components change as follows:

- Q_{wood} increases from 37.9 kWh/m^3 to 51.3 kWh/m^3 and from 75.9 kWh/m^3 to 89.0 kWh/m^3 both during steaming and cooking of beech logs;
- $Q_{construction}$ and $Q_{hot\ water}$ remain unchanged with values of 48.1 kWh/m^3 and 52.3 kWh/m^{3}respectively;
- $Q_{condensate}$ increases from 13.3 kWh/m^3 to 15.4 kWh/m^3 and from 16.6 kWh/m^3 to 18.5 kWh/m^3;
- $Q_{radiator}$ remains unchanged with a value of 1.4 kWh/m^3;
- $Q_{heat\ losses}$ increases from 5.6 kWh/m^3 to 6.6 kWh/m^3 and from 8.4 kWh/m^3 to 9.0 kWh/m^3 during steaming and cooking due to the increase in time τ_2 both with decreasing t_{h0} and with frozen wood.

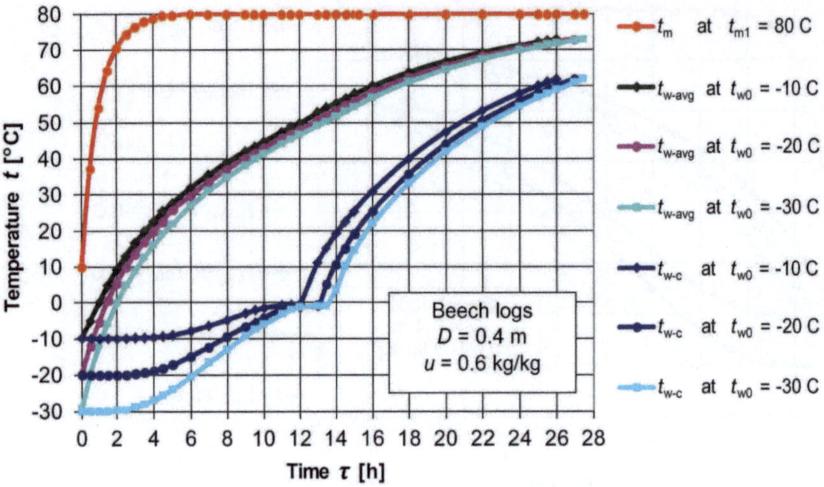

Figure 10.37 Change of t_m, $t_{w\text{-}z}$ and $t_{w\text{-}avg}$ in frozen beech logs with $u = 0.6$ kg/kg during their steaming or boiling, depending on t_{w0}

If the individual components of the heat balance of the pit Q_i are expressed as a percentage of the total energies $Q_{\Sigma\text{-steaming}}$ and $Q_{\Sigma\text{-boiling}}$ respectively, a reduction of t_{w0} from 20 °C to 0 °C and from –10 °C to –30 °C leads to the following changes in the shares of the individual components of this balance during steaming/boiling of non-frozen and frozen wood respectively:

- Q_{wood} increases from 35.7 % to 41.8 % and from 50.5 % to 53.6 % during steaming, and from 26.1 % to 32.1 % and from 40.8 % to 44.5 % during boiling;
- $Q_{construction}$ decreases from 45.2 % to 39.2 % and from 31.9 % to 29.1 % during steaming, and from 33.1 % to 30.1 % and from 25.8 % to 24.1 % during boiling;
- $Q_{hot\ water}$ decreases from 36.0 % to 32.8 %;
- $Q_{condensate}$ remains unchanged at 12.5 % for non-frozen and increases from 11.1 % to 11.2 % for frozen logs;
- $Q_{heat\ losses}$ increases from 5.3% to 5.4% and decreases from 5.6% to 5.4% when steaming; when cooking, the share of this energy increases from 3.9% to 4.1% and remains unchanged at 4.5% for the frozen logs;
- $Q_{radiator}$ decreases from 1.3% to 1.1% and from 0.9% to 0.8% when steaming; when cooking it remains unchanged at 0.9% for non-frozen logs and decreases from 0.8% to 0.7% for the frozen logs.

10.10.3 Influence of Wood Moisture Content on the Heat Balance of the Pit

Figure 10.38 and Figure 10.39 show the calculated change of $t_{w\text{-}z}$ and $t_{w\text{-}avg}$ of non-frozen wood with t_{h0} = 10 °C and frozen beech logs with t_{h0} = –10 °C during steaming or cooking at t_{m1} = 80 °C as a function of u. The duration of the treatment process τ_2 until reaching $t_{w\text{-}z}$ = 62 °C is:

- 19.5 h at u = 0.4 kg/kg, 20.0 h at u = 0.6 kg/kg, and 20.5 h at u = 0.8 kg/kg;
- 24.0 h at u = 0.4 kg/kg, 26.0 h at u = 0.6 kg/kg, and 27.5 h at u = 0.8 kg/kg.

At these values of τ_2 the average temperature of the steamed or cooked, both non-frozen and frozen logs, $t_{w\text{-}avg\text{-}end}$, a value of 72.8 °C. Using the equivalents of these values for the temperature $t_{w\text{-}avg\text{-}end}$ in Kelvin, the energy Q_{wood} and its components required for steaming or cooking the logs were calculated and plotted graphically (Figure 10.40–Figure 10.43).

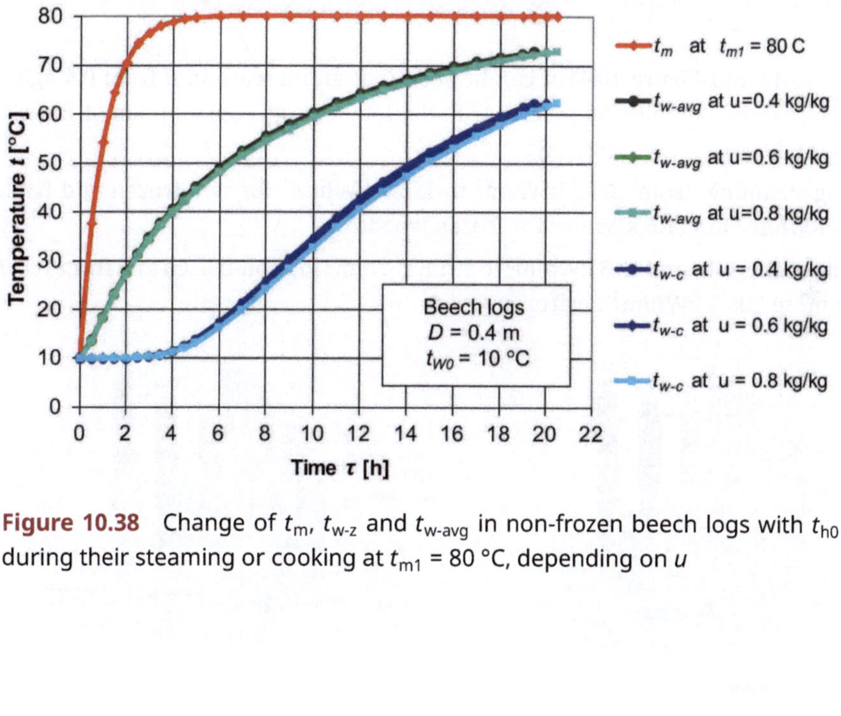

Figure 10.38 Change of t_m, $t_{w\text{-}z}$ and $t_{w\text{-}avg}$ in non-frozen beech logs with t_{h0} = 10 °C during their steaming or cooking at t_{m1} = 80 °C, depending on u

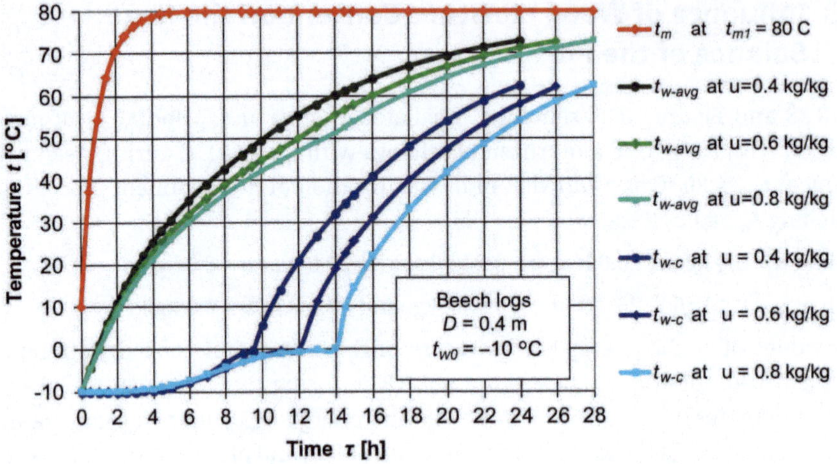

Figure 10.39 Change of t_m, t_{w-z} and t_{w-avg} in frozen beech logs with t_{w0} = –10 °C during their steaming or cooking at t_{m1} = 80 °C, depending on u (source: [41])

In Figure 10.40 and Figure 10.41 it can be seen that an increase in u from 0.4 kg/kg to 0.8 kg/kg causes the following increase in the total energy consumption of the pit, $Q_{\Sigma\text{-steaming}}$ and $Q_{\Sigma\text{-cooking}}$, respectively:

- during steaming: from 104.3 kWh/m³ to 125.2 kWh/m³ for non-frozen and from 127.2 kWh/m³ to 171.6 kWh/m³ for frozen wood;
- during cooking: from 143.5 kWh/m³ to 161.8 kWh/m³ for non-frozen and from 164.7 kWh/m³ to 205.5 kWh/m³ for frozen wood.

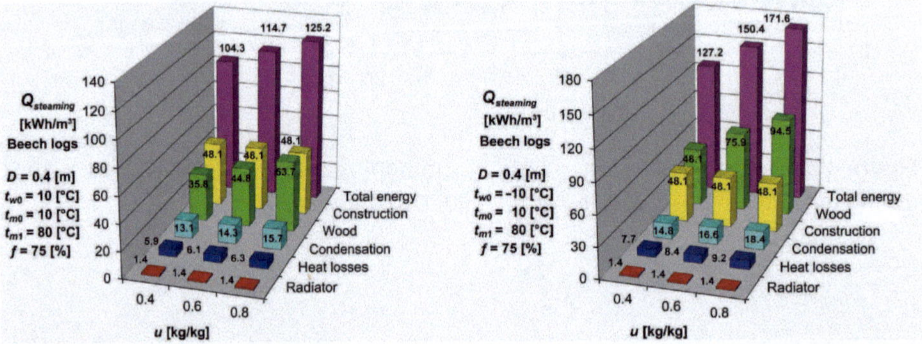

Figure 10.40 Change of components of the heat balance and total energy (in kWh/m³) during steaming of beech logs with t_{w0} = 10 °C (left) and t_{w0} = –10 °C (right), depending on u (source: [own illustration-left, 41])

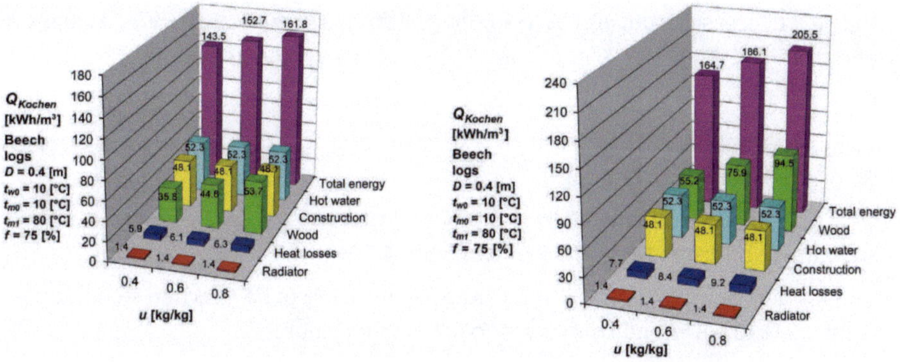

Figure 10.41 Change of components of the heat balance and total energy (in kWh/m^3) during cooking of beech logs with t_{w0} = 10 °C (left) and t_{w0} = –10 °C (right), depending on u (source: [own illustration-left, 41])

Figure 10.42 and Figure 10.43 show the change of the individual components of the heat balance Q_i of the pit at t_{m1} = 80 °C and f = 75 % as percentages of the total energy consumption, $Q_{\Sigma\text{-steaming}}$ and $Q_{\Sigma\text{-cooking}}$, depending on u.

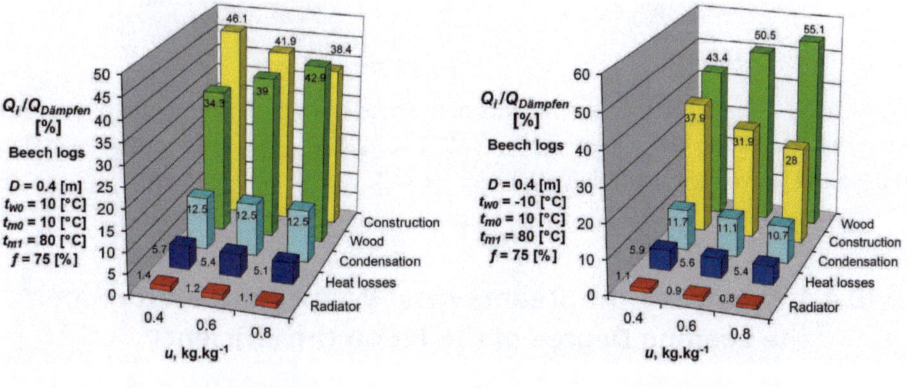

Figure 10.42 Change of the components of the heat balance of the pit in % of the total energy during steaming of beech logs with t_{w0} = 10 °C (left) and t_{w0} = –10 °C (right), depending on u (source: [own illustration-left, 41])

If the individual components of the heat balance of the pit Q_i are expressed as a percentage of the total energies $Q_{\Sigma\text{-steaming}}$ and $Q_{\Sigma\text{-cooking}}$ if we consider this, an increase in u from 0.4 kg/kg to 0.8 kg/kg leads to the following changes in the proportions of the components of this balance when steaming/cooking non-frozen and frozen wood (see Figure 10.42 and Figure 10.43):

- Q_{wood} increases from 34.3 % to 42.9 % and from 43.4 % to 55.1 % when steaming, and from 24.9 % to 33.2 % and from 33.5 % to 46.0 % when cooking;

- $Q_{structure}$ decreases from 46.1 % to 38.4 % and from 37.9 % to 28.0 % when steaming, and from 33.5 % to 29.7 % and from 29.1 % to 23.4 % when cooking;
- $Q_{hot\ water}$ decreases from 36.5 % to 32.4 % and from 31.8 to 25.4 %;
- $Q_{condensate}$ remains unchanged at 12.5 % for the non-frozen and decreases from 11.7 % to 10.7 % for the frozen logs;
- $Q_{heat\ losses}$ decreases from 5.7 % to 5.1 % and decreases from 5.9 % to 5.4 % when steaming; and from 4.1 % to 3.9 % and from 4.7 % to 4.5 % when cooking;
- $Q_{radiator}$ decreases from 1.4 % to 1.1 % and from 1.1 % to 0.8 % when steaming; and from 1.0 % to 0.8 % and from 0.9 % to 0.7 % when cooking.

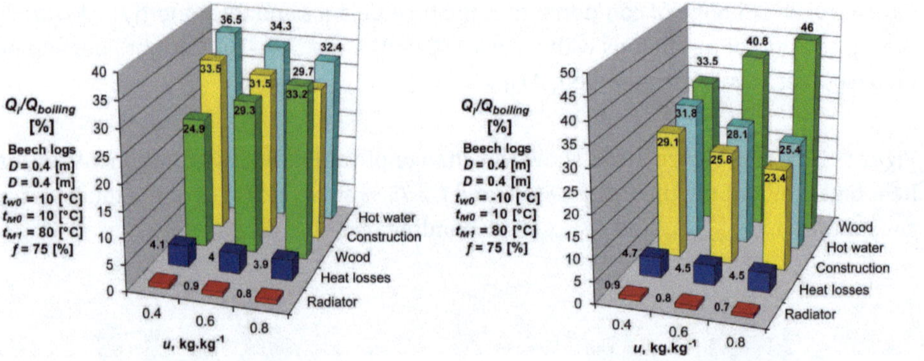

Figure 10.43 Change in the components of the heat balance of the pit in % of the total energy when cooking beech logs with t_{w0} = 10 °C (left) and t_{w0} = –10 °C (right), depending on u (source: [own illustration-left, 41])

10.10.4 Influence of the Steam or Hot Water Temperature and the Loading Degree of the Pit on the Efficiency

Figure 10.44 and Figure 10.45 show the change in the thermal efficiency of the steaming and cooking modes, calculated according to Equation 10.98 and Equation 10.99, $\eta_{steaming}$ and $\eta_{cooking}$, for non-frozen and frozen beech logs with u = 0.6 kg/kg depending on the steam or hot water temperature t_{m1} and on the loading factor of the pit with logs f.

In these figures it can be seen that the increase of t_{m1} from 70 °C to 90 °C leads to an insignificant decrease in the thermal efficiency η, which is attributable to the minor change in the effective specific heat capacity of wood:

- When steaming logs with t_{w0} = 10 °C: by 0.2 % for the three investigated values of f;

- When steaming logs with t_{w0} = –10 °C: by 3.2 % at f = 45 %, by 3.4 % at f = 60 % and by 3.5 % at f = 75 %;
- When boiling logs with t_{w0} = 10 °C: by 1.1 % at f = 45 %, by 1.0 % at f = 60 % and by 1.1 % at f = 75 %;
- When boiling logs with t_{w0} = –10 °C: by 2.4 % at f = 45 %, by 2.8 % at f = 60 % and by 3.0 % at f = 75 %.

At the maximum practically possible value of f = 75% causes this change in t_{M1} a reduction in efficiency $\eta_{steaming}$ and $\eta_{cooking}$ from 39.1% to 38.9% and from 29.9% to 28.8% respectively at t_{w0} = 10 °C and from 52.2% to 48.7% and from 42.4% to 39.4% respectively at t_{w0} = –10 °C.

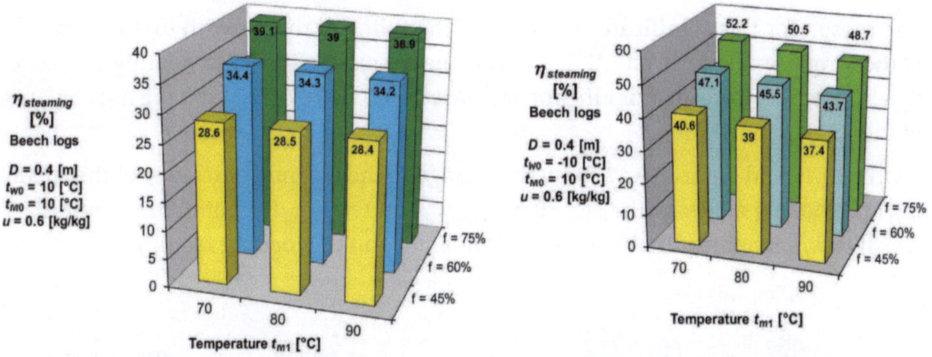

Figure 10.44 Change in the thermal efficiency of the steaming modes for non-frozen (left) and frozen (right) logs, depending on t_{m1} and f

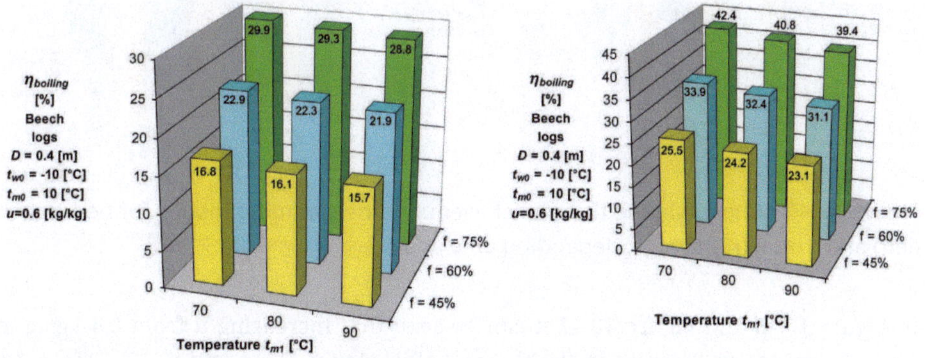

Figure 10.45 Change in the thermal efficiency of the cooking modes for non-frozen (left) and frozen (right) logs, depending on t_{m1} and f

The loading factor of the pit with logs has a significantly greater influence on the thermal efficiency, and when f is reduced from 75% to 45%, the values of $\eta_{steaming}$ and $\eta_{cooking}$ decrease as follows:

- When steaming logs with t_{w0} = 10 °C: from 39.1% to 28.6% at t_{m1} = 70 °C, from 39.0% to 28.5% at t_{m1} = 80 °C, and from 38.9% to 28.4% at t_{m1} = 90 °C;
- When steaming logs with t_{w0} = –10 °C: from 52.2 % to 40.6 % at t_{m1} = 70 °C, from 50.5 % to 39.0 % at t_{m1} = 80 °C, and from 48.7 % to 37.4 % at t_{m1} = 90 °C;
- When cooking logs with t_{w0} = 10 °C: from 29.9 % to 16.8 % at t_{m1} = 70 °C, from 29.3 % to 16.1 % at t_{m1} = 80 °C, and from 28.8 % to 15.7 % at t_{m1} = 90 °C;
- When cooking logs with t_{w0} = –10 °C: from 42.4 % to 25.5 % at t_{m1} = 70 °C, from 40.8 % to 24.2 % at t_{M1} = 80 °C, and from 39.4 % to 23.1 % at t_{m1} = 90 °C.

The reason for this significant influence of the filling of the pit with logs f on $\eta_{cooking}$ is the strong increase in the share of the energy expenditure for heating the water $Q_{hot\ water}$, which leads to a significant decrease in the share of Q_{wood} in the heat balance of the pit Q_{pit} contributes.

Figure 10.46 and Figure 10.47 show the change in the thermal efficiency of the steaming and cooking modes with t_{m1} = 80 °C, $\eta_{steaming}$ and $\eta_{cooking}$ for non-frozen and frozen beech logs, depending on the moisture content u and the loading factor f.

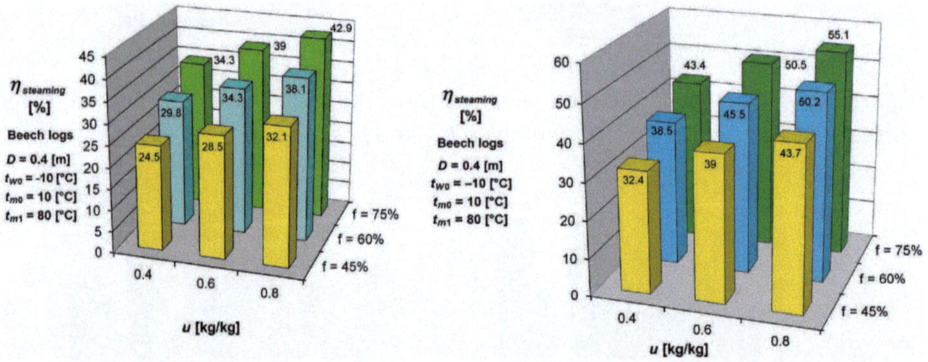

Figure 10.46 Change in the thermal efficiency of the steaming modes for non-frozen (left) and frozen (right) logs, depending on u and f

In Figure 10.46 and Figure 10.47 it can be seen that increasing u from 0.4 kg/kg to 0.8 kg/kg causes an increase in the thermal efficiency $\eta_{steaming}$ and $\eta_{cooking}$, since the specific heat capacity of wood increases more strongly. At the maximum possible value of f = 75 %, this change in u causes an increase in these efficiencies as follows:

- $\eta_{steaming}$ from 34.3% to 42.9% at t_{w0} = 10 °C and from 43.4% to 55.1% at t_{w0} = –10 °C;
- $\eta_{boiling}$ from 24.9% to 33.2% at t_{w0} = 10 °C and from 35.5% to 46.0% at t_{w0} = –10 °C.

In this case, the fill factor of the pit with logs f also has a particularly strong influence on the efficiency $\eta_{boiling}$, since under equal conditions, when reducing f = 75% to f = 45%, the efficiency decreases by approximately a factor of 2.

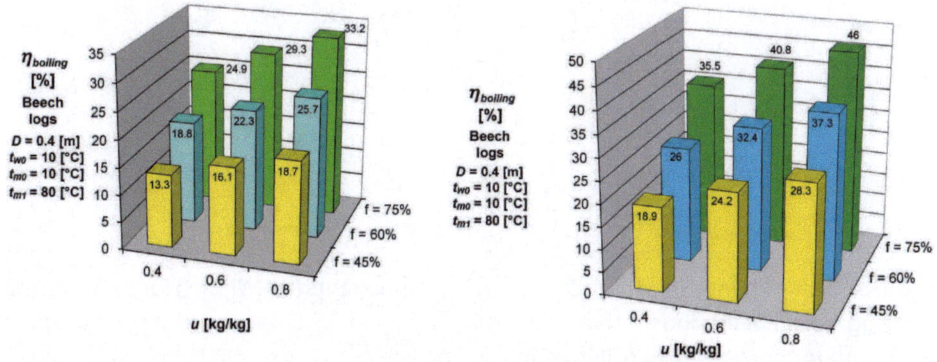

Figure 10.47 Change in the thermal efficiency of the cooking modes for non-frozen (left) and frozen (right) logs, depending on *u* and *f*

10.11 Calculation of the Energy Q_{wood} and Its Components During the Thermal Treatment of Frozen Beech Logs

Figure 10.48 shows the change in the total value of the energy Q_{wood} (in kWh/m^3) and its components (see Section 10.9.1.1), calculated according to Equation 10.70 – Equation 10.74, required for steaming or boiling frozen beech logs with an initial temperature of –10 °C, –20 °C and –30 °C and a moisture content of 0.6 kg/kg and 0.8 kg/kg [42].

According to Figure 10.48, the individual components of $Q_{H\text{-}total}$ change as follows:

- $Q_{ice\text{-}bw}$ increases from 4.2 kWh/m^3 to 7.3 kWh/m^3 at u = 0.6 kg/kg and from 3.8 kWh/m^3 to 6.5 kWh/m^3 at u = 0.8 kg/kg;
- $Q_{w\text{-}fr}$ increases from 5.1 kWh/m^3 to 15.0 kWh/m^3 at u = 0.6 kg/kg and from 5.8 kWh/m^3 to 17.3 kWh/m^3 at u = 0.8 kg/kg;
- $Q_{ice\text{-}fw}$ remains unchanged at a value of 14.3 kWh/m^3 at u = 0.6 kg/kg and 22.2 kWh/m^3 at u = 0.8 kg/kg due to the non-involvement of T_{w0} in Equation 10.73;
- $Q_{w\text{-}nfr}$ remains unchanged at a value of 52.3 kWh/m^3 at u = 0.6 kg/kg and 62.8 kWh/m^3 at u = 0.8 kg/kg due to the involvement of the same value of $T_{w\text{-}avg\text{-}end}$ = 72.9 °C in Equation 10.74, which is independent of T_{w0} (see Figure 10.37 and Figure 10.39).

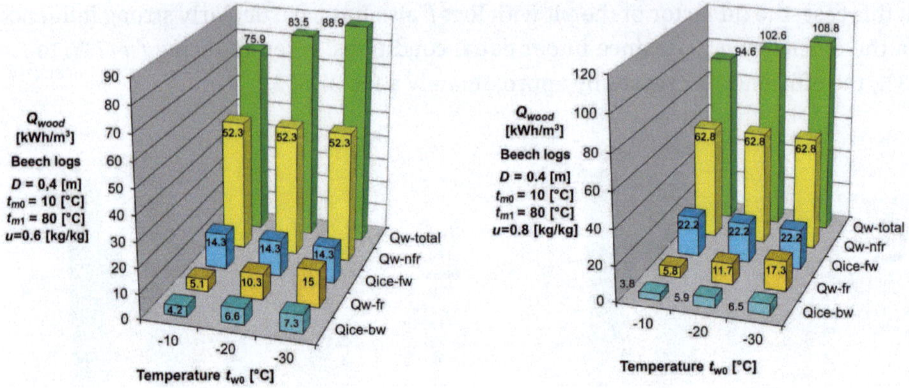

Figure 10.48 Change in energy Q_{wood} and its components (in kWh/m^3) for thawing and heating frozen beech logs with u = 0.6 kg/kg (left) and u = 0.8 kg/kg (right), depending on t_{w0} (data source: [42, own illustration-right])

Figure 10.49 shows the change in the individual components of the energy Q_{wood} in percentages of the total value $Q_{w\text{-}total}$, depending on the mentioned values of t_{w0}.

Expressing the individual components of the energy Q_{wood} as percentages of the total value, a decrease in t_{w0} from –10 °C to –30 °C the following change in the proportion of each component (Figure 10.49):

- $Q_{ice\text{-}bw}$ increases from 5.6 % to 8.2 % at u = 0.6 kg/kg and from 4.0 % to 6.0 % at u = 0.8 kg/kg;
- $Q_{w\text{-}fr}$ increases from 6.7 % to 16.9 % at u = 0.6 kg/kg and from 6.2 % to 15.9 % at u = 0.8 kg/kg;
- $Q_{ice\text{-}fw}$ decreases from 18.8 % to 16.1 % at u = 0.6 kg/kg and from 23.4 % to 20.4 % at u = 0.8 kg/kg;
- $Q_{w\text{-}nfr}$ decreases at u = 0.6 kg/kg from 68.9 % to 58.8 % and at u = 0.8 kg/kg from 66.4 % to 57.7 %.

The values of the energy Q_{wood} and its components given above apply to both steaming and cooking of frozen beech logs.

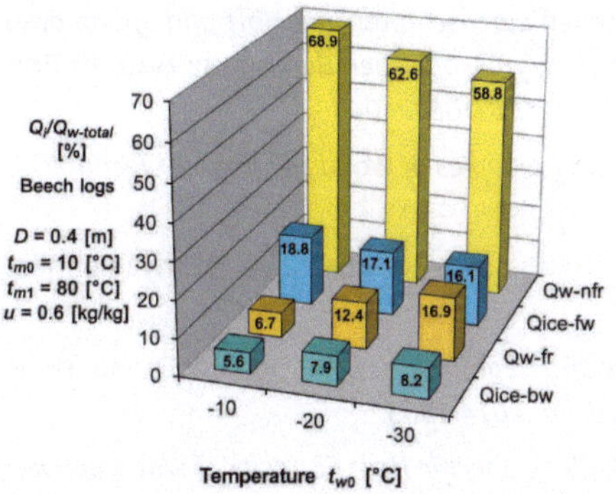

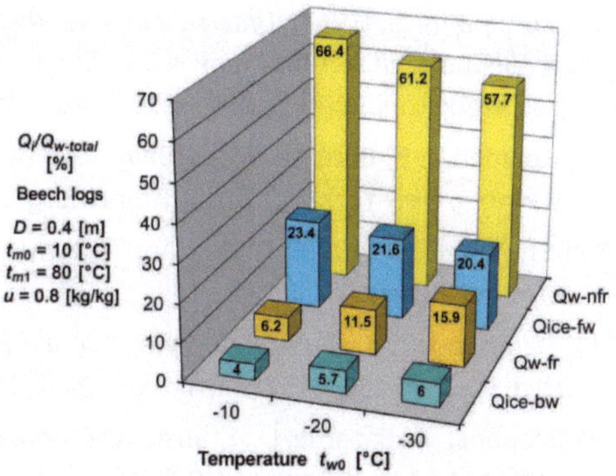

Figure 10.49 Change of the components of the energy Q_{wood} in % of the total energy for frozen beech logs with u = 0.6 kg/kg (left) and u = 0.8 kg/kg (right), depending on t_{w0} (data source: [42, own illustration-right])

References

[1] Deliiski, N. (1977): *Berechnung der instationären Temperaturverteilung im Holz bei der Erwärmung durch Wärmeleitung. Teil I: Mathematisches Modell für die Erwärmung des Holzes durch Wärmeleitung*. In: Holz als Roh- und Werkstoff, 35, p. 141 – 145

[2] Deliiski, N. (1979): *Berechnung der instationären Temperaturverteilung im Holz bei der Erwärmung durch Wärmeleitung. Teil I: Entwurf einer Technologie zum intensiven Dämpfen von Holz*. In: Holz als Roh- und Werkstoff, 37, p. 143 – 148

[3] Deliiski, N. (2003): *Modelling and Technologies for Steaming Wood Materials in Autoclaves*. Dissertation (DSc.), University of Forestry, Sofia, Bulgaria

[4] Deliiski, N.; Dzurenda, L. (2010): *Modelling of the Thermal Processes in the Technologies for Wood Thermal Treatment*. Technical University Zvolen, Zvolen, Slovakia

[5] Deliiski, N. (2013): *Modelling of the Energy Needed for Heating of Capillary Porous Bodies in Frozen and Non-Frozen States*. Lambert Academic Publishing / Scholars' Press, Saarbrücken, Germany

[6] Lawniczak, M. (1995): *Hydrothermal and Plasticizing Treatment of Wood. Part I. Boiling and Steaming of Wood*. Agricultural Academy, Poznań, Poland

[7] Trebula, P.; Klement, I. (2002): *Drying and Hydrothermal Treatment of Wood*. Technical University Zvolen, Zvolen, Slovakia

[8] Videlov, Ch. (2003): *Drying and Thermal Treatment of Wood*. University of Forestry, Sofia, Bulgaria

[9] Chudinov, B. S. (1966): *Theoretical Research of Thermo-Physical Properties and Thermal Treatment of Wood*. Dissertation (DSc.), SibLTI, Krasnoyarsk, Russia

[10] Vorreiter, L. (1949): *Holztechnologisches Handbuch. Band I*. Verlag Georg Fromme & Co., Wien, Austria

[11] Kühlmann, G. (1962): *Untersuchung der thermischen Eigenschaften von Holz und Spanplatten in Abhängigkeit von Feuchtigkeit und Temperatur im hygroskopischen Bereich*. In: Holz als Roh- und Werkstoff, 20(7), p. 259 – 270. *https://doi.org/10.1007/BF02604682*

[12] Kollman, F. F.; Côté, W. A., Jr. (1984): *Solid Wood*. In: *Principles of Wood Science and Technology*. Springer, New York/NY; Berlin/Heidelberg

[13] Shubin, G. S. (1990): Drying and Thermal Treatment of Wood. Lesnaya Promyshlennost, Moskau, Russia

[14] Radmanović, K.; Dukić, I.; Pervan, S. (2014): *Specific Heat Capacity of Wood*. In: Drvna Industrija, 65(2), p. 151 – 157. *https://doi.org/10.5552/drind.2014.1333*

[15] Nopens, M.; Sazama, U.; Krause, A.; Fröba, M. (2021): *Specific Heat Capacity of Wood Between –140 and 50 °C in Dry and Wet State*. In: Holzforschung, 75(9), p. 779 – 785

[16] Tumbarkova, N. (2019): *Modelling of the Freezing and Defrosting Processes of Logs and Their Energy Consumption*. Doctoral thesis, University of Forestry, Sofia, Bulgaria

[17] Siau, J. F. (1984): *Transport Processes in Wood*. Springer-Verlag, New York, USA

[18] Kanter, K. R. (1955): *Investigation of the Thermal Properties of Wood*. Doctoral Thesis, MLTI, Moskau, Russia

[19] Pervan, S. (2009): *Technology for Treatment of Wood with Water Steam*. University of Zagreb, Zagreb, Croatia

[20] Hrčka, R.; Babiak, M. (2017): *Wood Thermal Properties*. In , G. (Hrsg.), *Wood in Civil Engineering*. InTechOpen, London, UK, p. 25 – 43

[21] Steinhagen, H. P. (1977): *Thermal Conductive Properties of Wood, Green or Dry, from –40 to +100 °C: A Literature Review*. In: General Technical Report FPL-09, Forest Products Laboratory, Forest Service

[22] Khattabi, A.; Steinhagen, H. P. (1992): *Numerical Solution to Two-Dimensional Heating of Logs*. In: Holz als Roh- und Werkstoff, 50, p. 308 – 312

[23] Efimov, S. S. (1985): *Temperature Dependence of the Heat of Crystallization of Water*. In: Journal of Engineering Physics and Thermophysics, 49(4), p. 1229 – 1233. *https://doi.org/10.1007/BF00871924*

[24] Deliiski, N.; Dzurenda, L.; Tumbarkova, N.; Angelski, D. (2020): *Mathematical Description of the Latent Heat of Bound Water in Wood During Freezing and Defrosting*. In: Acta Facultatis Xylologiae Zvolen, 62(1), p. 41 – 53

[25] Niemz, P.; Sonderegger, W. (2017): *Holzphysik: Physik des Holzes und der Holzwerkstoffe*. Carl Hanser Verlag, Munich, Germany

[26] Niemz, P.; Teischinger, A.; Sandberg, D. (Hrsg.) (2023): *Springer Handbook of Wood Science and Technology*. Springer Nature Switzerland AG, Cham, Switzerland

[27] Niemz, P.; Sandberg, D. (2025): *General Data Related to Wood*. Chapter in *Wood Materials and Process Data. The Most Relevant Data, Tables and Figures*. Springer Nature Switzerland AG, Cham, Switzerland

[28] Campean, M. (2005): *Heat Treatments of Wood*. Transilvania University of Brasov, Brasov, Romania

[29] Sergovsky, P. S.; Rasev, A. I. (1987): *Hydrothermal Treatment and Preservation of Wood*. Lesnaya Promyshlennost, Moskau, Russia

[30] Dorn, W. S.; McCracken, D. D. (1972): *Numerical Methods with FORTRAN IV: Case Studies*. John Wiley & Sons, New York, USA

[31] Mörath, M. (1949): *Das Dämpfen und Kochen in der Furnier- und Sperrholzindustrie*. In: Holztechnik, Nr.7

[32] Deliiski, N. (2013): *3D Modeling and Visualization of the Non-Stationary Temperature Distribution During Heating of Frozen Wood*. In: Drvna Industrija, 64(4), p. 293 – 303

[33] Deliiski, N.; Vitchev, P.; Angelski, D.; Tumbarkova, N. (2022): *Change in the Duration of the Autoclave Steaming Regimes of Non-Frozen Prisms at Dispatching Interferences in the Production of Veneer*. In: Innovations in Woodworking Industry and Engineering Design, 11(1), p.25–32

[34] Deliiski, N.; Niemz, P.; Angelski, D.; Vitchev, P.; Tumbarkova, N. (2022): *Computing the Duration of Regimes for Autoclave Steaming of Frozen Wooden Prisms Under Variable Operating Conditions in Veneer Production*. In: Wood Material Science & Engineering, 17(6), p. 451 – 458

[35] Deliiski, N.; Niemz, P.; Angelski, D.; Vitchev, P.; Tumbarkova, N. (2023): *Computing the Thermal Efficiency of Autoclaves During Steaming of Frozen Wooden Prisms for Veneer Production at Changing Operational Conditions*. In: Processes, 11(3), 822

[36] Dzurenda, L.; Deliiski, N. (2019): *Thermal Processes in the Woodworking Technologies*. Technical University Zvolen, Zvolen, Slovakia

[37] Deliiski, N.; Dzurenda, L.; Niemz, P.; Câmpean, M.; Angelski, D.; Vitchev, P. (2023): *A Simulation Study of the Heat Balance of Concrete Pits During Boiling of Non-Frozen Logs*. In: Bulletin of the Transilvania University of Brasov, Series II: Forestry Wood Industry Agricultural Food Engineering, 16(65) (Special Edition 3), p. 67 – 82. *https://doi.org/10.31926/but.fwiafe.2023.16.65.2*

[38] Deliiski, N.; Niemz, P.; Dzurenda, L.; Vitchev, P.; Angelski, D. (2023): *An Approach for Computing the Thermal Balance and Energy Consumption of Concrete Pits During Boiling of Frozen Logs for Veneer Production*. In: Wood Material Science & Engineering, 18(6), p. 2153 – 2163. *https://doi.org/10.1080/17480272.2023.2275758*

[39] Dzurenda, L.; Deliiski, N.; Angelski, D.; Vitchev, P.; Atanasova, K. (2025): *Heat Balances of Concrete Pits When Steaming or Boiling Unfrozen Logs*. In: Acta Facultatis Xylologiae Zvolen, 67(1), p. 61 – 74. *https://doi.org/10.17423/afx.2025.67.1.06*

[40] Rivkin, S.; Aleksandrov, A. (1978): *Thermodynamic Properties of Water and Water Vapor*. Tehnika, Sofia, Bulgaria

[41] Deliiski, N.; Niemz, P.; Dzurenda, L.; Angelski, D.; Câmpean, M.; Vitchev, P. (2025): *Comparative Model-Based Study of the Thermal Balance and Efficiency of Concrete Pits During Steaming or Boiling Frozen Veneer Logs*. In: Wood Material Science & Engineering, online first. *https://doi.org/10.1080/17480272.2025.2555443*

[42] Deliiski, N.; Niemz, P.; Angelski, D.; Campean, M.; Vitchev, P. (2024): *An Engineering Approach for Calculating the Energy Required for Defrosting and Subsequent Heating of Frozen Wood During Thermal Treatment*. In: Wood Material Science & Engineering, 19(3), p. 811 – 820. *https://doi.org/10.1080/17480272.2024.2338567*

11

Appendix A: Drying Behavior of Various Native Wood Species

(based on [1–3])

Wood species	Physical properties	Drying behavior
Maple (Sycamore maple) *Acer pseudo-platanus* L.	ρ_0 = 480 - 590 - 750 kg/m^3 β_l = 0.5 % β_r = 3 % β_t = 8 % β_v = 11.5 - 11.8 %	Wood species with moderate difficulty in drying. Tendency to crack and warp. Excessively long storage in the green state leads to discoloration. Artificial drying should therefore be started as soon as possible. To preserve the light color, drying should be carried out at a moderate initial temperature (45 °C) and without humidification until the fiber saturation point is reached.
Aspen (trembling aspen) *Populus tremula* L.	ρ_0 = 360 - 450 - 560 kg/m^3 β_r = 3.5 % β_t = 6.7 - 8.5 % β_v = 11.0 - 12.8 %	Wood species with low difficulty level in drying. Fast drying. Low tendency to crack.
Birch *Betula verrucosa* Ehrh.	ρ_0 = 460 - 610 - 800 kg/m^3 β_l = 0.6 % β_r = 5.3 % β_t = 7.8 % β_v = 13.7 - 14.2 %	Wood species with moderate difficulty level in drying. Tendency to crack and warp, therefore drying must be carried out slowly.

Wood species	Physical properties	Drying behavior
Pear tree *Pyrus communis* L.	ρ_0 = 650 – 700 – 760 kg/m³ β_l = 0.4 % β_r = 4.6 % β_t = 9.1 % β_v = 13.6 – 14.7 %	Wood species with moderate difficulty level in drying. Tendency to warp. It is recommended to steam the material before drying in order to reduce internal stresses and even out the color.
Beech *Fagus sylvatica* L.	ρ_0 = 490 – 680 – 880 kg/m³ β_l = 0.3 % β_r = 5.8 % β_t = 11.8 % β_v = 14 – 17.9 – 21 %	Wood with high internal stresses, with strong shrinkage and low dimensional stability under moisture fluctuations. Tendency to crack and warp. A light steaming before drying opens the pores and equalizes the moisture content, which has a positive effect on both drying time and drying quality. Must be dried slowly and carefully. In the steamed condition it has very good bending properties.
Douglas fir *Pseudotsuga menziesii* Franco	ρ_0 = 320 – 470 – 730 kg/m³ β_l = 0.3 % β_r = 4 – 4.8 % β_t = 7 – 7.7 % β_v = 11.5 – 12.5 %	Wood species with moderate difficulty in drying. To avoid excessive resin exudation, drying must be carried out carefully (temperatures < 100 °C recommended). Low tendency to warp. Fine surface checks are common, especially in thick grades.
Yew *Taxus baccata* L.	ρ_0 = 610 – 640 – 740 kg/m³ β_r = 3.7 % β_t = 5.3 % β_v = 8.4 – 9.2 %	Wood species with moderate drying difficulty. Good dimensional stability.
Oak *Quercus robur* L. *Quercus petraea* (Matt.) Liebl	ρ_0 = 390 – 650 – 930 kg/m³ β_l = 0.4 % β_r = 4 – 4.6 % β_t = 7.8 – 10 % β_v = 12.6 – 15.6 %	Wood species with high drying difficulty, prone to cracking. Must therefore be dried slowly. Low initial temperature (30 °C) also helps to avoid discoloration. Pre-drying in the open air is beneficial, but often leads to case hardening. Contact with ferrous metals causes black stains due to the tannin content.

Wood species	Physical properties	Drying behavior
Alder *Alnus glutinosa* L. (Gaertn.)	ρ_0 = 450 – 510 – 600 kg/m^3 β_l = 0.5 % β_r = 4.4 – 4.8 % β_t = 7.7 – 9.3 % β_v = 12.3 – 14.2 %	Wood species with a low degree of difficulty in drying. Typical is the oxidative discoloration (initially orange, turning violet with prolonged storage in the green state). Artificial drying at a temperature of approximately 60 °C immediately after felling is recommended in order to remove the free water from the wood as quickly as possible.
Ash *Fraxinus excelsior* L.	ρ_0 = 410 – 650 – 820 kg/m^3 β_l = 0.2 % β_r = 4.6 – 5 % β_t = 8 – 8.4 % β_v = 12.8 – 13.6 %	Wood species with a low degree of difficulty in drying. In the steamed state it has very good bending properties.
Spruce *Picea abies* L.	ρ_0 = 300 – 430 – 640 kg/m^3 β_l = 0.3 % β_r = 3.5 – 3.7 % β_t = 7.8 – 8 % β_v = 11.6 – 12 %	Wood species with low drying difficulty. Fast drying. Low tendency to cracking and warping. Good dimensional stability.
European hornbeam *Carpinus betulus* L.	ρ_0 = 500 – 790 – 820 kg/m^3 β_l = 0.5 % β_r = 5.2 – 6.8 % β_t = 11.5 % β_v = 18.8 %	Wood species with high growth stresses, therefore with great drying difficulty. Strong tendency to cracking and warping. Slow drying is recommended. Natural drying leads to better results. End-grain coatings with paraffin or wax dispersions are recommended to avoid end checks.
Chestnut (Horse chestnut) *Aesculus hippocastanum* L.	ρ_0 = 460 – 510 – 550 kg/m^3 β_l = 0.9 % β_r = 3.3 % β_t = 6.8 % β_v = 11 %	Wood species with moderate difficulty in drying. Low tendency to crackβing and warping. Can be dried both in the open air and artificially. For technical drying, do not exceed an initial temperature of 60 °C.

Wood species	Physical properties	Drying behavior
Pine *Pinus sylvestris* L.	ρ_0 = 300 – 490 – 860 kg/m^3 β_l = 0.2 – 0.4 % β_r = 3.3 – 4.5 % β_t = 7.5 – 8.7 % β_v = 11.2 – 12.4 %	Wood species with low difficulty in drying. Tendency to swelling and resin exudation. Higher tendency to cracking compared to fir and spruce, therefore slow drying is recommended, but with initial temperatures above 45 °C, otherwise there is a risk of blue stain as long as the moisture content exceeds 25 %.
Cherry *Prunus avium* L.	ρ_0 = 490 – 550 – 670 kg/m^3 β_r = 3.5 – 5 % βt = 6.5 – 8.7 % β_v = 13.7 – 14 %	Wood species with moderate difficulty in drying. Can be dried both in the open air and artificially. For technical drying, do not exceed temperatures of 60–70 °C, otherwise it tends to crack and warp.
Larch *Larix decidua* L.	ρ_0 = 400 – 550 – 820 kg/m^3 β_l = 0.3 % β_r = 3.3 – 4.3 % β_t = 7.8 – 10.4 % β_v = 11.4 – 15 %	Wood species with moderate difficulty in drying. Tendency to swelling and resin exudation. Higher tendency to cracking compared to fir and spruce, but still low compared to hardwood.
Linden *Tilia cordata* Mill.	ρ_0 = 320 – 490 – 560 kg/m^3 β_l = 0.2 – 0.3 % β_r = 5.5 – 6.6 % β_t = 9.1 – 10.7 % β_v = 14.4 – 14.9 %	Wood species with moderate difficulty in drying. Tendency to discoloration and crack formation; therefore, slow drying at 60 °C and relative humidity < 65 % is recommended.
Walnut *Juglans regia* L.	ρ_0 = 450 – 640 – 750 kg/m^3 β_l = 0.5 % β_r = 5.4 % β_t = 7.5 % β_v = 13.4 – 14 %	Wood species with moderate difficulty in drying. Slow drying is recommended to avoid cracking. In steamed condition it has good bending ability.
Poplar (Black poplar) *Populus nigra* L.	ρ_0 = 370 – 410 – 520 kg/m^3 β_l = 0.3 % β_r = 3.3 – 5.2 % β_t = 7.1 – 9.8 % β_v = 10.7 – 14.3 %	Wood species with moderate difficulty in drying. Shrinks evenly. Hardly prone to cracking, but frequently prone to warping.

Wood species	Physical properties	Drying behavior
Plane tree *Platanus acerifolia* Willd.	ρ_0 = 380 - 580 - 650 kg/m^3 β_l = 0.5 % β_r = 4.5 - 5 % β_t = 8.4 - 8.7 % β_v = 13.7 - 14.2 %	Wood species with moderate difficulty in drying. Slight tendency to crack and warp. In steamed condition it has good bending ability.
Black locust *Robinia pseudoacacia* L.	ρ_0 = 540 - 740 - 870 kg/m^3 β_l = 0.1 % β_r = 3.2 - 4.6 % β_t = 5.4...7.2 % β_v = 11.4...12.2 %	Wood species with a high degree of difficulty in drying due to pores closed with tyloses and high density. Strong tendency to crack and warp. Slow drying is recommended. Natural drying leads to better results than artificial drying. End-grain coatings with paraffin or wax dispersions are recommended to prevent end checks.
Silver fir *Abies alba* Mill.	ρ_0 = 320 - 410 - 710 kg/m^3 β_l = 0.1 % β_r = 2.9 - 3.8 % β_t = 7.2 - 7.6 % β_v = 10.2 - 11.5 %	Wood species with a low degree of difficulty in drying. Fast drying. Low tendency to crack and warp. Good dimensional stability.
Elm (wych elm) *Ulmus glabra* Huds.	ρ_0 = 440 - 640 - 820 kg/m^3 β_l = 0.3 % β_r = 4.6 - 4.8 % β_t = 6.9 - 8.3 % β_v = 11.8 - 13.8 %	Wood species with moderate difficulty in drying. Tendency to crack and warp. Can be dried both outdoors and artificially. Technical drying should be carried out gently, with initial temperatures below 60 °C.
Willow *Salix alba* L.	ρ_0 = 270 - 330 - 380 kg/m^3 β_l = 0.5 - 0.9 - 1.1 % β_r = 1.9 - 2.4 - 3.9 % β_t = 5.4 - 6.3 - 8.7 % β_v = 7.9 - 9.6 - 11.3 %	Wood species with low difficulty in drying. Fast drying. Slight tendency to crack and warp.

ρ_0 = oven-dry density (value min–mean–max) [kg/m^3]
β_l = total shrinkage (from green to oven-dry condition) in the longitudinal direction [%]
β_r = total shrinkage (from green to oven-dry condition) in the radial direction [%]
β_t = total shrinkage (from green to oven-dry condition) in tangential direction [%]
β_v = total volumetric shrinkage (from green to oven-dry condition) [%]

References

[1] Wagenführ, A. (2008): *Holzatlas*: Carl Hanser Verlag, Munich, Germany

[2] Trübswetter, T. (2009): *Holztrocknung. Verfahren zur Trocknung von Schnittholz – Planung von Trocknungsanlagen* 2nd ed., Carl Hanser Verlag, Leipzig, Germany

[3] Campean, M. (2005): *Heat Treatments of Wood*: Publishing House of Transilvania University, Brasov, Romania

12 Appendix B: Determination of Relative Humidity and Equilibrium Moisture Content

Determination of relative humidity (φ) and equilibrium moisture content (*EMC*) based on the readings (t, Δt) of the psychrometer according to the following table (source: [4])

Reading examples:

1. The psychrometer of a drying kiln shows at a certain point in time during drying: t = 60 °C and t_{wet} = 52 °C. The psychrometric difference in this state is Δt = 8 °C and therefore the relative humidity is: φ = 65 % and the corresponding equilibrium moisture content *EMC* = 9.8 %.
2. An environment with a temperature of t = 20 °C and a relative humidity of φ = 42 % corresponds to an equilibrium moisture content of *EMC* = 8.2 %.

References

[1] Cividini, R. (2001): *Conventional kiln-drying of lumber*. NARDI S.p.A., Chiampo, Italy

t°	Δt°	1	1.5	2	3	4	5	6	7	8	9	10	11	12	13	14	15	16	18	20	22	24	26	28	30
0	***φ***	81	73	63	43	31	17	13	11																
	EMC	16.4	14.2	12	8.5	66.6	4	3.3	3																
10	***φ***	87	81	76	66	57	45	35	25	15	6														
	EMC	18.8	15.6	12.4	10.4	9.2	8.6	7	5.4	3.6	1.2														
20	***φ***	89	85	82	70	64	55	49	42	35	29	24	18	12	6										
	EMC	20	18.2	17	14	12	10	9.2	8.2	7	6	5.2	4.6	3.4	1.6										
30	***φ***	90.6	88.8	87	75.3	69	64	59	54	48	43	38.5	35	30	24	20	17	13	8						
	EMC	20.7	19.4	17.9	14	13	11.4	10.4	9.6	8.6	8	7.3	6.8	5.9	5	4.3	3.7	2.3	1.4						
35	***φ***	91.3	90	88	78.5	73	67.5	63	57.5	52	47	43	38	35	29	27	24	20	13	6					
	EMC	21	19.6	18	14.6	13	11.8	11	9.7	9	8.4	7.7	7	6.5	5.7	5.2	4.7	4.2	3	1.4					
40	***φ***	92	90.5	89	80.5	75.5	70	65	61	55	51	46	43	40	34	32	29	25	20	15	8				
	EMC	21	19.9	18.1	16.1	14.2	12	11	10.2	9.3	8.6	8.2	7.6	6.8	6.3	5.9	5.4	4.9	4	3	1.8				

t°	Δt°	1	1.5	2	3	4	5	6	7	8	9	10	11	12	13	14	15	16	18	20	22	24	26	28	30
45	***φ***	92.5	90.9	89.3	82	77	73	68	62	58	54	50	47	43	39	36	33	30	24	20	14	9			
	EMC	21	19.9	18.2	15.4	14	12.4	11.4	10.6	9.8	9.1	8.4	7.8	7.1	6.6	6.2	5.8	5.4	4.6	3.9	3	2			
50	***φ***	93	91.3	89.5	83	77	74	68	64	60	57	54	50	47	44	40	37	35	29	23	18	14	10	5	
	EMC	20.8	19.4	18	15.8	14.1	12.1	11.5	10.6	9.8	9.2	8.6	8	7.5	7	6.5	6.1	5.6	4.9	4.2	3.5	2.8	2.1	1.2	
55	***φ***	93.5	91.6	89.7	83.8	79	75	71	67	63	59	56	52	48	44	43	40	37	33	27	23	19	14	10	7
	EMC	20.5	18.9	17.8	15.6	14	12.7	11.6	10.7	9.9	9.2	8.6	8.2	7.6	7	6.7	6.3	5.9	5.2	4.5	4	3.4	2.9	2	1.5
60	***φ***	94	92	90	84.5	80.5	76.5	73.5	69	65	62	59	54	51	48	45	43	40	35	30	27	22	19	15	12
	EMC	20.2	18.8	17.5	15.3	13.9	12.5	11.5	10.6	9.8	9.2	8.7	8.1	7.7	7.2	6.7	6.4	6	5.4	4.8	4.2	3.7	3.1	2.6	2.1
65	***φ***	94.3	92.4	90.5	85	82	77	69	68	66	63	59	56	53	50	47	45	43	38	33	29	25	22	18	15
	EMC	19.5	18.4	17	15	13.6	12.3	11.3	10.5	9.7	9.1	8.6	8	7.6	7.1	6.7	6.4	6	5.4	4.9	4.3	3.9	3.4	2.9	2.4
70	***φ***	94.5	92.8	91	85	80.5	77	73.5	70	66.5	63	60	57	54	51	49	46	44	39	35	32	27	24	21	17
	EMC	19	17.6	16.2	14.3	13	11	10.8	10	9.4	9	8.5	8	7.5	7	6.7	6.4	6	5.4	4.9	4.4	4	3.5	3.1	2.7

t°	Δt°	1	1.5	2	3	4	5	6	7	8	9	10	11	12	13	14	15	16	18	20	22	24	26	28	30
75	***φ***	94.7	93.1	91.4	87	83	79.5	76	72.5	69	65.5	62	59	56	53	50	48	45	41	37	33	30	26	24	19
	EMC	18.7	17.6	16.4	14.4	13.1	12	11	10.1	9.5	8.8	8.3	7.8	7.3	7	6.6	6.3	5.9	5.4	4.9	4.4	4	3.6	3.2	2.8
80	***φ***	95	93.4	91.8	86.5	83	79.5	76	73	70	67	59	60	57	54	52	49	47	44	39	35	31	27	25	21
	EMC	18.1	17.1	15.8	14.2	12.8	11.6	10.7	9.9	9.3	8.7	8.2	7.7	7.2	6.8	6.5	6.2	5.9	5.4	4.9	4.4	4	3.7	3.3	2.9
85	***φ***	95.2	93.6	92.1	87.5	83.5	80	77	74	71	68	64.5	61.5	58.5	56	54	52	49	45	41	37	33	30	27	23
	EMC	17.7	16.4	15.5	13.8	12.5	11.4	10.5	9.7	9.1	8.5	8	7.5	7.1	6.7	6.4	6.1	5.8	5.3	4.9	4.4	4	3.7	3.3	2.9
90	***φ***	95.3	93.8	92.4	87.5	84	81	78	75	72	69	66	63	61	57	55	53	49	46	42	39	35	32	29	26
	EMC	17.2	16.3	15.2	13.6	12.3	11.1	10.2	9.5	8.9	8.4	7.9	7.4	7	6.6	6.3	6	5.7	5.2	4.8	4.4	4	3.7	3.3	3
95	***φ***	95.4	94	92.7	88.5	85	82	79	76	74	70	67	65	63	59.5	57	55	51	48	44.5	41	38	33	30	28
	EMC	16.5	15.2	14.9	13.2	12	11	10.1	9.3	8.8	8.2	7.7	7.3	6.9	6.5	6.2	5.9	5.6	5.1	4.8	4.4	4	3.7	3.3	3
100	***φ***	95.5	94.3	93	88.5	85	82	79	76	73	70	67.5	65	62.5	60	57.5	55	53	48.5	45	42	38	35	33	29
	EMC	16	15.2	13.8	12.4	11.4	10.6	9.9	9.1	8.6	8.1	7.5	7.1	6.7	6.3	6	5.8	5.5	5	4.7	4.3	3.9	3.6	3.3	3

13

Appendix C: State Parameters of Moist Air at Atmospheric Pressure (p = 1 Bar)

(Source: [1])

t [°C]	Relative humidity ϕ [%]										
	0	10	20	30	40	50	60	70	80	90	100
0	p_D	0.61	1.22	1.83	2.44	3.06	3.67	4.28	4.89	5.50	6.11
	x	0.38	0.76	1.14	1.52	1.91	2.29	2.67	3.06	3.44	3.82
	h	0.95	1.90	2.85	3.80	4.75	5.70	6.65	7.60	8.55	9.55
5	p_D	0.87	1.74	2.61	3.48	4.36	5.23	6.10	6.97	7.84	8.72
	x	0.54	1.08	1.63	2.17	2.72	3.27	3.82	4.37	4.92	5.47
	h	6.35	7.71	9.09	10.4	11.8	13.2	14.6	16.0	17.3	18.7
10	p_D	1.23	2.45	3.68	4.91	6.14	7.36	8.59	9.82	11.04	12.27
	x	0.77	1.53	2.30	3.07	3.84	4.61	5.39	6.17	6.94	7.73
	h	11.9	13.9	15.8	17.7	19.7	21.6	23.6	25.5	27.5	29.5
15	p_D	1.70	3.40	5.11	6.81	8.52	10.2	11.9	13.6	15.3	17.0
	x	1.06	2.12	3.19	4.26	5.34	6.41	7.49	8.58	9.66	10.8
	h	17.7	20.4	23.1	25.8	28.5	31.2	33.9	36.7	39.4	42.0
20	p_D	2.34	4.67	7.01	9.35	11.7	14.0	16.4	18.7	21.0	23.4
	x	1.46	2.92	4.39	5.87	7.36	8.83	10.4	11.9	13.3	14.9
	h	23.7	27.4	31.1	34.9	38.7	42.4	46.4	50.2	53.7	57.8
25	p_D	3.17	6.33	9.50	12.7	15.8	19.0	22.2	25.3	28.5	31.7
	x	1.98	3.96	5.97	8.00	9.99	12.1	14.1	16.2	18.3	20.4
	h	30.0	31.5	40.2	45.4	50.4	55.8	60.9	66.3	71.6	76.9

t [°C]	Relative humidity ϕ [%]										
	0	10	20	30	40	50	60	70	80	90	100
30	p_D	4.24	8.48	12.7	17.00	21.2	21.4	29.7	33.9	38.2	42.4
	x	2.65	5.32	8.00	10.8	13.5	16.2	19.0	21.8	24.7	27.5
	h	36.8	43.6	50.4	57.6	64.5	71.4	78.6	85.7	93.1	100.3
35	p_D	5.62	11.2	16.9	22.5	28.1	33.7	39.4	45.0	50.6	56.2
	x	3.52	7.05	10.7	14.3	18.0	21.7	25.5	29.3	33.2	37.0
	h	44.0	53.1	62.4	71.7	81.2	90.7	100.4	110.2	120.2	129.9
40	p_D	7.38	14.8	22.1	29.5	36.9	44.3	51.6	59.0	66.4	73.8
	x	4.62	9.34	14.1	18.9	23.8	28.8	33.8	39.0	44.2	49.5
	h	51.9	64.0	76.3	88.7	101.3	114.1	127.0	140.4	153.8	167.7
45	p_D	9.58	19.2	28.7	38.3	47.9	57.5	67.1	76.7	86.2	95.8
	x	6.02	12.2	18.4	24.8	31.3	38.0	44.7	51.7	58.7	65.9
	h	60.6	76.5	92.5	109.1	125.9	143.2	160.5	178.6	196.7	215.3
50	p_D	12.3	24.7	37.0	49.3	61.7	74.0	86.3	98.7	111.0	123.3
	x	7.75	15.8	23.9	32.3	40.9	49.7	58.8	68.1	77.7	87.5
	h	70.1	91.0	112.0	133.8	156.1	178.9	202.5	226.6	251.5	276.9

p_D = partial pressure of water vapor [mbar]
x = moisture content [g/kg$_{dry\ air}$]
h = enthalpy [kJ/(1+x) kg]

References

[1] Frühwald, A., Ressel, J. (1982): *Holztrocknung. Teil 1: Grundlagen der Holztrocknung*. Universität Hamburg, Germany

14 Appendix D: Drying Schedules

Drying Schedules for Spruce Wood

Drying Schedule for Spruce Wood, 24 mm

Phase	Wood moisture content [%]	Temperature [°C]	Equilibrium moisture content [%]	Time [h]
Heating phase	>50	60	18	9
Pure drying	50 – 40	60	15	determined by the rate of moisture reduction
	40 – 30	60	12.3	
	30 – 20	60	11.5	
	20 – 15	62	9.8	
	15 – 10	65	6.7	
Conditioning	10	55	8	5
Cooling	10	30	–	6
Total duration: approximately 3 days				

Drying schedule for spruce wood, 48 mm

Phase	Wood moisture content [%]	Temperature [°C]	Equilibrium moisture content [%]	Time [h]
Heating phase	>50	60	18	15
Pure drying	50 – 40	60	15	determined by the rate of moisture reduction
	40 – 30	60	13.4	
	30 – 20	60	12.5	
	20 – 15	62	10.6	
	15 – 10	65	7.2	
Conditioning	10	55	8	6
Cooling	10	30	–	10
Total duration: approximately 5 days				

Drying schedule for spruce wood, 75 mm

Phase	Wood moisture content [%]	Temperature [°C]	Equilibrium moisture content [%]	Time [h]
Heating phase	>50	60	18	19
Pure drying	50 – 40	60	16	determined by the rate of moisture reduction
	40 – 30	60	14.8	
	30 – 20	60	13.7	
	20 – 15	62	11.4	
	15 – 10	65	7.6	
Conditioning	10	55	8	7
Cooling	10	30	–	12
Total duration: approximately 9 days				

Drying Schedules for Fir Wood

Drying schedule for fir wood, 24 mm

Phase	Wood moisture content [%]	Temperature [°C]	Equilibrium moisture content [%]	Time [h]
Heating phase	>50	50	18	9
Pure drying	50 - 40	50	15	determined by the rate of moisture reduction
	40 - 30	50	10.3	
	30 - 20	50	8.9	
	20 - 15	55	7.1	
	15 - 10	65	5.0	
Conditioning	10	55	8	5
Cooling	10	30	–	8
Total duration: approximately 4 days				

Drying schedule for fir wood, 48 mm

Phase	Wood moisture content [%]	Temperature [°C]	Equilibrium moisture content [%]	Time [h]
Heating phase	>50	48	18	15
Pure drying	50 - 40	48	15	determined by the rate of moisture reduction
	40 - 30	48	11	
	30 - 20	50	9.5	
	20 - 15	54	7.8	
	15 - 10	63	5.3	
Conditioning	10	53	8	6
Cooling	10	30	–	10
Total duration: approximately 6 days				

Drying schedule for fir wood, 75 mm

Phase	Wood moisture content [%]	Temperature [°C]	Equilibrium moisture content [%]	Time [h]
Heating phase	>50	45	18	19
Pure drying	50 - 40	45	15	determined by the rate of moisture reduction
	40 - 30	45	12	
	30 - 20	47	10.2	
	20 - 15	50	8.4	
	15 - 10	60	5.8	
Conditioning	10	53	8	7
Cooling	10	30	–	12
Total duration: approximately 10 days				

Drying Schedules for Poplar and Linden Wood

Drying schedule for poplar- and linden wood, 25 mm

Phase	Wood moisture content [%]	Temperature [°C]	Equilibrium moisture content [%]	Time [h]
Heating phase	>50	60	18	9
Pure drying	50 – 40	60	15	determined by the rate of moisture decrease
	40 – 30	60	11.9	
	30 – 20	62	10.2	
	20 – 15	64	7.8	
	15 – 10	65	4.4	
Conditioning	10	55	8	5
Cooling	10	30	–	6
Total duration: approximately 4 days				

Drying schedule for poplar and linden wood, 50 mm

Phase	Wood moisture content [%]	Temperature [°C]	Equilibrium moisture content [%]	Time [h]
Heating phase	>50	60	18	15
Pure drying	50 – 40	60	16	determined by the rate of moisture decrease
	40 – 30	60	13.2	
	30 – 20	62	11	
	20 – 15	64	8.1	
	15 – 10	65	4.8	
Conditioning	10	55	8	6
Cooling	10	30	–	10
Total duration: approximately 5.5 days				

Drying Schedules for Alder Wood

Drying schedule for alder wood, 25 mm

Phase	Wood moisture content [%]	Temperature [°C]	Equilibrium moisture content [%]	Time [h]
Heating phase	>50	60	18	11
Pure drying	50 – 40	60	15	determined by the rate of moisture decrease
	40 – 30	60	12.5	
	30 – 20	62	10.2	
	20 – 15	64	7.9	
	15 – 10	65	4.5	
Conditioning	10	55	8	5
Cooling	10	30	–	6
Total duration: approximately 5 days				

Drying schedule for alder wood, 50 mm

Phase	Wood moisture content [%]	Temperature [°C]	Equilibrium moisture content [%]	Time [h]
Heating phase	>50	60	18	17
Pure drying	50 – 40	60	15	determined by the rate of moisture reduction
	40 – 30	60	13.1	
	30 – 20	62	11	
	20 – 15	64	8.2	
	15 – 10	65	4.7	
Conditioning	10	55	8	6
Cooling	10	30	–	10
Total duration: approximately 8.5 days				

Drying Schedules for Cherry Wood

Drying schedule for cherry wood, 25 mm

Phase	Wood moisture content [%]	Temperature [°C]	Equilibrium moisture content [%]	Time [h]
Heating phase	>50	55	18	13
Pure drying	50 – 40	55	16.5	determined by the rate of moisture reduction
	40 – 30	55	14.4	
	30 – 20	58	11.0	
	20 – 15	60	7.6	
	15 – 10	63	4.8	
Conditioning	10	53	8	5
Cooling	10	30	–	6
Total duration: approximately 5 days				

Drying schedule for cherry wood, 50 mm

Phase	Wood moisture content [%]	Temperature [°C]	Equilibrium moisture content [%]	Time [h]
Heating phase	>50	60	18	21
Pure drying	50 – 40	60	16.5	determined by the rate of moisture decrease
	40 – 30	60	15.4	
	30 – 20	62	11.9	
	20 – 15	64	8.2	
	15 – 10	65	5.4	
Conditioning	10	55	8	6
Cooling	10	30	–	10
Total duration: approximately 10 days				

Drying Schedules for Maple Wood

Drying schedule for maple wood, 25 mm

Phase	Wood moisture content [%]	Temperature [°C]	Equilibrium moisture content [%]	Time [h]
Heating phase	>50	50	18	11
Pure drying	50 – 40	50	15	determined by the rate of moisture decrease
	40 – 30	50	14.2	
	30 – 20	52	11.8	
	20 – 15	54	8.5	
	15 – 10	64	5.6	
Conditioning	10	54	8	5
Cooling	10	30	–	6
Total duration: approximately 5 days				

Drying schedule for maple wood, 50 mm

Phase	Wood moisture content [%]	Temperature [°C]	Equilibrium moisture content [%]	Time [h]
Heating phase	>50	50	18	17
Pure drying	50 – 40	50	15.7	determined by the rate of moisture decrease
	40 – 30	50	15.3	
	30 – 20	52	13.9	
	20 – 15	54	9.3	
	15 – 10	64	6.5	
Conditioning	10	54	8	6
Cooling	10	30	–	10
Total duration: approximately 9 days				

Drying Schedules for Birch Wood

Drying schedule for birch wood, 25 mm

Phase	Wood moisture content [%]	Temperature [°C]	Equilibrium moisture content [%]	Time [h]
Heating phase	>50	50	18	13
Pure drying	50 – 40	50	13.7	determined by the rate of moisture reduction
	40 – 30	50	12.6	
	30 – 20	52	10.5	
	20 – 15	54	7.6	
	15 – 10	60	4.8	
Conditioning	10	50	8	5
Cooling	10	30	–	6
Total duration: approximately 6 days				

Drying schedule for birch wood, 50 mm

Phase	Wood moisture content [%]	Temperature [°C]	Equilibrium moisture content [%]	Time [h]
Heating phase	>50	50	18	21
Pure drying	50 – 40	50	15.0	determined by the rate of moisture reduction
	40 – 30	50	13.8	
	30 – 20	51	11.5	
	20 – 15	53	8.2	
	15 – 10	60	5.0	
Conditioning	10	50	8	6
Cooling	10	30	–	10
Total duration: approximately 12 days				

Drying Schedules for Walnut Wood

Drying schedule for walnut wood, 25 mm

Phase	Wood moisture content [%]	Temperature [°C]	Equilibrium moisture content [%]	Time [h]
Heating phase	>50	50	18	13
Pure drying	50 – 40	50	16.5	determined by the rate of moisture reduction
	40 – 30	50	14.4	
	30 – 20	60	11.0	
	20 – 15	61	7.6	
	15 – 10	63	4.0	
Conditioning	10	53	8	5
Cooling	10	30	–	6
Total duration: approximately 7 days				

Drying schedule for walnut wood, 50 mm

Phase	Wood moisture content [%]	Temperature [°C]	Equilibrium moisture content [%]	Time [h]
Heating phase	>50	50	18	21
Pure drying	50 – 40	50	16.5	determined by the rate of moisture reduction
	40 – 30	50	15.4	
	30 – 20	55	11.9	
	20 – 15	60	8.2	
	15 – 10	62	5.4	
Conditioning	10	52	8	6
Cooling	10	30	–	10
Total duration: approximately 12 days				

Drying Schedules for Acacia Wood

Drying schedule for acacia wood, 25 mm

Phase	Wood moisture content [%]	Temperature [°C]	Equilibrium moisture content [%]	Time [h]
Heating phase	>50	50	18	13
Pure drying	50 – 40	50	18	determined by the rate of moisture reduction
	40 – 30	50	13.6	
	30 – 20	51	9.3	
	20 – 15	60	3.5	
	15 – 10	63	3.2	
Conditioning	10	53	8	5
Cooling	10	30	–	6
Total duration: approximately 13 days				

Drying schedule for acacia wood, 50 mm

Phase	Wood moisture content [%]	Temperature [°C]	Equilibrium moisture content [%]	Time [h]
Heating phase	>50	50	18	19
Pure drying	50 – 40	50	18	determined by the rate of moisture decrease
	40 – 30	50	14.8	
	30 – 20	51	9.3	
	20 – 15	57	3.6	
	15 – 10	60	3.4	
Conditioning	10	50	8	6
Cooling	10	30	–	10
Total duration: approximately 26 days				

Drying Schedules for Ash Wood

Drying schedule for ash wood, 25 mm

Phase	Wood moisture content [%]	Temperature [°C]	Equilibrium moisture content [%]	Time [h]
Heating phase	>50	44	18	13
Pure drying	50 – 40	44	16	determined by the rate of moisture reduction
	40 – 30	44	14.0	
	30 – 20	45	11.1	
	20 – 15	47	8.4	
	15 – 10	56	7.0	
Conditioning	10	46	8	5
Cooling	10	30	–	6
Total duration: approximately 8 days				

Drying schedule for ash wood, 50 mm

Phase	Wood moisture content [%]	Temperature [°C]	Equilibrium moisture content [%]	Time [h]
Heating phase	>50	44	18	21
Pure drying	50 – 40	44	16.2	determined by the rate of moisture reduction
	40 – 30	44	15.2	
	30 – 20	45	12.2	
	20 – 15	47	9.1	
	15 – 10	54	7.7	
Conditioning	10	44	8	6
Cooling	10	30	–	10
Total duration: approximately 15 days				

Drying Schedules for Beech Wood

Drying schedule for beech wood, 25 mm

Phase	Wood moisture content [%]	Temperature [°C]	Equilibrium moisture content [%]	Time [h]
Heating phase	>50	44	18	13
Pure drying	50 - 40	44	16.5	governed by the rate of moisture decrease
	40 - 30	44	16.4	
	30 - 20	46	12.8	
	20 - 15	54	8.7	
	15 - 10	62	4.8	
Conditioning	10	52	8	12
Cooling	10	30	–	6
Total duration: approximately 9 days				

Drying schedule for beech wood, 50 mm

Phase	Wood moisture content [%]	Temperature [°C]	Equilibrium moisture content [%]	Time [h]
Heating phase	>50	44	18	21
Pure drying	50 - 40	44	16.5	given by the rate of moisture decrease
	40 - 30	44	16.1	
	30 - 20	46	13.9	
	20 - 15	52	10.0	
	15 - 10	58	6.0	
Conditioning	10	48	8	16
Cooling	10	30	–	10
Total duration: approximately 16 days				

Drying Schedules for Oak Wood

Drying schedule for oak wood, 25 mm

Phase	Wood moisture content [%]	Temperature [°C]	Equilibrium moisture content [%]	Time [h]
Heating phase	>50	44	18	15
Pure drying	50 – 40	44	17	given by the rate of moisture decrease
	40 – 30	44	14.1	
	30 – 20	46	11.5	
	20 – 15	54	9.0	
	15 – 10	60	7.0	
Conditioning	10	50	8	12
Cooling	10	30	–	6
Total duration: approximately 16 days				

Drying schedule for oak wood, 50 mm

Phase	Wood moisture content [%]	Temperature [°C]	Equilibrium moisture content [%]	Time [h]
Heating phase	>50	43	18	23
Pure drying	50 – 40	43	17	determined by the rate of moisture reduction
	40 – 30	43	15.2	
	30 – 20	44	12.5	
	20 – 15	46	9.8	
	15 – 10	56	7.7	
Conditioning	10	46	8	15
Cooling	10	30	–	10
Total duration: approximately 30 days				

Drying Schedules for Turkey Oak

Drying schedule for Turkey oak, 25 mm (Source: [1], with permission)

Phase	Wood moisture content [%]	Temperature [°C]	Equilibrium moisture content [%]	Time [h]
Heating phase	>50	30	18	15
Pure drying	50 – 40	30	17	determined by the rate of moisture decrease
	40 – 30	30	15.5	
	30 – 24	35	12.6	
	24 – 20	50	11.5	
	20 – 15	52	7.0	
	15 – 10	55	5.0	
Conditioning	10	45	8	12
Cooling	10	30	–	6
Total duration: approximately 21 days				

Drying schedule for Turkey oak, 50 mm (source: [1], with permission)

Phase	Wood moisture content [%]	Temperature [°C]	Equilibrium moisture content [%]	Time [h]
Heating phase	>50	30	18	25
Pure drying	50 – 40	30	17	determined by the rate of moisture decrease
	40 – 30	30	16	
	30 – 24	35	14.5	
	24 – 20	40	12.5	
	20 – 15	45	9.0	
	15 – 10	50	5.3	
Conditioning	10	40	8	18
Cooling	10	30	–	10
Total duration: approximately 38 days				

References

[1] Deaconu, I. (2025): *Cercetări privind comportamentul la uscare al cherestelei din lemn de cer (Quercus cerris L.)*. Doktorthesis, Transilvania University Brasov, Romania

15 The Authors

Dr. Mihaela Campean is a professor at Transilvania University in Brasov, Romania, in the Faculty of Furniture Design and Wood Engineering, where she teaches wood drying and heat treatment of wood. Since 2009, she has been coordinating doctoral dissertations in the field of wood engineering.

Dr. Nencho Deliiski, DSc., is a professor at the Forestry University in Sofia, Bulgaria, specializing in the automation of woodworking processes, modeling and model-based control of processes, and energy consumption in the heat treatment of wood.

Dr. Bogdan Bedelean is a lecturer at Transilvania University in Brasov, Romania, in the Faculty of Furniture Design and Wood Engineering. His main subjects are applied computer science, technological transport in the wood industry, and automation of processing in the wood industry.

Index

R

S

T